Modern Practice in

Stress and Vibration Analysis

Pergamon Titles of Related Interest

BAUTISTA et al
Theory of Machines & Mechanisms

DU
Boundary Elements

GHOSH & NIKU-LARI
CAD/CAM & FEM in Metal Working

HAYWOOD
Analysis of Engineering Cycles, 3rd Edition
Analysis of Engineering Cycles: Worked Problems

JAPAN SOCIETY OF MECHANICAL ENGINEERS
Visualized Flow

NIKU-LARI
Structural Analysis Systems, Volumes 1–6

TANAKA & CRUSE
Boundary Element Methods in Applied Mechanics

Pergamon Related Journals *(free specimen copy gladly sent on request)*

Chinese Journal of Mechanical Engineering

Computers & Fluids

Computers & Industrial Engineering

Computers & Structures

International Journal of Applied Engineering Education

International Journal of Engineering Science

International Journal of Solids and Structures

International Journal of Machine Tools & Manufacture

International Journal of Mechanical Sciences

Journal of Applied Mathematics & Mechanics

Mechanism & Machine Theory

Modern Practice in
Stress and Vibration Analysis

*Proceedings of the Conference Held at
the University of Liverpool, 3–5 April 1989*

Edited by

J. E. MOTTERSHEAD
University of Liverpool, UK

PERGAMON PRESS

OXFORD · NEW YORK · BEIJING · FRANKFURT
SÃO PAULO · SYDNEY · TOKYO · TORONTO

U.K.	Pergamon Press plc, Headington Hill Hall, Oxford OX3 0BW, England
U.S.A.	Pergamon Press, Inc., Maxwell House, Fairview Park, Elmsford, New York 10523, U.S.A.
PEOPLE'S REPUBLIC OF CHINA	Pergamon Press, Room 4037, Qianmen Hotel, Beijing, People's Republic of China
FEDERAL REPUBLIC OF GERMANY	Pergamon Press GmbH, Hammerweg 6, D-6242 Kronberg, Federal Republic of Germany
BRAZIL	Pergamon Editora Ltda, Rua Eça de Queiros, 346, CEP 04011, Paraiso, São Paulo, Brazil
AUSTRALIA	Pergamon Press Australia Pty Ltd., P.O. Box 544, Potts Point, N.S.W. 2011, Australia
JAPAN	Pergamon Press, 5th Floor, Matsuoka Central Building, 1-7-1 Nishishinjuku, Shinjuku-ku, Tokyo 160, Japan
CANADA	Pergamon Press Canada Ltd., Suite No. 271, 253 College Street, Toronto, Ontario, Canada M5T 1R5

First edition 1989

Library of Congress Cataloging in Publication Data

Modern practice in stress and vibration analysis: proceedings of the conference held at the University of Liverpool, 3–5 April 1989/ edited by J. E. Mottershead.—1st ed.
p. cm.
Conference organized by the Stress Analysis Group of the Institute of Physics, University of Liverpool, and others.
1. Structural analysis (Engineering)—Congresses.
2. Strains and stresses—Congresses.
3. Vibration—Congresses.   I. Mottershead, J. E.
II. University of Liverpool. Institute of Physics.
Stress Analysis Group.
TA645.M63   1989   624.1'71—dc20   89–8753

British Library Cataloguing in Publication Data

Modern practice in stress and vibration analysis.
1. Stress analysis.   2. Mechanical vibration.
Analysis
I. Mottershead, J. E.
620.1'123
ISBN 0-08-037522-7

In order to make this volume available as economically and as rapidly as possible the authors' typescripts have been reproduced in their original form. This method has its typographical limitations but it is hoped that they in no way distract the reader.

Printed in Great Britain by BPCC Wheatons Ltd, Exeter

Modern Practice in Stress and Vibration Analysis

Organized by

The Stress Analysis Group of the Institute of Physics

in collaboration with

The British Society for Strain Measurement
The Institute of Mechanics, Chinese Academy of Sciences
The Institution of Mechanical Engineers
The Japan Society of Mechanical Engineers
The Joint British Committee for Stress Analysis
The Society for Experimental Mechanics (USA)

Stress Analysis Group Committee

J. E. Mottershead (Chairman)
University of Liverpool

T. K. Hellen (Hon. Secretary)
CEGB Berkeley Nuclear Laboratories

R. D. Adams
University of Bristol

D. Burrows
GEC Research Ltd

J. Clark
Alcan International Ltd

M. R. Goldthorpe
University of Sheffield

A. C. Pickard
Rolls Royce plc

J. L. Wearing
University of Sheffield

A. L. Yettram
Brunel University

CONTENTS

FOREWORD

The papers in this volume constitute the complete documentation of the meeting on MODERN PRACTICE IN STRESS AND VIBRATION ANALYSIS organised by the Stress Analysis Group of the Institute of Physics at the University of Liverpool, 3-5 April 1989.

The Stress Analysis Group was formed some forty years ago and since its inception it has always taken the broad view which allows for <u>motion</u> in stress analysis. The Group has been known in the U.K. for its contribution in providing meetings with an emphasis on application, covering topics which range widely to include modern numerical techniques and advanced experimentation.

It is hoped that the contributions published here will be of value to the broad community of practitioners in stress and vibration analysis whom the Stress Analysis Group exists to serve.

J. E. Mottershead,
Liverpool, April 1989.

Structural Modification Analysis using Rayleigh Quotient Iteration

W. M. TO and D. J. EWINS

Imperial College of Science, Technology and Medicine, London, UK

ABSTRACT

This paper highlights the application to structural dynamics of the sensitivity analysis methods developed by numerical analysts and presents a historical development of first- and higher-order eigenvalue and eigenvector sensitivities. Different formulae for an eigenvalue sensitivity are presented and it is shown that all of these are implicitly the same. A condition number is presented to give the limited bound of application for the first-order eigenvalue and eigenvector sensitivities. An alternative structural modification method based on Rayleigh Quotient Iteration is presented. A lumped spring-mass system with 7 degrees-of-freedom (7DoF) is used to show the applicability of the Rayleigh quotient iteration method.

KEYWORDS

Structural modification; sensitivity analysis; condition number; Rayleigh quotient iteration

1. INTRODUCTION

Sensitivity analysis has been applied by several workers to the general eigenvalue problem [1-8] and, more specifically, to applications of structural modification in references [9-11]. In this area, both first- and higher-order eigenvalue and eigenvector sensitivities have been investigated with a view to predicting the dynamics of a modified structure from knowledge of its properties in an original, or unmodified, state. As the sensitivity analysis of a mechanical structure is based on a Taylor's expansion of the eigenvalues and eigenvectors of the unmodified structure, and the computation of the higher-order terms of this series is difficult and time consuming, the effectiveness of this method is limited to small modifications. However, it is not easy to determine what is 'small'. In this paper, a condition number is presented to indicate how sensitive the eigenvalues and eigenvectors of a mechanical structure are to small modifications. The value of this condition number is used to determine a limit of applicability for the first-order eigenvalue and eigenvector sensitivities.

The Rayleigh quotient provides a well-known procedure for the approximate evaluation of the eigenvalue of a structure. It is shown that under some conditions the Rayleigh quotient and first-order eigenvalue sensitivity are equivalent.

Structural modification problems possess an inherent advantage in that some initial conditions are known and this information is very useful for an iteration method of analysis. Accordingly, an iterative procedure using the Rayleigh quotient would seem to offer advantages for the solution of the modification analysis problem, especially in view of the cubic convergence property of this method.

1

NOMENCLATURE

[A], [dA]   System matrix of the eigenvalue problem $[A]\{x\} = \lambda\{x\}$ and the change in [A]
[K], [M]   System stiffness and mass matrices
[dK], [dM]   Modifications in system stiffness and mass matrices
$\{y_i\}$, $\{x_i\}$, λ_i Left-hand & right-hand mass-normalised eigenvectors and eigenvalue of ith mode
$\{x'\}_o$, $\{x'\}_m$ Orthonormalised eigenvectors of the original system, modified system
λ_o, λ_m Eigenvalues of the original system, modified system
α Design parameter of system matrices
$\| \{ \} \|_2$ 2-norm (Euclidean norm) of a vector i.e. $\| \{x\} \|_2 \equiv (\{x\}^H\{x\})^{1/2}$
$\| [ ] \|_2$ 2-norm of a matrix   i.e. $\| [A] \|_2 = $ (max. eigenvalue of $[A]^H[A])^{1/2}$
$\{ \}^H, [ ]^H$ Hermitian (complex-conjugate) transpose of a vector and a matrix

2. THE ORIGINAL PURPOSE OF DEVELOPING THE FIRST-ORDER PERTURBATION

First-order perturbation estimates (i.e. eigenvalue and eigenvector sensitivities) have long been used by numerical analysts and scientists to investigate the stability of the eigenvalue problem $[A]\{x\}=\lambda\{x\}$. These estimates have the advantage that they provide a quick, nonrigorous look at how eigenvalues and eigenvectors change when the elements of matrix [A] vary within the limits of permissible error. In most matrix computational textbooks, first-order perturbation theory is used to obtain the error bounds in computing eigenvalues and eigenvectors on a digital computer.

By translating the above-mentioned feature to suit structural analysts, a first-order sensitivity analysis is used to determine the "order of importance" ranking of a structure's degrees of freedom for structural modification, for each mode of vibration.

3. HISTORICAL DEVELOPMENT OF FIRST- (AND HIGHER-) ORDER SENSITIVITY

In 1846, Jacobi [1] published a result on eigenvalue sensitivities for an eigenvalue problem $[A]\{x\}=\lambda\{x\}$. Wilkinson [2] presented clear derivations for the first-order perturbation equation of an eigenvalue in terms of the normalised left-hand and right-hand eigenvectors. The equation presented in ref.2 is :

$$\frac{\partial \lambda_i}{\partial \alpha} = \frac{\{y_i\}^T [\frac{\partial A}{\partial \alpha}] \{x_i\}}{\{y_i\}^T \{x_i\}} \tag{1}$$

In the 1960s, some methods were developed for sensitivity analysis of electronic networks and are notable for their nonreliance on eigenvectors in the eigenvalue sensitivity formula. Rosenbrock [3] and Reddy [4] developed a formula for eigenvalue sensitivities in terms of the matrix [A] and its eigenvalues.

$$\frac{\partial \lambda_i}{\partial \alpha} = \frac{\text{trace } \{adj([A] - \lambda_i[I]) [\frac{\partial A}{\partial \alpha}]\}}{\text{trace } \{adj([A] - \lambda_i[I])\}} \tag{2}$$

If λ_i is a simple eigenvalue of the matrix [A], then $adj([A] - \lambda_i[I])$ can be expressed in terms of the corresponding left-hand and right-hand eigenvectors of λ_i :

$$\text{adj}([A] - \lambda_i[I]) = \tau_i \{x_i\} \{y_i\}^T \tag{3}$$

where τ_i is a constant. Therefore,

$$\text{trace} \{\text{adj}([A] - \lambda_i[I])\} = \text{trace} (\tau_i\{x_i\}\{y_i\}^T) = \tau_i\{y_i\}^T\{x_i\} \tag{4}$$

Bodewig [12] showed that if [Q] is square and rank [Q] = 1 so that [Q] is a simple product of a column $\{x_i\}$ and a row $\{y_i\}^T$, i.e. $[Q] = \tau_i\{x_i\}\{y_i\}^T$, and if $[\partial A/\partial \alpha]$ is arbitrary, but square, then

$$\text{trace} ([Q][\frac{\partial A}{\partial \alpha}]) = \text{trace} ([\frac{\partial A}{\partial \alpha}][Q]) = \tau_i \{y_i\}^T [\frac{\partial A}{\partial \alpha}] \{x_i\} \tag{5}$$

By substituting eqns.(4) and (5) in eqn.(2), eqn.(1) is obtained. Thus, in the computation of $\text{adj}([A] - \lambda_i[I])$, both left-hand and right-hand eigenvectors are implicitly computed, in view of eqn.(4).

The trace of the $\text{adj}([A] - \lambda_i[I])$ can also be expressed as:

$$\text{trace} \{\text{adj}([A] - \lambda_i[I])\} = \prod_{i,\ i \ne r}^{n} (\lambda_r - \lambda_i) \tag{6}$$

In structural analysis [A] is related to the stiffness matrix (usually written as [K]) and [I] is the unit mass matrix. If a point Single Degree of Freedom (SDoF) stiffness modification is made at location j, the numerator of eqn.(2) can be written as :

$$\text{trace} \left\{ \begin{bmatrix} \Delta_{11}(\lambda_i) & \Delta_{21}(\lambda_i) & ... & \Delta_{n1}(\lambda_i) \\ \vdots & & & \vdots \\ \Delta_{1n}(\lambda_i) & \Delta_{2n}(\lambda_i) & ... & \Delta_{nn}(\lambda_i) \end{bmatrix} \begin{bmatrix} 0 & 0 & ... & 0 \\ \vdots & & 1_{(j,j)} & \vdots \\ 0 & 0 & ... & 0 \end{bmatrix} \right\}$$

$$= \Delta_{jj}(\lambda_i)$$
$$= \text{the (j,j) cofactor of } ([A] - \lambda_i[I]) \tag{7}$$

The equation of motion for a linear system is :

$$([A] - \lambda [I]) \{x\} = \{f\} \tag{8}$$

It has been shown [13] that any frequency response function (FRF) of a grounded undamped structure can be described completely in terms of its poles or eigenvalues (squares of natural frequencies), its zeros (squares of anti-resonance frequencies), and a constant .

$$H_{jj}(\lambda) = \frac{C_{jj} \prod_{r=1}^{n-1} (1 - \frac{\lambda}{jj\beta_k})}{\prod_{r=1}^{n} (1 - \frac{\lambda}{\lambda_r})} \tag{9}$$

where $H_{jj}(\lambda)$ = receptance of point measurement at location j
C_{jj} = static flexibility
λ_r = r^{th} pole
$jj\beta_k$ = k^{th} zero for H_{jj}

The FRF $H_{jj}(\lambda)$ can also be obtained from eqn.(8).

$$H_{jj}(\lambda) = \frac{x_j}{f_j}(\lambda) = \frac{\text{the (j,j) cofactor of } ([A] - \lambda[I])}{\det([A] - \lambda[I])} \qquad (10)$$

since $\quad \det([A] - \lambda[I]) = \prod_{r=1}^{n}(\lambda - \lambda_r) = \prod_{r=1}^{n}(1 - \frac{\lambda}{\lambda_r})\prod_{r=1}^{n}\lambda_r \qquad (11)$

Using eqns.(9), (10) and (11), the (j,j)th cofactor of ([A] - λ[I]) can be expressed as :

$$\text{the (j,j)th cofactor of } ([A] - \lambda[I]) = \frac{C_{jj}\prod_{r=1}^{n}\lambda_r\prod_{k=1}^{n-1}(^{jj}\beta_k - \lambda)}{\prod_{k=1}^{n-1}{}^{jj}\beta_k} \qquad (12)$$

Substituting eqns.(6) and (12) in eqn.(2), we obtain :

$$\frac{\partial\lambda_i}{\partial\alpha} = \frac{C_{jj}\prod_{r=1}^{n}\lambda_r\prod_{k=1}^{n-1}(^{jj}\beta_k - \lambda_i)}{\prod_{k=1}^{n-1}{}^{jj}\beta_k\prod_{i,\ i\neq r}^{n}(\lambda_r - \lambda_i)} \qquad (13)$$

and this is the equation for a point SDoF stiffness modification obtained independently by Skingle [11].

The formula shown above for calculating the sensitivity of an eigenvalue to changes in the matrix (eqn.(13)) is implicitly the same as these shown in eqns.(1) & (2). This is to be expected if one begins with equation $[A]\{x\}=\lambda\{x\}$. As the system properties do not vary under a harmonic excitation, the difference between eqn.(13) and eqns.(1) & (2) is the number of design parameters used to calculate the first-order eigenvalue sensitivity.

Fox and Kapoor [5] considered the special case of symmetric stiffness [K] and mass [M] matrices but developed techniques applicable to more general cases. For eigenvalues their formula is :

$$\frac{\partial\lambda_i}{\partial\alpha} = \{x_i\}^T([\frac{\partial K}{\partial\alpha}] - \lambda_i[\frac{\partial M}{\partial\alpha}])\{x_i\} \qquad (14)$$

in which it is assumed that the eigenvectors are normalised such that

$$\{x_i\}^T[M]\{x_i\} = 1 \qquad (15)$$

The first-order eigenvector sensitivity is :

$$\frac{\partial\{x_i\}}{\partial\alpha} = -\frac{1}{2}\{x_i\}^T[\frac{\partial M}{\partial\alpha}]\{x_i\}\{x_i\} + \sum_{k=1,k\neq i}^{n}\frac{1}{(\lambda_i - \lambda_k)}\{x_i\}^T([\frac{\partial K}{\partial\alpha}] - \lambda_i[\frac{\partial M}{\partial\alpha}])\{x_k\}\{x_k\} \qquad (16)$$

Rogers [6] derived sensitivity formulae for eigenvalues and eigenvectors of a general problem and stated the need for two sets of normalisation conditions for non-self-adjoint system. Vanhonacker [9] derived some formulae calculating the differential sensitivities of a mechanical structure subjected to parameter changes. The formulae obtained were based on an nDoF system under a sinusoidal excitation. Comparing his derivations with the equations shown above, one can observe that the differential sensitivity is a special case in the classical methods of sensitivity analysis.

Many papers [7,8] have presented the higher-order eigenvalue and eigenvector sensitivities. Wang *et al.* [10] have investigated the accuracy of structural modification by calculating the first- and second-order eigenvalue and eigenvector sensitivities and showed the divergence phenomenon for large structural modification. It has been shown [10] that including some higher-order terms does not always ensure a more accurate prediction.

4. CONDITION NUMBERS

For an eigenvalue problem $[A]\{x'\} = \lambda\{x'\}$ Stewart [14] has shown that if $\{y'\}$ and $\{x'\}$ are orthonormalised left-hand and right-hand eigenvectors of $[A]$ with simple eigenvalue λ, then $\|\{y'\}\|_2$ is a condition number for λ (this condition number was first observed by Wilkinson [2]). Let $[U] \in \mathbb{C}^{n \times (n-1)}$ be a matrix such that $(\{x'\},[U])$ is unitary, then $\|(\lambda[I] - [U]^H[A][U])^{-1}\|_2$ is a condition number for the eigenvector $\{x'\}$. The inverse of $\|(...)^{-1}\|_2$ is a measure the separation of λ from its neighbours. In Stewart's book, the following theorem is given :- Let λ be a simple eigenvalue of $[A] \in \mathbb{C}^{n \times n}$ with right-hand eigenvector $\{x'\}$ and left hand eigenvector $\{y'\}$. Suppose $\{x'\}$ has been scaled so that $\| \{x'\} \|_2 = 1$ and $\{y'\}$ has been scaled so that $\{y'\}^H\{x'\} = 1$. Let $[U] \in \mathbb{C}^{n \times (n-1)}$ be chosen so that $(\{x'\}, [U])$ is unitary and set

$$(\{x'\},[U])^H [A] (\{x'\},[U]) = \begin{bmatrix} \lambda & \{x'\}^H[A][U] \\ 0 & [U]^H[A][U] \end{bmatrix} \qquad (17)$$

Let $[dA] \in \mathbb{C}^{n \times n}$ be given and let

$$\varepsilon = \| [dA] \|_2 \ , \quad \eta = \| [U]^H[A]^H\{x'\} \|_2 \ , \quad \gamma = \| [U]^H[dA]\{x'\} \|_2$$
$$\text{and} \quad \mu = \| (\lambda[I] - [U]^H[A][U])^{-1} \|_2^{-1} \qquad (18)$$

Then if, $\quad \dfrac{\gamma (\eta + \varepsilon)}{(\mu - \varepsilon)^2} < \dfrac{1}{4} \ , \qquad (19)$

an eigenvalue λ_m of $[A+dA]$ with eigenvector $\{x'\}_m$ can be predicted accurately by first-order perturbation theory.

This theorem is modified to deal with the generalised eigenvalue problem $[K]\{x'\}=\lambda[M]\{x'\}$. The system matrix $[A]$ is substituted by the matrix $[M]^{-1}[K]$ if $[M]$ is invertible, the perturbation matrix $[dA]$ is approximated by $[M+dM]^{-1}([dK]-\lambda[dM])$ if $[M+dM]$ is invertible.

5. RAYLEIGH QUOTIENT

The Rayleigh quotient provides an approximation of the eigenvalues based on trial vectors that are often approximate eigenvectors of a self-adjoint system. Assume that the eigenvector of a modified system $\{x'\}_m$ is approximated by the eigenvector of the original system $\{x'\}_0$, then :

$$\lambda_R = \dfrac{\{x'\}_0^T [K+dK] \{x'\}_0}{\{x'\}_0^T [M+dM] \{x'\}_0} \qquad (20)$$

where λ_R is called the Rayleigh quotient.

For a stiffness modification, the Rayleigh quotient is expressed as :

$$\lambda_R = \lambda_o + \cfrac{1}{\{x'\}_o^T [M] \{x'\}_o} \{x'\}_o^T [dK] \{x'\}_o \qquad (21)$$

Comparing the results obtained by Rayleigh quotient (eqn.(21)) with the results obtained by perturbation theory (eqn.(14)), it is noticed that the Rayleigh quotient and first-order perturbation theory yield identical results for a self-adjoint system when there is no mass perturbation.

For a mass modification, the Rayleigh quotient can be expanded to a power series:

$$\lambda_R = \lambda_o + \cfrac{1}{\{x'\}_o^T [M] \{x'\}_o} (-\lambda_o \{x'\}_o^T [dM] \{x'\}_o)$$
$$\lambda_o \frac{(\{x'\}_o^T [dM] \{x'\}_o)^2}{(\{x'\}_o^T [M] \{x'\}_o)^2} - \lambda_o \frac{(\{x'\}_o^T [dM] \{x'\}_o)^3}{(\{x'\}_o^T [M] \{x'\}_o)^3} + \dots \qquad (22)$$

Hence, the first-order term in the Rayleigh quotient is the first-order eigenvalue perturbation $((\delta\lambda_i/\delta\alpha)\Delta\alpha)$. The second-order term is one of the mass perturbation terms which appear in the second-order eigenvalue sensitivity. However, most of the mass perturbation terms for the second-order eigenvalue sensitivity are not accounted for in the Rayleigh quotient expression. Some higher-order mass perturbations terms are present because mass terms appear in the denominator of the Rayleigh quotient.

6. RAYLEIGH QUOTIENT ITERATION

If $\{x'\}_o$ is an approximate eigenvector for the modified system, then λ_R (as defined above in eqn.(20)) is a reasonable choice for the corresponding eigenvalue. On the other hand, if λ_R is an approximate eigenvalue, inverse iteration theory shows that the solution to $([K+dK] - \lambda_R[M+dM])\{x'\}_m = [M+dM]\{x'\}_o$ will almost always be a good approximate eigenvector.

Combining these two ideas in the natural way gives rise to Rayleigh Quotient Iteration. For a self-adjoint system $[K+dK]$ is symmetric and $[M+dM]$ is symmetric and positive definite, and the Rayleigh Quotient Iteration is expressed as follows [15] :

$\{x'\}_o$ given, $\quad \| \{x'\}_o \|_2 = 1$
For $k = 0, 1, \dots$

$$\lambda_k = \frac{\{x'\}_k^T [K+dK] \{x'\}_k}{\{x'\}_k^T [M+dM] \{x'\}_k}$$

Solve $([K+dK] - \lambda_k[M+dM])\{z\}_{k+1} = [M+dM]\{x'\}_k \quad$ for $\{z\}_{k+1}$

$$\{x'\}_{k+1} = \frac{\{z\}_{k+1}}{\| \{z\}_{k+1} \|_2}$$

7. NUMERICAL EXAMPLES

In order to evaluate the effectiveness of the first-order sensitivity and the Rayleigh quotient iteration methods, a modification study was made on a lumped spring-mass model with 7DoF shown in Fig.1. Figure 2 shows the point frequency response function for this model when the excitation was applied on point 4 (damping loss factors 0.001 were used to produce the FRF).

For several parameter changes at different locations of the model the first-order sensitivity and the Rayleigh quotient iteration have been applied to the original data to predict the consequent changes to the system's natural frequencies. The results are compared with the exact solution obtained by complete reanalysis.

In Figs.3.1-3.3, the shifts of eigenvalues are plotted with respect to a mass change at point 6. From these figures, it is noticed that the Rayleigh quotient iteration method gives accurate results for modes 1 to 6 although some numerical difficulties are observed for mode 7. Figure 3.3 shows that mode 7 converges to mode 6 if the mass change is greater than 60%. The first-order sensitivity method is a linear approximation based on an infinitesimal change of a parameter, and it is seen that the accuracy of prediction based on this parameter decreases with increasing magnitude of the mass change. Figure 3.4 shows the condition number (LHS of inequality (25)) against the percentage of mass change. Modes 3 and 4 are close modes and for these the first-order sensitivity is applicable only for small mass changes (≈ 2.5 %). Modes 5, 6 and 7 are also ill-conditioned for the mass perturbation at point 6.

The accuracy of the mode shapes predicted by the two techniques is assessed by using the Modal Assurance Criterion (MAC). Tables 1 and 2 contain the MAC values between eigenvectors from the first-order sensitivity analysis and the exact solution and between the Rayleigh quotient iteration and the exact solution. Examining the tables, it is seen that the Rayleigh quotient iteration yields consistently better correlation with the exact solution except when the mass change is greater than 60%, mode 7 converges to mode 6.

In Figs.4.1-4.3, the shifts of eigenvalues are plotted for the case of a stiffness change made between points 3 and 4. It is seen that the Rayleigh quotient iteration method gives accurate prediction for all modes. The condition number against the percentage of stiffness modification change is shown in Fig.4.4. From this figure, it is noticed that modes 2, 3, 5 and 6 are ill-conditioned for this stiffness modification. The limit of application for the first-order perturbation theory for all modes is approximately equal to 4%. The MAC values between eigenvectors from those two techniques and those from the exact solution are shown in Tables 3 and 4. From the results, it is observed that the Rayleigh quotient iteration again yields better results even when the magnitude of stiffness modification is large.

8. POSSIBLE USE OF THESE MODIFICATION TOOLS

An analytical or Finite Element model of a complex structure normally has a large number of degrees-of-freedom and design variables. The primary analysis yields some of eigenvalues in a given frequency range, together with the corresponding eigenvectors.

Usually, the model needs to be progressively modified during an automated optimum design process. The *first-order eigenvalue sensitivity* ranks possible structural modification sites in their order of effectiveness at influencing each particular mode. The *Rayleigh quotient iteration* produces a more accurate prediction. The *condition number* monitors the behaviour of the modes in structural modification. By combining these three tools, an iterative program can be written to calculate the optimum condition for a structure specified by the user.

9. SUMMARY

First-order eigenvalue sensitivities are very useful to rank the order of importance for the sites during structural modification. First-order eigenvector sensitivities are sometimes impossible to compute because to do so requires the complete set of right-hand eigenvectors for a self-adjoint system. Calculations of higher-order eigenvalue and eigenvector sensitivities are difficult and expensive. Summation of a truncated eigensystem sensitivities series does not converge rapidly to give satisfactory results.

8

A condition number has been presented to provide information about the sensitivities of the eigenvalues and eigenvectors of the generalised eigenvalue problem $[K]\{x\} = \lambda[M]\{x\}$ to small perturbations of $[K]$ and/or $[M]$.

A Rayleigh quotient iteration method has been presented to compute the modified eigenvalues and their associated eigenvectors of a large analytical model. It has long been known that if the first prediction of an eigenvalue by using the unperturbed eigenvector is closer to the modified one than its neighbour eigenvalues, this iteration method converges globally and the convergence is ultimately cubic.

ACKNOWLEDGEMENT

The authors gratefully acknowledge the financial support of the Croucher Foundation in Hong Kong.

REFERENCES

1. Jacobi, C.G.J., *"Uber ein leichtes Verfahren die in der Theorie der Saecularstoerungen vorkommenden Gleichungen numerisch aufzuloesen"*, Zeritshrift fur Reine und Angewandte Mathematik, Vol.30, 1846, pp.51-95
2. Wilkinson, J.H., *The Algebraic Eigenvalue Problem* Oxford University Press, London, 1963, pp.62-109
3. Rosenbrock, H.H., *"Sensitivity of an Eigenvalue to Changes in the Matrix"* Electronics Letters, Vol.1, 1965, pp.278-279
4. Reddy, D.C., *"Sensitivity of an Eigenvalue of a Multivariable Control System"* Electronics Letters, Vol.2, 1966, pp.446
5. Fox, R.L. and Kapoor, M.P., *"Rates of Changes of Eigenvalues and Eigenvectors"* AIAA Journal, Vol.6, 1968, pp.2426-2429
6. Rogers, L.C., *"Derivatives of Eigenvalues and Eigenvectors"* AIAA Journal, Vol.8, 1970, pp.943-944
7. Rudisill, C.S., *"Derivatives of Eigenvalues and Eigenvectors for a General Matrix"* AIAA Journal, Vol.12, 1974, pp.721-722
8. Belle, H.V., *"Higher Order Sensitivities in Structural Systems"* AIAA Journal, Vol.20, 1982, pp.286-288
9. Vanhonacker, P., *"Differential and Difference Sensitivities of Natural Frequencies and Mode Shapes of Mechanical Structure"*, AIAA Journal, Vol.18, 1980, pp.1511-1514
10. Wang, J., Heylen, W. and Sas, P., *"Accuracy of Structural Modification Techniques"* Proc. of the 5th International Modal Analysis Conference, 1987, pp.65-71
11. Skingle, G.W., *"Resonance and Anti-Resonance Sensitivities for SDoF Mass and Stiffness Modification"*, Imperial College, Dynamics Section, Int.Rep.No.8802, Jan.1988
12. Bodewig, E., *Matrix Calculus* North Holland, Amsterdam, 1959, pp.35-36
13. Duncan, W.J., *"Mechanical Admittances and their Applications to Oscillation Problems"* HMSO., London, 1947
14. Stewart, G.W., *Introduction to Matrix Computations* Academic Press, New York, 1973, pp.289-307
15. Golub, G.H. and Van Loan, C.F., *Matrix Computations* North Oxford Academic, Oxford, 1983, pp.317-318

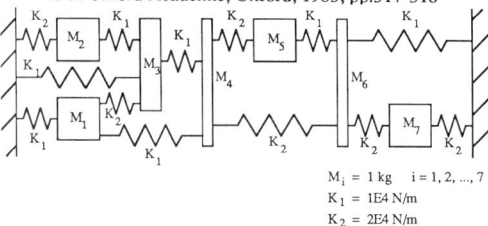

$M_i = 1 \text{ kg} \quad i = 1, 2, ..., 7$
$K_1 = 1E4 \text{ N/m}$
$K_2 = 2E4 \text{ N/m}$

Figure 1. A lumped spring-mass model with 7 degrees-of-freedom

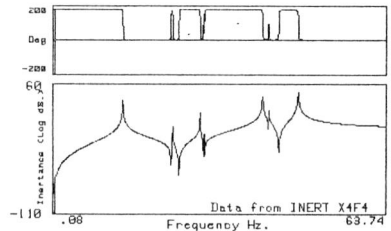

Figure 2. A point FRF of the analytical model

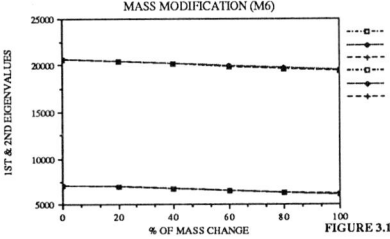

FIGURE 3.1

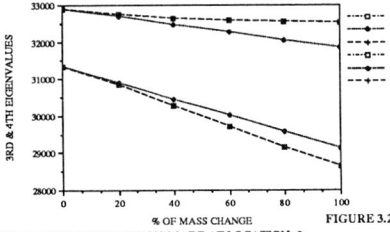

FIGURE 3.3

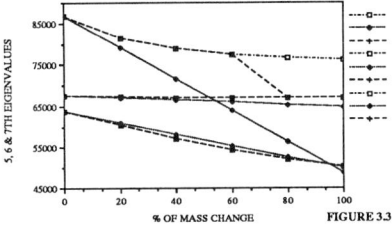

FIGURE 3.2

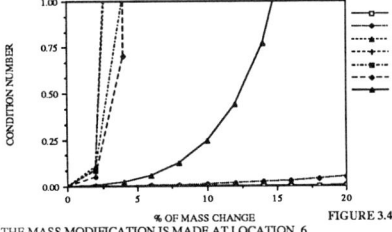

FIGURE 3.4

THE MASS MODIFICATION IS MADE AT LOCATION 6

THE MASS MODIFICATION IS MADE AT LOCATION 6

60% PERTURBATION : (MAC NO.S)
PREDICTED (First-Order Sensitivity)

EXACT						
1.000	.003	.002	.002	.005	.002	.021
.004	1.000	.000	.003	.001	.001	.011
.006	.003	.959	.035	.000	.004	.018
.000	.000	.031	.964	.000	.000	.001
.016	.007	.019	.003	.812	.094	.061
.000	.000	.000	.000	.157	.834	.000
.004	.002	.004	.000	.012	.082	.938

100% PERTURBATION : (MAC NO.S)
PREDICTED (First-Order Sensitivity)

EXACT						
1.000	.008	.004	.005	.007	.003	.037
.012	.999	.001	.009	.000	.003	.022
.016	.007	.855	.097	.001	.008	.035
.001	.000	.111	.879	.000	.000	.001
.034	.016	.049	.025	.625	.130	.101
.000	.000	.000	.000	.312	.667	.000
.005	.003	.005	.001	.009	.203	.821

TABLE 1

60% PERTURBATION : (MAC NO.S)
PREDICTED (Rayleigh Quotient Iteration)

EXACT						
1.000	.004	.006	.000	.016	.000	.004
.004	1.000	.002	.000	.007	.000	.002
.006	.002	1.000	.000	.010	.000	.003
.000	.000	.000	1.000	.001	.000	.000
.016	.007	.010	.001	1.000	.000	.006
.000	.000	.000	.000	.000	1.000	.000
.004	.002	.003	.000	.006	.000	1.000

100% PERTURBATION : (MAC NO.S)
PREDICTED (Rayleigh Quotient Iteration)

EXACT						
1.000	.014	.017	.001	.034	.000	.000
.014	1.000	.007	.000	.014	.000	.000
.017	.007	1.000	.000	.017	.000	.000
.001	.000	.000	1.000	.001	.000	.000
.034	.014	.017	.001	1.000	.000	.000
.000	.000	.000	.000	.000	1.000	.000
.005	.002	.003	.000	.005	.000	1.000

TABLE 2

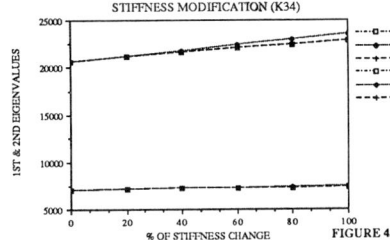

FIGURE 4.1

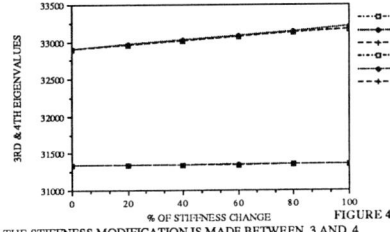

FIGURE 4.3

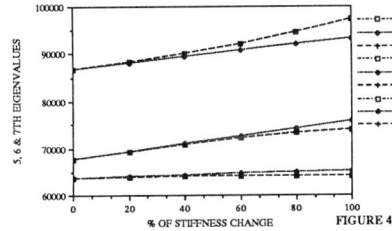

FIGURE 4.2

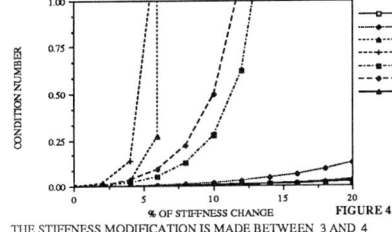

FIGURE 4.4

THE STIFFNESS MODIFICATION IS MADE BETWEEN 3 AND 4

THE STIFFNESS MODIFICATION IS MADE BETWEEN 3 AND 4

100% PERTURBATION : (MAC NO.S)
PREDICTED (First-Order Sensitivity)

EXACT						
.979	.017	.000	.000	.001	.001	.000
.014	.909	.001	.023	.016	.014	.005
.000	.001	.994	.005	.000	.000	.000
.001	.017	.004	.964	.004	.003	.001
.000	.001	.000	.000	.255	.697	.004
.002	.020	.000	.003	.532	.125	.506
.004	.035	.000	.005	.192	.161	.483

TABLE 3

100% PERTURBATION : (MAC NO.S)
PREDICTED (Rayleigh Quotient Iteration)

EXACT						
1.000	.000	.000	.000	.000	.000	.000
.000	1.000	.000	.000	.000	.000	.000
.000	.000	1.000	.000	.000	.000	.000
.000	.000	.000	1.000	.000	.000	.000
.000	.000	.000	.000	1.000	.000	.000
.000	.000	.000	.000	.000	1.000	.000
.000	.000	.000	.000	.000	.000	1.000

TABLE 4

Vibration Control in Spacecraft

D. J. INMAN and A. SOOM

Department of Mechanical and Aerospace Engineering,
State University of New York at Buffalo, Buffalo, NY 14260, USA

ABSTRACT

This paper examines both passive and active vibration control for use in vibration suppression in spacecraft. Proposed spacecraft designs are very lightweight and hence extremely flexible when compared with earthbound structures. To further complicate matters, spacecraft are intended to function in a vacuum void of external viscous damping mechanisms. For these reasons, the vibration suppression problem for spacecraft presents new challenges to the vibration community. The unique problems of controlling the structural vibration in flexible spacecraft are examined and discussed in this paper.

KEYWORDS

control, flexible structure, vibration

INTRODUCTION

This paper considers vibration control problems common to flexible spacecraft. The type of spacecraft considered are those typified by the US space station model and common to large satellites. Such structures and/or their components consist of large, low mass and consequently highly flexible devices. An example familiar to most is the Solar Array Flight Experiment (SAFE), performed during the Shuttle flight of August 1984 (Mission 41-D). This test forms the sole base of public domain information on the vibration behavior of large flexible structures in space. This device is 31.5m x 4m x 1.6mm extended in a cantilevered fashion out of the shuttle bay and experienced substantial vibration as the result of motion aboard the shuttle (Brumfield et al, 1985). The SAFE structure conceptually consists of a flat beam or plate like structure which is often modeled in ground tests as a cantilevered beam for the sake of performing proof of concept experiments.

Mechanism for controlling the vibrations of flat structures, such as solar arrays, consist of added passive damping treatments or added vibration absorbers, and active control in the form of layered piezoelectric devices or lumped actuators such as proof-mass actuators. Passive damping treatments have the advantage of high reliability, no supporting electronics and no moving parts, all of which are disadvantages of active control. However, active control is capable of much better response performance than passive control.

PASSIVE VS ACTIVE CONTROL

In general the mathematical models of large flexible spacecraft are distributed parameter models which in principle are infinite dimensional and are at best large order finite element models in practice. Both design and control methods work best for low order systems, hence the spacecraft models are usually reduced in size before control and design calculations are made. This unfortunately leads to difficulty known as spillover (Balas, 1982). Spillover occurs any time unmodeled dynamics are excited by a control law constructed based on only the modeled dynamics. Passive control and redesign cause less (if any) of this sort of problem. It is natural at this point to ask why bother with active control.

To answer this, a comparison is made between active and passive devices. Comparisons are difficult to make because of the physical differences between the two approaches. Inman and Horner (1989) chose to make the comparison between active and passive damping based on comparing a layered damping treatment (passive) with a layered piezoelectric material (active). Their choice of comparison was based on the fact that both the active and passive device are of identical geometry (occupy the same space) and both are distributed parameter elements, negating the effects of truncated modeling. In addition, if a pinned-pinned Euler-Bernoulli beam is used for the model, a closed form solution for the loss factor, and hence damping ratio is available (Ross et al, 1959). Using the closed form expression, Inman and Horner (1989) were able to show that the piezoelectric active control device is capable of providing up to 5 times more damping to the beam than a passive damping treatment of the same mass and thickness. The point of view taken here is that active control is used to provide damping, and hence vibration suppression, in those circumstances where more than 20% critical damping is required in the system.

ACTIVE CONTROL FORMULATION

Next, a formulation for active control is presented. To make the presentation simple and to the point as possible, several simplifying assumptions are made. The systems considered here are those with open loop dynamics that decouple, and are subject only to velocity feedback. Both of these assumptions can be relaxed (Cudney and Inman, 1989). Consider a structure modeled by

$$M \ddot{\mathbf{x}}(t) + D \dot{\mathbf{x}}(t) + K\mathbf{x}(t) = \mathbf{f}(t) \tag{1}$$

where M, D and K are positive definite, symmetric matrices of size nxn, representing the mass damping and stiffness in the system respectively, such that the matrix produce $DM^{-1}K$ is symmetric. Here $\mathbf{x}(t)$, $\dot{\mathbf{x}}(t)$ and $\ddot{\mathbf{x}}(t)$ are nx1 vectors of displacement, velocity and acceleration respectively and $\mathbf{f}(t)$is an nx1 vector of external forces (both disturbances and/or control forces). Alternately, the system of equation (1) can be written in state space form by defining the (augmented) vectors

$$\mathbf{q} = \begin{bmatrix} \mathbf{x} \\ \dot{\mathbf{x}} \end{bmatrix} \quad , \quad \mathbf{F} = \begin{bmatrix} \mathbf{0} \\ \mathbf{f} \end{bmatrix} \tag{2}$$

and the $2n$x$2n$ state matrix, A, defined by

$$A = \begin{bmatrix} 0 & I \\ -M^{-1}K & -M^{-1}D \end{bmatrix} \tag{3}$$

The vector $\mathbf{q}$ is called the state vector and in state space form equation (1) becomes

$$\dot{\mathbf{q}} = A\mathbf{q} + \mathbf{F} \tag{4}$$

with output equation defined by the sx1 vector $\mathbf{q}$

$$y = Cq \tag{5}$$

where C is an $s \times 2n$ matrix defining which states q_i are measured or observed.

The state matrix A may also contain actuator dynamics as well as structural dynamics. If the state matrix A has a diagonal Jordon form, then there exists a nonsingular matrix P such that

$$P^{-1}AP = \Lambda \tag{6}$$

where Λ is a diagonal matrix of eigenvalues of the matrix A. With this transformation in mind, the state equations can be decoupled by letting $q = Pz$ and premultiplying equation (4) by P^{-1}. This yields the decoupled system

$$\dot{z} = \Lambda z + P^{-1}F \tag{7}$$
$$y = CPz$$

If output feedback is used to control the response, then F takes the form By, where B is the control gain matrix of dimension $2nxs$. For single-input single-output systems y is a scalar, and C and B become the vectors c^T and b respectively. Making the appropriate substitution, the control system (7) reduces to the equivalent open loop system given by

$$\dot{z} = (\Lambda + P^{-1}bc^TP)z \tag{8}$$

Here the matrix Λ is diagonal while the matrix $P^{-1}bc^TP$ will be fully populated and not diagonal.

This illustrates the coupling effect of feedback control. The control matrix $P^{-1}bc^TP$ effectively recouples the previously decoupled open loop equations.

A major problem associated with controlling large space structures is illustrated by a partitioned form of equation (8). Typically the order of the system ($2n$) is large. Since control computations work best for small order systems, and because of computational constraints, it is common to consider only the first k states in the design process. This choice also fits well with the philosophy that a structure behaves like a low pass filter and hence there exist some value of the index on the element of the state vector z, say k such that the first k elements are a good approximation of the structure. With this in mind, partition z into two vectors z_k consisting of the first k modes and z_{2n-k} consisting of the remaining modes. Then the system of equation (8) becomes partition as

$$
\begin{bmatrix} \dot{z}_k \\ \dot{z}_{2n-k} \end{bmatrix} = \left(\begin{bmatrix} \Lambda_k & 0 \\ 0 & \Lambda_{2n-k} \end{bmatrix} + \begin{bmatrix} b_k c_k^T & b_k c_{2n-k}^T \\ b_{2n-k} c_k^T & b_{2n-k} c_{2n-k}^T \end{bmatrix} \right) \begin{bmatrix} z_k \\ z_{2n-k} \end{bmatrix} \tag{9}
$$

Here 0 is a rectangular matrix of zeros of appropriate dimension, Λ_k are the first k eigenvalues, Λ_{2n-k} is a diagonal submatrix of the of the last $2n-k$ eigenvalues, and where the vectors b and c^T have been partitioned as indicated by the dimensions listed in equation (8).

The partitioned form of equation (9) lends itself to an explanation of the flexible structure control problem. If Λ_{2n-k} is left out of the control design the effects of the terms $b_k c_{2n-k}^T$,

$b_{2n-k} c_n^T$ and $b_{2n-k} c_{2n-k}^T$ are to couple the neglected dynamics back into the closed loop system. If the higher modes of the sensor (c_{2n-k}^T) and actuator (b_{2n-k}) are zero then, the control problem for the truncated modal z_k is effectively decoupled and no special difficulties are encountered. However, if the sensors are such that c_{2n-k}^T is not zero, then the term $b_k c_{2n-k}^T$ recouples the higher order dynamics to the equation for z_k. This is called *observation spillover* and is known to cause instabilities if neglected (Balas, 1982). Likewise if the actuator locations or gains are such that b_{2n-k} is not zero then the term $b_{2n-k} c_k^T$ couples the higher order dynamics to z_k. This is called *control spillover*, and can cause serious degradation of the closed loop performance.

ACTUATOR/STRUCTURE INTERACTION

The problems encountered in actively suppressing the vibration of flat appendages, such as solar panels can be understood by examining the control of a cantilevered beam with a single proof-mass actuator. In the previous section the phenomena of spillover was illustrated as a possible source of instability. Here it is shown that instabilities may also result due to the interaction between the structure and the actuator. This is significant because the possibility of instability arises from including the actuator dynamics which are normally left out of the design.

This point is easily made by considering a simple single degree of freedom model of the structure. Let m_s, c_s and k_s denote the mass, damping and stiffness of the structure. Included in m_s is the non-moving mass of the actuator fixed to the structure as well as any changes in stiffness that occurs as a result of the presence of the actuators dead mass. Let x_1 be the coordinate associated with the vibration of the structure and let x_2 be the coordinate of the actuator moving mass, m_p. Let c_p, and k_p represent the actuators in inherent damping and stiffness respectively. Then the equation of motion of the combined structure/actuator system is

$$\begin{bmatrix} m_s & 0 \\ 0 & m_p \end{bmatrix} \ddot{x} + \begin{bmatrix} c_s+c_p & -c_p \\ -c_p & c_p \end{bmatrix} \dot{x} + \begin{bmatrix} k_s+k_p & -k_p \\ -k_p & k_p \end{bmatrix} x = \begin{bmatrix} -f_g \\ f_g \end{bmatrix} \qquad (10)$$

where $x = [x_1 \ x_2]^T$ and $-f_g$ is the control force applied by the actuator to the structure and f_g is same force applied from the structure to the actuator. The quantity f_g and the coordinate x_2 are often left out of control of flexible structure problems.

For simplicity consider a simple velocity feedback law of the form

$$f_g = g\dot{x}_1 \qquad (11)$$

where g is an electronic gain adjusted to achieve the best performance. This form is often desirable because it effectively adds damping to the structure directly suppressing vibration. Substitution of (11) into (10) yields the equivalent open loop system

$$\begin{bmatrix} m_s & 0 \\ 0 & m_p \end{bmatrix} \ddot{x} + \begin{bmatrix} c_s+c_p+g & -c_p \\ -g-c_p & c_p \end{bmatrix} \dot{x} + \begin{bmatrix} k_s+k_p & -k_p \\ -k_p & k_p \end{bmatrix} x = 0 \qquad (12)$$

Note the asymmetric appearance of the gain g in the damping matrix. This presents a source of potential instability.

Recall that systems of the form given in (1), with $f = 0$, are asymptotically stable if the coefficient matrices M, D and K are symmetric and positive definite. If one or more of these matrices is not symmetric or positive definite, then the system may or may not be asymptotically stable. In the case of the control system modeled by equation (12) the damping matrix is asymmetric. As such a fairly strong statement can be made regarding the systems stabilities. In particular, if K and M are symmetric and positive definite, then the equilibrium of equation (12) and its corresponding response is asymptotically stable if and only if the symmetric part of the damping matrix is positive definite. This is an immediate consequence of the Kelvin-Tait-Chetaev theorem (or KTC theorem) first reported by Zajac (1964, 1965) and summarized by Inman (1989) for gyroscopic systems. Applying this result to the damping matrix of equation (12) yields the desired stability criteria for the combined structure actuator model.

Recall that the symmetric part of a matrix D, denoted D_s is given by $(D^T+D)/2$ and that the skew symmetric part, denoted by D_{ss}, is given by $(D - D^T)/2$. Applying this to the damping matrix of equation (12) yields

$$D_s = \begin{bmatrix} c+c_p+g & \frac{-g}{2}-c_p \\ \frac{-g}{2}-c_p & c_p \end{bmatrix} \tag{13}$$

and

$$D_{ss} = \begin{bmatrix} 0 & \frac{-g}{2} \\ \frac{g}{2} & 0 \end{bmatrix} \tag{14}$$

If the KTC theorem states that if D_s is positive definite, then the system will be asymptotically stable. The matrix D_s is positive definite if and only if the determinants of its principal minors are all positive. Applying this to equation (13) yields the two inequalities

$$c + c_p + g > 0 \tag{15}$$
$$4cc_p > g^2 \tag{16}$$

Interpreting this in terms of the control problem provides a limitation on the gain, g. As long as g satisfies these two inequalities stable performance results.

In the control problem, the quantities c and c_p are fixed by the design of the structure and actuator. The gain g however is free to vary. The larger g is the more damping the actuator supplies to the structure and the faster and unwanted disturbance is rejected. Without considering the actuator dynamics there is no reason to limit the gain. However, it has been observed experimentally that the system becomes unstable for high gains (Inman and Zimmerman). This is well explained by equation (15) and (16) which states that if g does not satisfy these inequalities, the response may be unstable (for negative feedback (15) is always satisfied). To see that these inequalities are also sufficient conditions consider the following theorems.

If D is not positive definite, if M and K are symmetric and positive definite, then the free response of (1) is unstable. To see that this is true note that the eigenvalue problem for (1) is

$$(\lambda_i^2 M + \lambda_i D + K) \, \mathbf{u}_i = \mathbf{0} \tag{17}$$

Premultiplying by $\mathbf{u}_i^*$, the complex conjugate of the vector $\mathbf{u}_i$ and solving the λ_i yields

$$\lambda_i = -\frac{\mathbf{u}_i{}^*D\mathbf{u}_i}{2\mathbf{u}_iM\mathbf{u}_i} \pm \left(\frac{(\mathbf{u}_i{}^*D\mathbf{u}_i)^2 - 4\mathbf{u}_i{}^*M\mathbf{u}_i\mathbf{u}_i{}^*K\mathbf{u}_i}{\mathbf{u}_iM\mathbf{u}_i}\right)^{1/2} \tag{18}$$

For underdamped systems the real part of the system eigenvalue is

$$2Re\lambda_i = -\mathbf{u}_i{}^*D\mathbf{u}_i/\mathbf{u}_i{}^*M\mathbf{u}_i \tag{19}$$

as derived by Inman and Andry (1980). The system is asymptotically stable if and only if $Re\lambda_i < 0$ for all values of i. Since $\mathbf{u}_i{}^*M\mathbf{u}_i > 0$ this requires that

$$\mathbf{u}_i{}^*D\mathbf{u}_i > 0 \tag{20}$$

for all $\mathbf{u}_i$. Substitution of $D = D_s + D_{ss}$ yields that

$$\mathbf{u}_i{}^*D_s\mathbf{u}_i + \mathbf{u}_i{}^*D_{ss}\mathbf{u}_i > 0 \tag{21}$$

Since D_{ss} is skew symmetric $\mathbf{u}^*D_{ss}\mathbf{u}$ is zero for any vector $\mathbf{u}$ this yields

$$\mathbf{u}_i{}^*D_s\mathbf{u}_i > 0 \tag{22}$$

as a necessary and sufficient condition for asymptotic stability. If D_s is not positive definite, then there exists a vector $\mathbf{u}_i$ which violates inequality (22). Hence the system is asymptotically stable if and only if the matrix D_s is positive definite.

Applying this result to the control problem yields that if the inequalities given in (15) and (16) are violated, the velocity feedback system will become unstable. This provides a theoretical explanation for an experimentally observed phenomena. In addition it provides a design constraint useful for control law development.

Note that while this result is illustrated by a simple two degree of freedom example the theory is not limited by n and the result holds for any number of degrees of freedom.

CONCLUSION

This paper summarizes two sources of instability present when active control is used to suppress vibrations in flexible structures. The first is the well known problem of spillover the second is a relatively unknown result of actuator/structure interaction. A stability result and design condition is presented which prevents high gain instability in such systems.

REFERENCES

Balas, M.J. (1982), Trends in large space structure control theory: fondest hopes, wildest dreams, *IEEE on Automatic Control*, 27, 552-535.

Brumfield, M.L., P.S. Pappa, J.B. Miller and R. R. Adams (1985), Orbital dynamics of the OAST-1 solar array using video measurements, Proc. 26, AIAA, SDM Conference.

Cudney, H.H. and D.J. Inman (1989), Control formulation for flexible structures using physical coordinates, *Mechanics of Structures and Machines*, 16(4), 111-130.

Inman, D.J. (1989), *Vibration with Control Measurement and Stability*, Prentice Hall, New Jersey.

Inman, D.J. and G.C. Horner (1989), A survey of damping in flexible structures, *Damping 89*, HBA-1-HBA-5.

Inman, D.J. and D.C. Zimmerman (1988), Actuator issues in controlling flexible structures, 2nd NASA/DOD CSI Tech Conf. AWAL-TR-88-3052, 376-388.

Zajac, E.E. (1964), The Kelvin-Tait-Chetaev theorem and extensions, *J. Astronautical Science*, 11, 46-49.

Zajac, E.E. (1965), Comments on stability of damped mechanical systems and further extension, *AIAA J*, 3, 1749-1750.

A Finite Element Stress Analysis of Porous Media

R. W. LEWIS, D. V. TRAN and W. K. SZE

*Department of Civil Engineering, University College of Swansea,
Swansea SA2 8PP, UK*

1. INTRODUCTION

A stress analysis of a porous medium is normally a complex problem due to
the fact that there always exist two or more phases of material. The pores
are partially or fully saturated with a fluid phase and this strongly influences the induced stresses of the solid phase. The deformation is usually accompanied by a flow of pore fluid through the medium and vice versa.
The fluid flow is driven either by the imposed external forces in the soil
consolidation process or by the temperature gradients in timber drying and
ground freezing processes. In many cases the solid phase is strongly coupled with the fluid phase and a coupled theory is usually necessary for analysing problems of porous media.

In this paper particular attention will be focussed on three relevant processes: soil consolidation, timber drying, and ground freezing processes.

1.1 Soil consolidation

Finite element formulation of Biot's theory has been used to predict the
time dependent deformations of soil domains by Zienkiewicz et al [1], Lewis
et al [2,4]. The formulation incorporates the total stress equilibrium and
strain compatibility during consolidation. Pore fluid effects are incorporated into the equilibrium equation using the principle of effective stress.
The stress-strain relationship for the solid phase is generally nonlinear
and often time dependent. Tran [3] has studied different constitutive models
for soils in which various forms of nonlinear elasticity and plasticity were
utilised. Viscous soil properties such as creep or secondary consolidation
have also been considered using a Kelvin rheological model [5,6].

1.2 Timber drying process

The coupled heat and mass transfer processes appearing in the porous medium
can be described by Luikov's equation [7]. These are derived using the
principle of irreversible thermodynamics which provide a systematic account

17

of the phenomenon of heat and mass transfer in such material. Since these equations are extremely complex any analytical solution is restricted to simple geometry with constant material properties [8]. For realistic problems it is necessary to employ numerical techniques such as the finite element method to solve the resulting equations. Some classical papers describing the use of finite element methods for the Luikov coupled heat and mass transfer equations include those presented by Comini and Lewis [9], Thomas et al [10,11].

The change in temperature and moisture content from their initial values is often associated with the development of shrinkage stresses. The deformation and stress can be calculated using an appropriate constitutive model. The simplest relationship is that of isotropic linear elasticity and this has been examined by Lewis et al [12]. The model has also been extended to an orthotropic linear elastic model for a more realistic representation of timber behaviour by Johnson et al [13]. Later, the elastic-viscoplastic rheological model was proposed by Lewis et al [14] and Morgan et al [15,16]. This model assumes that timber behaves purely elastically before the plastic state is attained and becomes plastic/viscoplastic if that limit is exceeded. Furthermore, Lewis et al [17] proposed a combination of linear viscoelastic and nonlinear elastic-viscoplastic models to describe the physical behaviour of timber and this model will be discussed in this paper.

1.3 Ground freezing process

In addition to the drying analysis the migration of water moisture in a freezing soil system can also be described by the Luikov coupled heat and mass transfer equations. A numerical model is subsequently developed to simulate the heaving mechanism associated with soil freezing. The proposed frost heave model assumed that the movement of water moisture into the frozen region ceased to propagate on the formation of an ice lens. The corresponding accumulation of water flux and its consequent phase change volume increase generated the heaving velocity of the frozen soil mass. Below the ice lensing front is found a region where water and ice co-exist in a state of thermodynamic equilibrium. It is assumed that water flux is the only contributing source to the calculation of the magnitude of frost heave. A detailed description of this model, particularly the mathematical formulation is well documented in the papers presented by Lewis and Sze [18,19] and Lewis et al [20].

It is well known that the freezing induced mechanism (heave) in soil causes a vast amount of damage to structures placed upon it. The structure may be the foundation of a building, underground services such as gas pipes, etc., or the top of road or runway surfaces. An example is presented in this paper to demonstrate the devastating power associated with frost heave.

2.   GOVERNING EQUATIONS

The finite element formulations of the governing equations have been proposed in detail elsewhere [1,2,4,7,10] and are only briefly discussed in the following.

2.1 Soil consolidation

The governing equations are the equilibrium of total stress derived using

the principle of virtual work and the continuity equation of fluid flow
using the compatibility condition of volume change.

The equilibrium equations in incremental form are as follows

$$K_T \frac{du}{dt} - L \frac{dp}{dt} = \frac{df}{dt} + c \tag{1}$$

The continuity equations are given by

$$HP + S \frac{dp}{dt} + L^T \frac{du}{dt} = \bar{f} \tag{2}$$

where u,p are the nodal displacement and pore fluid pressure

K_T the stiffness matrix of solid phase

L   the coupled matrix between solid and fluid phase

f   the external force vector including initial effects such as stress, strain

c   the force vector due to creep effects of solid phase

H   the permeability matrix

S   the compressibility matrix

$\bar{f}$ the outflow (or inflow) of fluid phase

t   the time.

2.1.1 Initial and boundary conditions:

At $t = 0$, any prescribed u,p

and any prescribed displacement $u = U_p$ on Γ_u

any prescribed pore fluid pressure $P = P_p$ on Γ_p

2.1.2 Neuman boundary conditions:

Any imposed external forces on boundary Γ_F

and the continuity of flow across the boundary Γ_q

$$-n^T K \nabla (P + \gamma_w h) - q = 0$$

where q is outflow (or inflow) on the boundary Γ_q

2.2 Luikov coupled heat and mass transfer equations

The partial differential equations for heat and mass fluxes may be derived
by applying the methods of thermodynamics of irreversible processes. In
addition, the basis of the analysis of heat and mass transfer are two fundamental laws of nature, the law of conservation and transformation of energy
and the law of mass conservation. Based on these considerations, the coupled heat and mass transfer equations may be expressed as:

$$C_g \frac{\partial T}{\partial t} = \nabla \cdot [K_g \; \nabla T + K_\varepsilon \; \nabla U] \qquad (3)$$

for heat transfer and

$$C_m \frac{\partial U}{\partial t} = \nabla \cdot [K_\delta \; \nabla T + K_m \; \nabla U] \qquad (4)$$

for mass transfer

where T = temperature

U = moisture potential

C_g = generalised heat capacity

C_m = generalised mass capacity

K_g, K_ε, K_δ & K_m = generalised conductivities.

These coefficients may be constants or functions of the prime variables depending on three different finite element formulations, namely fully non-linear, partially nonlinear and fully linear. Detailed descriptions of these three formulations are given by Lewis et al [16].

2.2.1 <u>Boundary conditions.</u> A general set of boundary conditions for the system of equations (3) and (4) is given by:

$T = \bar{T}$ on boundary Γ_1

$u = \bar{u}$ on boundary Γ_2

$K_g \dfrac{\partial T}{\partial n} + J_g^* = 0$ on boundary Γ_3 subjected to heat flux boundary conditions

$K_m \dfrac{\partial u}{\partial n} + J_m^* = 0$ on boundary Γ_4 subjected to moisture flux

J_g^* and J_m^* represent the generalised heat and mass fluxes respectively.

3.   STRESS DETERMINATION OF POROUS MEDIA

3.1 Stress determination for soil consolidation process

The effective stresses within a soil element are determined at the Gauss points using an appropriate constitutive model. A linear elastic model can give a reasonable solution for a single-load path problem. However, in practical situations involving local failure of the soil, variable soil strength with depth and nonrecoverable strains upon unloadings, it is more appropriate to use either a variable elasticity approach or an elasto-plastic relationship [3,4].

For a general nonlinear material the constitutive equation can be written in tangential form allowing plasticity to be incorporated

$$d\sigma' = D_T(d\varepsilon - d\varepsilon^c - d\varepsilon^*)$$

where $d\varepsilon^c$ is the creep strain of the solid phase

$d\varepsilon^*$ is the hydrostatic strain caused by uniform compression of solid particles under pore fluid pressure

$d\varepsilon$ represents total strain of soil skeleton

and $d\varepsilon = d\varepsilon^e + d\varepsilon^p$ e,p denotes elastic and plastic respectively

$d\sigma'$= effective stress responsible for deformation and failure in soil.

The elasto-plastic matrix D_T can be written as:

D_T = De - Dp where De is the elasticity matrix

and $D_p = \dfrac{De\ ba^T De}{A + a^T De\ b}$ is the plasticity matrix

a,b are the yield direction and plastic flow direction respectively

A is the hardening modulus.

Depending on the type of soil and structure in geotechnical problems, an appropriate form of yield surface and plastic potential surface is required. The most popular forms are those of Von Mises, Tresca, and Mohr-Coulomb yield criterion [3]. Creep strain of solid phase is also considered using the nonlinear Kelvin rheological model by Lewis et al [5,6].

3.2   Stress determination for timber drying process

After calculating the distribution of temperature and moisture in the timber body, the resulting deformation and stresses can be determined using an appropriate constitutive relationship. The stress-strain behaviour of timber undergoing drying may be characterised by two distinct stages: in the first the short term viscoelastic creep effects are modelled through a viscoelastic rheological model. In the second stage an elasto-viscoplastic model is used to characterise the behaviour of timber for the remaining drying period.

The viscoelastic model may be expressed as:

$$\sigma = D(o)\varepsilon - \int_o^t \frac{\partial}{\partial \xi'} [D(\xi - \xi')]\varepsilon(t)\ d\xi'$$

$$-D(o)(\varepsilon_T + \varepsilon_U) + \int_o^t \frac{\partial}{\partial \xi'} [D(\xi - \xi')](\varepsilon_T + \varepsilon_U)\ d\xi'$$

where σ = the stress vector

ε = the strain vector

ξ = shifted time = $\int_o^t \phi(T,U)\ dt$

ξ' = shifted time prior to ξ

$D(o)$ = initial elastic matrix

$D(\xi - \xi')$ = viscoelastic matrix

$\varepsilon_T, \varepsilon_U$ = initial strain vector due to temperature and moisture changes respectively.

The constitutive relationship for the elasto-viscoplastic model may be expressed as:

$$\dot{\varepsilon} = \underset{\text{(elastic)}}{D^{-1}\dot{\sigma}} + \underset{\text{(plastic)}}{\langle\Psi(F)\rangle\frac{dQ}{d\sigma}} + \underset{\text{(initial)}}{\dot{\varepsilon}_i}$$

where D = elastic matrix

F = yield function such that

$$\langle\Psi(F)\rangle = \begin{cases} 0 & \text{if } F \leqslant 0 \\ \Psi(F) & \text{if } F > 0 \end{cases}$$

Q = plastic potential

$\dot{\varepsilon}_i$ = initial strain rate.

In order to obtain a solution the following general relations must be specified: (1) the elastic constitutive relation, (2) a yield criterion to indicate the onset of plastic flow, (3) a relationship to determine plastic strain. All these points are fully documented in a paper presented by Lewis et al [16].

4.   APPLICATIONS

Applications of the coupled theories which were presented previously for porous media problems have been successfully carried out on a number of occasions [4,5,6,9,11,16,18]. Due to the limited space available for this paper, several typical examples will be presented.

4.1 Soil-structure interaction

There are two ways of analysing the soil-structure interaction during consolidation. One is termed superpositions and the other a true-time dependent analysis. The superposition concept first predicts the final shape of the soil surface then calculates the interaction of the superimposed flexural member with the distorted surface. This method is simpler than the second but problems could arise due to the simplified assumptions at the interface boundary, namely perfectly smooth or rough contact. Also, in propagating solutions in the time domain these interface boundaries, with the originally assumed contacts, are no longer realistic. The second method steps forward in time and includes the interaction effects at each time step. This method is more costly than the superpositions method in terms of computer memory storage. However, it is claimed to be justified for a better physical simulation and more confident interpretation of the interaction effects [3,6].

4.1.1 Soil-pile interaction. Soil-pile interaction during consolidation has been extensively investigated by Tran [3] involving axially and horizontally loaded piles, negative skin friction, bearing capacity, and a non-linear soil. A simple problem is presented here.

A concrete pile embedded in soft clay is axially loaded with 200 kN. The problem is analysed assuming axisymmetric conditions. The mesh configuration is shown in Figure 1. The pile is assumed to have linear elastic properties since the pile stiffness is very high compared to that of the

soft clay. The clay is assumed to have an elasto-plastic stress-strain relationship. The interface between the pile and the soil is modelled using a
nonlinear hyperbolic relation. Detailed descriptions of material properties
can be obtained from Reference [3]. Results of the pile tip settlement are
shown in Figure 2. Pore water pressures generated within the soil mass and
effective stress plots for a typical time step are shown in Figures 3 and 4.
In many of the cases which have been studied [3], the interface conditions
between the pile and soil play an important role in pile analysis and design.

4.1.2 Framework-soil interactions. Superstructure frames are convention-
ally analysed assuming their bases to be either completely rigid or hinged.
However, the foundation resting on deformable soil (clay) also undergoes
time dependent settlements, rotations depending on the relative rigidity of
the foundation, the superstructure and the soil. Interactivity is therefore
necessary for correctly assessing the member forces in the superstructure.
The problem of a single bay frame on soft soil is presented. The mesh configuration is shown in Figure 5. Details of material properties can again
be obtained from Reference [3]. The frame is loaded vertically or a combination of vertical and horizontal loads. The results of a vertically loaded
frame for some time steps are shown in subsequent Figures 6, 7 and 8.

In general, the differential settlements due to soil consolidation often
provoke an increase of critical stresses and lead to a redistribution of
stresses in the members of the frame.

4.2   Heave induced stress during ground freezing

A practical example is presented here to demonstrate the effect of a concrete footing subject to differential heave. The structure is assumed to be
located at the centre between two different types of soils: namely sand and
silt which are known to have different frost susceptible behaviours. In
order to simplify the problem the location of the ice lens is assumed to be
at a fixed level below the concrete footing. As a result the soils sand-
wiched between the footing and the ice lens are already frozen from the onset of the simulation and the problem analysed is illustrated in Figure 9.

The frost heave mechanism of sand and silt is determined by a new ground
heave model which is fully documented in the following papers presented by
Lewis and Sze [19], Lewis et al [20] and Sze [21]. According to previous
analyses of frost heave for the above types of soils, the heaving magnitude
at two hours, during the freezing period, is 0.4 mm and 8.0 mm for the sand
and silt respectively.

The example analysed is treated as an elastic plane strain problem with the
geometry and material properties illustrated in Figure 9. The corresponding
constitutive relationship and finite element solution is well presented by
Zienkiewicz [22] and will not be repeated here. The stress induced in the
concrete footing during the ground freezing process is illustrated in Figures 10 and 11. These are respectively the minor and major principle stress
contours evaluated at the Gauss points. The result has clearly illustrated
the weak point in the concrete footing caused by the differential heave
resulting from the ground freezing phenomenon.

4.3   Shrinkage stress in timber drying

The problem analysed is a 200 mm x 50 mm section of timber, as shown in
Figure 12, subjected to drying conditions on its four faces. This section
is chosen since it is considered representative of the size and type of problem generally dealt with by the timber drying industry. Hence, the results
obtained for the heat and mass transfer distribution in the section may be
considered relevant when compared with experimental results. Considering
the symmetry of the problem, only one quarter of the section need be analysed. The finite element mesh of the resulting 100 mm x 25 mm quarter section is also shown in Figure 12.

The overall picture of the stress development for the case of an orthotropic
elastic and plastic body, with the extension of the isotropic Von Mises
theory to include orthotropic properties, is shown in Figures 13, 14, 15 and
16 in terms of the principle stresses at 2.4 hr, 9.5 hr, 21.3 hr and 59.5 hr
respectively. The results obtained clearly show the phenomenon of stress
reversal taking place during timber drying

5.   GENERAL CONCLUSION AND FUTURE DEVELOPMENT

A finite element stress analysis of porous media has been presented. Typical processes such as timber drying, soil consolidation and ground freezing
are analysed. In many cases the solid phase is strongly coupled with the
fluid phase and the coupled theories proposed previously are necessary for
three stress analyses of porous media.

For soil consolidation problems a true time interaction analysis using the
coupled equations is very useful in design practice. Important factors such
as critical stresses, contact pressure, displacement and pore water pressure
field distribution can be obtained explicitly from the finite element program.

Other complexities, namely the initial stress effects, the material nonlinearities, creep and temperature effects in the soil mass or structure can be
easily incorporated into the analysis. In principle, it is now possible to
carry out a fully three-dimensional analysis but the cost of such analyses
is generally so high as to preclude their application to practical design
problem associated with the majority of structures. Some recent developments have been investigated in order to overcome the limitation. Soil
regions of large of infinite extent can be represented by a boundary of infinite elements or boundary elements.

Interaction techniques for solving a large system of equations, such as the
"staggered" method, element by element method have been investigated and the
possibility of their applications in parallel machines has been suggested.
Other physical processes such as oil recovery and the stability of oil plat-
forms can be studied by incorporating the concept of multi-phase fluid flow
into the consolidation theory.

For frost heave problems the results obtained from the stress contours
clearly show the behaviour of a raft foundation subjected to the differential heave in a ground freezing problem. The physical characteristics of
the frozen soil are not as simple as assumed in the example. However, this
demonstrates the cornerstone of a research program currently under investigation by the authors for the stress behaviour within frozen soil. Particular interest is focussed on the stability in the design of a frozen

excavation in conjunction with artificial ground freezing techniques. In addition, the overall picture of the stress developed in a timber drying example is given and where the important physical phenomenon of stress reversal is clearly depicted by the proposed model.

REFERENCES

1. Zienkiewicz, O.C., Lewis, R.W. and Humpheson, C. A unified approach to soil mechanics problems including plasticity and viscoplasticity. Finite Elements in Geomechanics, Ed. Gudehas, Wiley and Sons, Chapt.4, 151-177.

2. Lewis, R.W., Roberts, G.K. and Zienkiewicz, O.C. A nonlinear flow and deformation analysis of consolidated problems. Numer. Meth. in Geomech., Ed. Desai, C.S., 1106-1118.

3. Tran, D.V. A finite element analysis of soil-structure interaction during consolidation. Ph.D. thesis, University College of Swansea, U.K., 1988.

4. Lewis, R.W. and Schrefler, B.A. A fully coupled consolidation model of subsidence of Venice. Water Resources Research, Vol. 14, No. 2, 223-229, 1978.

5. Lewis, R.W. and Schrefler, B.A. The finite element method in the deformation and consolidation of porous media. John Wiley and Sons, 1987.

6. Lewis, R.W. and Tran, D.V. Numerical simulations of secondary consolidation of soil : Finite element application, Int. J. Num. Ana. Meth. Geomechanics, Vol. 13, 1-18, 1989.

7. Luikov, A.V. Systems of differential equations of heat and mass transfer in capillary-porous bodies (Review), Int. J. Heat Mass Transfer, Vol. 18, 1-14, 1976.

8. Mikhailov, M.G. Exact solution for freezing of humid porous half-space, Int. J. Heat Mass Transfer, Vol. 19, 651-655, 1976.

9. Comini, G. and Lewis, R.W. A numerical solution of two-dimensional problems involving heat and mass transfer, Int. J. Heat Mass Transfer, Vol. 19, 1387-1392, 1976.

10. Thomas, H.R., Lewis, R.W. and Morgan, K. An application of the finite element method to the drying of timber, Wood and Fibre, Vol. 11, 237-243, 1980.

11. Thomas, H.R., Morgan, K. and Lewis, R.W. A fully nonlinear analysis of heat and mass transfer problems, Int. J. Num. Meth. Engng., Vol. 15, 1381-1393, 1980.

12. Lewis, R.W., Strada, M. and Comini, G. Drying induced stresses in porous bodies, Int. J. Num. Meth. Engng., Vol. 11, 1175-1184, 1977.

13. Johnson, K.H., White, I.R., Lewis, R.W. and Morgan, K. The analysis of drying-induced stresses in wood, Proc. 2nd Int. Conf. Appl. Num. Modelling, Madrid, 1978.

14. Lewis, R.W., Morgan, K. Thomas, H.R. and Strada, M. Drying induced stresses in porous bodies - an elasto-viscoplastic model, Comp. Meth. Appl. Mech. Engng., Vol. 20, 291-301, 1979.

15. Morgan, K., Lewis, R.W. and Thomas, H.R. Numerical modelling of drying induced stresses in porous bodies, in "Developments in Drying", Science Press, New York, 1979.

16. Morgan, K., Thomas, H.R. and Lewis, R.W. The numerical modelling of stress reversal in timber drying, Wood Science, Vol. 15, 139-149,1982.

17. Lewis, R.W., Srinatha, H.R. and Thomas, H.R. A finite element study of the drying stresses in timber using viscoelastic and elasto-viscoplastic rheological models, Chapter 5 in "Numerical Methods in Coupled Systems", Ed. Lewis et al, John Wiley and Sons, 1984.

18. Lewis, R.W., Sze, W.K. and Roberts, P.M. A finite element investigation into heat and mass transfer in porous materials with particular reference to ground freezing, Keynote lecture presented at Drying 86, Boston, August, 1986, Vol. 1, 21-29, Hemisphere Press.
19. Lewis, R.W. and Sze, W.K. A finite element simulation of frost heave in soils, Proc. 5th Int. Symp. on Ground Freezing, Nottingham, England, 1988.
20. Lewis, R.W., Sze, W.K. and Huang, H.C. Some novel techniques for the finite element analysis of heat and mass transfer problems, Int. J. Num. Meth. Engng., Vol. 25, 611-624, 1988.
21. Sze, W.K. Finite element method in ground freezing problems, Ph.D. thesis, University College of Swansea, U.K., 1989.
22. Zienkiewicz, O.C. The Finite Element Method (3rd Edition), McGraw-Hill, 1977.

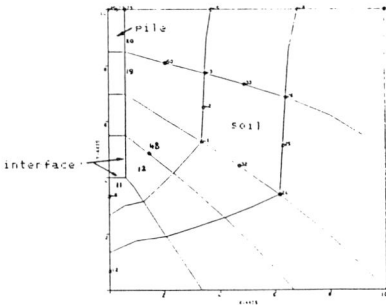

Figure 1

The original mesh, 24 elements,
85 nodes (15 soil els., 5 interface els., 4 pile els.)

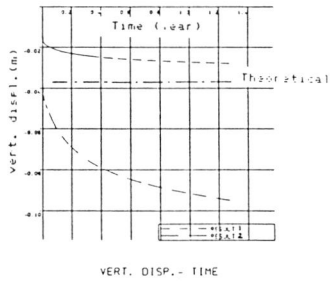

Figure 2

Vert. displacement of pile
head comparison
1 : rough pile   2 : smooth pile

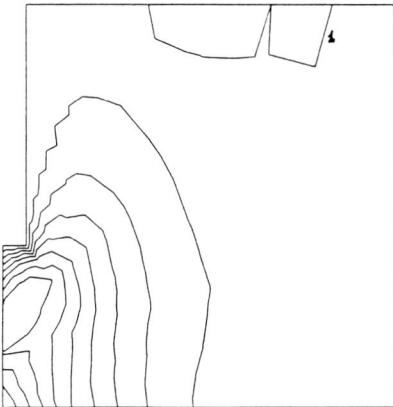

10 contours
interval 0.93
1^{sf} contour at 0.0

Figure 3

P.W.P. contours in soil mass at
time step 30 rough interface pile

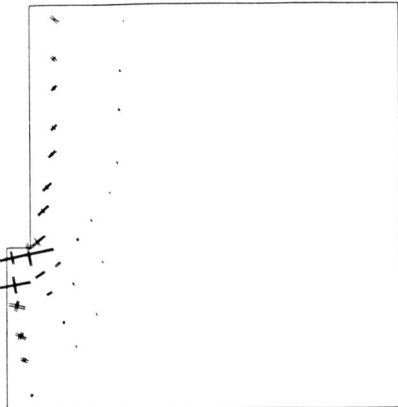

PRINCIPAL STRES. AT STEP 30

Figure 4

Effective principal stresses in
soil at time step 30 (1.3 years),
rough pile, scale = 50 kPa/mm

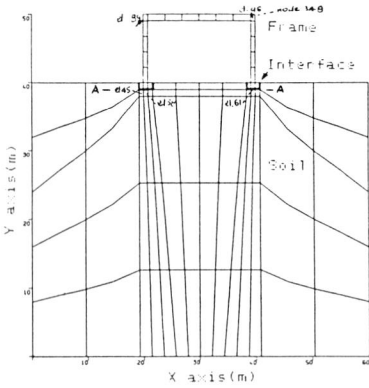

Figure 5

The original mesh
Total 96 elements, 348 nodes

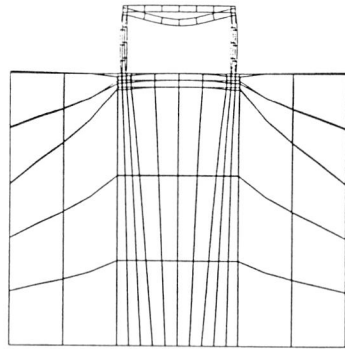

Figure 6

The deformed mesh at time step 16
(t = 2 years), Disp. scale = 0.004/mm

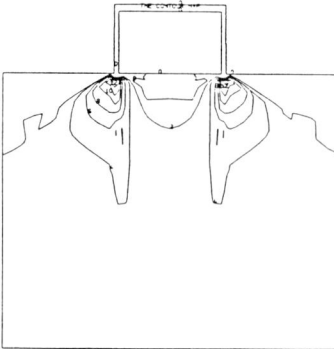

Figure 7

The P.W.P. contours at
time step 16 (t = 2 years)

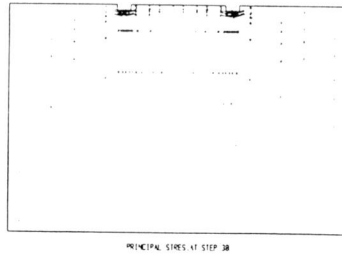

Figure 8

Effective principal stresses
in soil mass at time step 30
scale = 50 kPa/mm

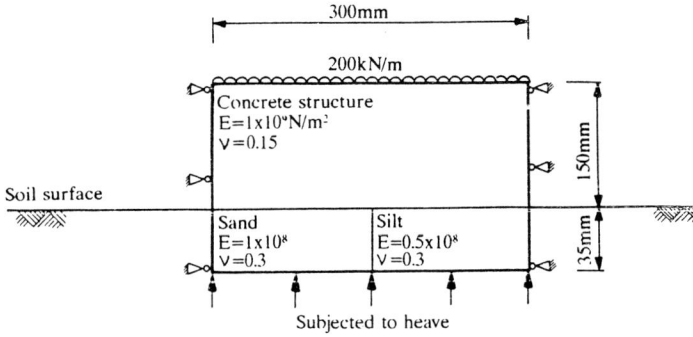

Figure 9(a)   Soil/structure interaction due to frost heave

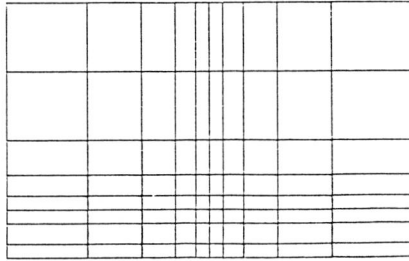

Figure 9(b)   Finite element mesh

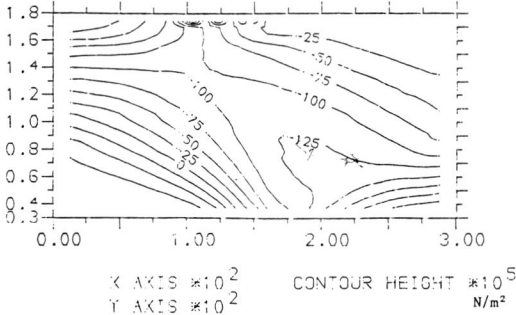

X AXIS $*10^2$ CONTOUR HEIGHT $*10^5$
Y AXIS $*10^2$ N/m^2

Figure 10   Minor principal stress contour for
the concrete foundation

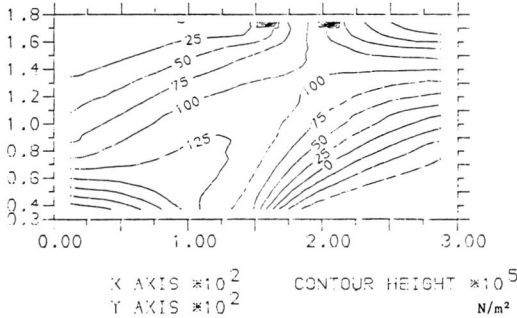

X AXIS $*10^2$ CONTOUR HEIGHT $*10^5$
Y AXIS $*10^2$ N/m^2

Figure 11   Major principal stress contour for
the concrete foundation

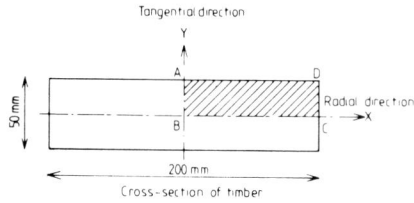

Figure 12 Finite element discretization of timber drying problem.

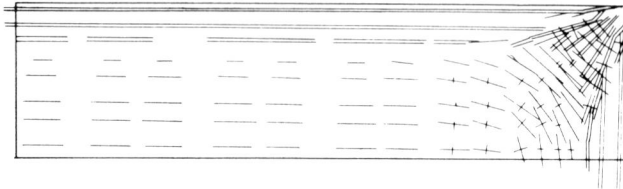

Figure 13 Principal stress distributions at 2.4 h – orthotropic elasticity and plasticity.
Scale: 1 mm = 200 kN/m². ═══, tension; ——, compression.

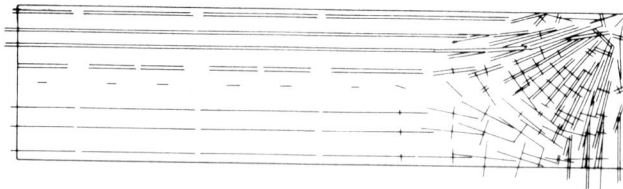

Figure 14 Principal stress distributions at 9.5 h – orthotropic elasticity and plasticity.

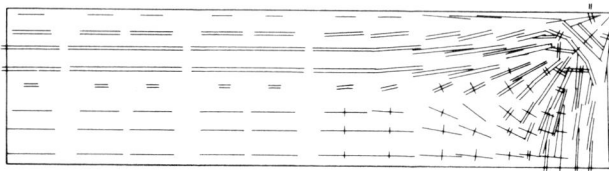

Figure 15 Principal stress distributions at 21.3 h – orthotropic elasticity and plasticity.

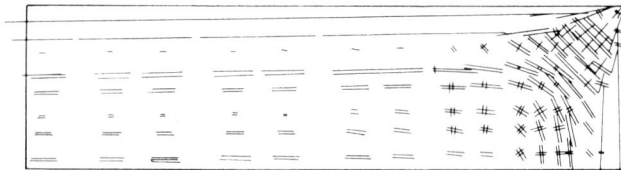

Figure 16 Principal stress distributions at 59.5 h – orthotropic elasticity and plasticity.

Stress Analysis using Boundary Elements

R. T. FENNER

Department of Mechanical Engineering,
Imperial College of Science, Technology and Medicine,
London SW7 2BX, UK

ABSTRACT

Although finite element techniques remain the most commonly used methods of numerical stress analysis, boundary element methods offer significant advantages for at least some types of problems, particularly linear ones. Perhaps the most important of these advantages is the ease of mesh data preparation. In two-dimensional problems, only simple line elements are required, for which data are easily generated. In three-dimensional problems, only the two-dimensional boundary surfaces have to be discretised into elements, a task very similar to that encountered in two-dimensional finite element analysis. Also, using boundary element methods, much less unwanted information is generated. In the vast majority of stress analysis problems, the maximum stresses occur on the boundary, and it is unnecessary to compute internal values. Boundary element methods are usually substantially more economical of both computing time and storage than other numerical methods. In order to achieve this economy, however, it is important that boundary element methods are carefully programmed, that efficient types of element are used, and that the mesh design is appropriate. For example, a boundary element mesh should not be simply a finite element mesh with the internal nodes and elements removed: a much coarser boundary element mesh is capable of giving results of the same accuracy. Another major advantage of boundary element methods is that they offer continuous and accurate modelling within the region of interest, giving high resolution of stresses and displacements. This makes them particularly good for solving stress concentration and crack problems.

The boundary element approach, which eliminates the need to solve for unknowns within the region of interest, represents a significant refinement of domain methods such as finite difference and finite element techniques. In many practical problems, however, it is not necessary to determine even all the unknowns on the boundary, but only those at particular geometric features. This can be done very efficiently using a local boundary element technique based on series expansions of the unknowns. A further important application of the boundary element method is to design optimisation. As the shape of a component is changed, the rates of change of the boundary stresses can be directly computed and the optimum shape determined in a small number of iterative steps.

KEYWORDS

Boundary elements, boundary integral equations, numerical methods, stress analysis

INTRODUCTION TO BOUNDARY ELEMENTS

The boundary element method (BEM) is an efficient numerical solution technique for a wide range of continuum mechanics problems, although attention is confined here to static stress analysis of solids. The BEM, as its name suggests, is a *boundary* method, in which only the boundary of the region of interest is discretised. This

31

is in contrast to *domain* techniques such as finite difference and finite element methods (FEM), where not only the boundary but also the interior of the region must be discretised. It is worth reviewing briefly how this simplification is possible.

Theoretical Formulation

The usual basis for developing the formulation for boundary element stress analysis is Betti's reciprocal theorem for a linearly elastic system, which can be derived from the principle of virtual work. Suppose that σ_{ij} and e_{ij} represent one set of distributions of stresses and strains which satisfy equilibrium, compatibility and Hooke's law (but no particular boundary conditions), and that $\sigma_{ij}*$ and $e_{ij}*$ represent another independent set which satisfy the same conditions. Then

$$\int_V \sigma_{ij} e^*_{ij} \, dV = \int_V \sigma^*_{ij} e_{ij} \, dV$$

where V is the volume of the region of interest. It is convenient to let the unstarred quantities represent the as yet unknown solution to the problem in hand, and the starred quantities a particular known elasticity solution, often referred to as the *fundamental* solution. The particular solution which is usually selected is that associated with a point force in an *infinite* domain, for which analytical expressions exist (see, for example, Brebbia and Dominguez, 1989, or Cruse, 1988). A necessary feature of the fundamental solution is that it is singular (stresses become infinite at the point of application of the load). Provided the unknown solution satisfies the reciprocal relationship, it also exactly satisfies equilibrium, compatibility and Hooke's law everywhere within the chosen region, and must be made to satisfy the boundary conditions.

After some manipulation, which takes advantage of the singular nature of the fundamental solution, the following result is obtained (Cruse, 1988)

$$u_i(p) = \int_S U_{ij}(p,Q) \, t_j(Q) \, dS - \int_S T_{ij}(p,Q) \, u_j(Q) \, dS$$

In this equation, which is known as the Somigliana identity for displacements, the u_i are the components of displacement in the problem solution, and the t_i are the components of traction acting on the surface S of the region of interest (traction is the resultant force per unit surface area). The symbol p (lower case) indicates a point within the region of analysis, while Q (upper case) indicates one on the boundary. The *kernel* functions $U_{ij}(p, Q)$ and $T_{ij}(p, Q)$ representing the fundamental solution are respectively the displacement and traction components in the x_i coordinate direction at point Q due to a unit point load in the x_j direction at point p.

The significance of the Somigliana identity for displacements is that it defines the displacements at any internal point in terms of boundary data (displacements and tractions) only, and a similar form of identity can be obtained for the stresses at internal points. Now, if the point p is taken to the boundary, the following *boundary integral equation* (BIE) is obtained

$$C_{ij} u_j(P) + \int_S T_{ij}(P,Q) u_j(Q) \, dS = \int_S U_{ij}(P,Q) \, t_j(Q) \, dS$$

The constant C_{ij} depends on the local geometry of the boundary at P (whether it is smooth there or whether P is at a sharp corner). This is now an equation linking boundary data only.

Boundary Discretisation

The boundary can be discretised in some way, such that the distributions of displacements and tractions are represented in terms of values at a finite number of points, together with appropriate interpolation functions for intermediate positions. If P is located at each of the points in turn, the BIE allows a set of linear algebraic equations involving the displacements and tractions at these points to be generated by numerical integration. At any given point either displacements or tractions are known from the boundary conditions of the problem, or there may be a linear relationship between them. The remaining unknowns are found by solving the equations numerically.

The BIE provides a *direct* way for calculating unknown boundary data from known boundary data, working in terms of the physical quantities displacement and traction. There are also *indirect* methods, which work in terms of fictitious density functions over the boundary, but which are generally unsuitable for typical engineering problems involving corners (and possibly cracks).

The usual way of discretising the boundary is by means of *boundary elements*, with characteristics similar to finite elements. For example, a commonly-used type of boundary element for two-dimensional problems is a three-noded line element which employs quadratic interpolation functions to represent both boundary geometry and the variables (displacements and tractions). The required integrations over the elements are usually performed by Gaussian quadrature. Careful programming is important to ensure that the singular kernel functions are accurately integrated, and that the calculations are arranged in the most efficient manner.

Solution of Equations

The resulting set of linear algebraic equations is rather different in form from those obtained in finite element or finite difference methods. In general it is fully populated with non-zero coefficients, and it is not symmetric. Both of these properties make a set of equations of a given size more expensive to store and solve. On the other hand, since only the boundary of the domain is discretised and the number of nodes at which unknowns must be found is relatively small, the number of equations is much reduced. A further point in favour of boundary elements is that because of their greater inherent accuracy it is not necessary to discretise even the boundary as finely as with, say, finite elements. In other words, a boundary element mesh should not be simply a finite element mesh with the internal nodes and elements removed: a much coarser boundary element mesh is capable of giving results of the same accuracy. In most cases, the overall result of these interacting influences on relative storage requirements and solution time is that boundary element methods are substantially more economical than finite element methods (Fenner, 1983). Although savings do depend on the type of problem involved, practical experience suggests that storage requirements and solution times can be cut by a factor of the order of three.

Choice of Elements

These savings are only achieved, however, if efficient types of boundary element are used. For example, straight line or flat surface elements, over which the variables are assumed either to be constant or to vary linearly with position, are generally more expensive to use than finite elements. Elements with shape functions which are at least quadratic in form should be used. Also, there has in the past been a tendency to use "discontinuous" elements (Brebbia & Dominguez, 1989). Instead of locating common nodes on inter-element boundaries, the nodes are duplicated and located just inside each of the elements involved. The main purpose of doing this is to deal with traction discontinuities at sharp corners. If these elements are used generally, however, the total number of nodes, and hence equations to be solved, is dramatically increased. For example, for quadratic line elements, the number of nodes is increased by 50%, while for quadratic quadrilateral surface elements the increase is 167%. With solution costs proportional to at least the square of the number of nodes, the potential advantage of boundary elements over finite elements is turned into a disadvantage.

Post-Processing of Results

Having solved the linear equations for the boundary nodal point values of displacement and traction, the stresses and displacements can be found at any internal point (except very close to the boundary) using the Somigliana identities. This means that internal values, which are determined very accurately, can be freely explored in a post-processing operation. On the other hand, since boundary integrations are required for each internal point, the cost of obtaining the data can become substantial if information at a large number of points is required.

Special Features of Boundary Elements

There are several features of boundary element methods which have no counterpart in finite element or finite difference methods. Boundary elements can be applied to problems with boundaries at infinity without having to introduce artificial finite boundaries for discretisation purposes. This is because the fundamental solution is valid for an infinite domain, which ensures that the reciprocal solution is similarly valid. The ability to model infinite or semi-infinite domains is very important in many mining and civil engineering applications involving, for example, structures and the ground on which they stand. It is also important in problems of fluid/structure interaction, such as ships in water.

Another unique feature of boundary element methods is that it is not necessary to discretise lines or planes of symmetry. Figure 1 shows the familiar problem of a flat plate in tension with a hole at its centre. By invoking symmetry, only one quarter of the plate need be modelled: AE and CD are lines of symmetry. Boundary elements need only be placed on the other boundaries AB, BC and DE. Integrations are still carried out over the full boundaries of the original plate using boundary information which is mirrored in the lines of symmetry. Alternatively, AE and CD can be discretised into boundary elements, subjected to symmetry boundary conditions, and integration confined to the region ABCDE. The choice between these two approaches for a given problem depends on the relative lengths of symmetry lines and other boundaries. If the symmetry lines form only a small proportion of the total boundaries, the saving in the number of elements, and hence nodes and equations, by not discretising them is more than offset by the extra computation involved in integrating over the entire boundary.

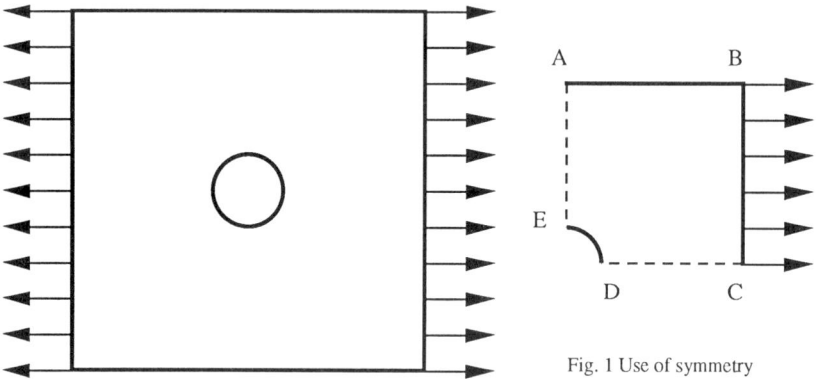

Fig. 1 Use of symmetry

In axisymmetric problems, where the solution domain is a body of revolution about an axis of symmetry it is again unnecessary to model this axis when analysing a typical radial plane through the domain. Figure 2 shows an extreme example of the simplifications which are possible due to both symmetry and infinite boundaries. Two (quadratic) boundary elements are used to model the three-dimensional problem of a spherical cavity in an infinite body, subject to, say, uniform internal pressure. FG and GH are the line elements, each subtending an angle of 45^0 at the centre O of the sphere. OF may be regarded as the axis of symmetry, and OH is another line of symmetry in a typical radial plane through the axis.

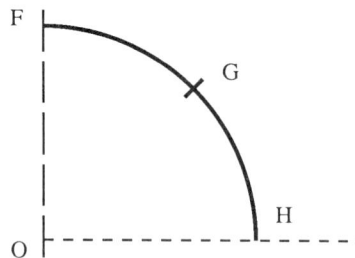

Fig. 2 A one-element mesh

Another feature of boundary element methods, which is not true of conventional displacement-based finite element methods, is that they can solve problems involving incompressible or nearly incompressible materials without difficulty (Fenner & Remzi, 1983). This is particularly relevant to rubber materials.

Limitations of Boundary Elements

A major limitation of boundary element methods is that they are unsuitable for regions with high aspect ratios (ratios of length to width). This is because of the difficulties of maintaining adequate accuracy of integration when the load point (P in the BIE) is close to, but not on, a boundary. As a general rule, no such point should be within about one boundary element length of the boundary. This means that for a long slender region, the elements along the sides should not be longer than the width of the region, which can demand a large number of elements. Under these circumstances, the cost advantages over finite element methods are lost. Boundary element methods are unsuitable for regions with aspect ratios greater than about 10 to 1, and are therefore unsuitable for many structural problems.

One way in which this limitation can be overcome to some extent is by the use of subregions of relatively small aspect ratio. BIEs are applied to each subregion independently, which are then linked along their interfaces by conditions of continuity of displacements and tractions. While the use of subregions further increases the number of boundary elements required, it does offer the advantages of being able to define different material properties for each of the regions, and of introducing a substantial proportion of zeros in the equation coefficient matrix. It is also particularly beneficial for some types of crack problems.

Another feature of boundary element methods which can sometimes prove to be inconvenient in practice is the fact that point forces, which imply infinite stresses, cannot be applied. This means not only that all applied forces have to be distributed over finite regions of the boundary, but also that it is not possible to impose a displacement constraint at a point which would require a point force there to maintain.

Application to Fracture Mechanics Problems

The fact that the boundary element methods offer continuous and accurate modelling within the region of interest, giving high resolution of stresses and displacements, makes them particularly suitable for stress concentration and crack problems. Indeed, crack problems are one of their most important application areas, which has received a lot of attention (Cruse, 1988). At first sight, boundary element methods are unsuitable for crack problems which are not symmetrical about the crack, because both sides of the crack must then be modelled and their coincidence or very close proximity would cause difficulties due to load points on one side being very close to the other side of the crack. Various ways of overcoming this difficulty have been devised (Cruse, 1988), the most obvious of which is to make the crack form part of the boundary between two subregions.

Mixing Boundary Elements and Finite Elements

There is considerable scope for using a mixture of boundary elements and finite elements in certain types of problems. For example, a structure built on the ground: the structure can be modelled using finite elements, and the ground using boundary elements. This takes advantage of boundary elements' ability to easily treat semi-infinite domains, and finite elements' ability to analyse long slender structural members. A further example is provided by the problem of a crack in a large complex structure such as an oil rig. While finite elements can be used for the structure as a whole, the region around the crack is better modelled using boundary elements.

ADVANTAGES AND DISADVANTAGES OF BOUNDARY ELEMENTS

The main advantages and disadvantages of boundary element methods when compared with domain methods of the finite element or finite difference types, may be summarized as follows.

Advantages

(i)   Perhaps the most important practical advantage is the ease of mesh data preparation. In two-dimensional (linear) problems, only a relatively small number of simple line elements are required, for which data are easily generated. In three-dimensional problems, only the two-dimensional boundary surfaces have to be modelled with relatively few elements. This makes boundary elements particularly attractive for use in computer aided design systems, where geometric data is most readily available in terms of surface geometry.

(ii)   Using boundary elements, much less unwanted information is obtained. For example, in the vast majority of stress analysis problems, the maximum stresses occur on the boundary, so that it is unnecessary to compute internal stresses. If internal stresses are required, however, they can be found at any points, which can be selected after the boundary solutions have been obtained.

(iii)   Boundary element methods are usually significantly more economical of both computing time and storage. This is particularly important for three-dimensional problems.

(iv)   The boundary element approach offers continuous interior modelling within the solution domain of the problem, giving high resolution of stresses and displacements. This is particularly beneficial for stress concentration and fracture mechanics problems.

(v)   Boundary element methods are particularly attractive for problems involving infinite or semi-infinite regions.

(vi)   Incompressible and nearly incompressible material behaviour can be accommodated without difficulty, in sharp contrast to displacement-based finite element methods.

Disadvantages

(i)   Very little good boundary element software is commercially available.

(ii)   Boundary element methods require a relatively sophisticated level of mathematics for their formulation, of a type which is unfamiliar to most stress analysts.

(iii)   While boundary elements are good for physically compact problems having small surface-to-volume ratios, slender structural type components may be better modelled by other methods.

(iv)   Boundary element methods are unsuitable if information is required at a large number of internal points. At least for stress analysis problems, however, such information is rarely required.

(v)   Boundary element methods are, in general, much less suitable for nonlinear problems than for linear ones.

NONLINEAR PROBLEMS

The suitability of boundary element methods for nonlinear problems depends on the type of nonlinearity involved. Nonlinearities can be introduced by geometric, material property or boundary condition effects. Geometric nonlinearities arise in problems in which the strains are sufficiently large either to invalidate the small strain assumptions on which conventional elasticity theory is based, or to give rise to displacements which significantly affect the size and shape of the solution domain. Nonlinear material behaviour can take the form of nonlinear elasticity or of plasticity. Nonlinearities due to boundary conditions arise, particularly in regions of high local contact stresses between two bodies, when the extent of the part of a boundary subjected to applied load varies with the magnitude of the load.

While boundary element methods are particularly suitable for nonlinear contact problems (see, for example, Karami & Fenner, 1986), the same is not necessarily true of other nonlinear problems. This is because in elastic contact problems it is still only necessary to integrate over the boundaries of the bodies concerned. In problems involving material or geometric nonlinearities, however, the BIEs retain terms involving integration not only over the boundaries but also over the interior of the domain. Although it is still not necessary to solve for values of the unknowns at interior points, the need to integrate over the interior significantly increases the cost of the analysis. Although a great deal of effort has gone into developing boundary element methods for many types of nonlinear problems, the relative advantages over finite element methods are much less clearcut than for linear problems. At the very least, they do, however, provide a means of independently checking finite element results.

CURRENT AND FUTURE DEVELOPMENTS

Much current work on boundary element development is aimed at further improving the accuracy and efficiency of the method. For some time the methods of boundary discretisation have been very strongly influenced by the approaches adopted for finite elements, when it is not clear that this is the necessarily the best way to go. With BIEs offering highly accurate interior modelling, the accuracy of boundary element methods is crucially dependent upon the treatment of the boundaries. For example, isoparametric elements with polynomial shape functions, although quite satisfactory for general purpose interpolation, do suffer from some significant disadvantages. In particular, they result in boundary geometries (especially circular arcs which are very common in practical problems) which are known exactly being unnecessarily approximated, and artificial discontinuities of slope are introduced at the inter-element boundaries. These approximations are one of the major sources of error in boundary element analysis, but are comparitively straightforward to correct. Apart from modelling the exact geometry, the slope discontinuity problem can be overcome by using elements which equate slopes at inter-element boundaries on smooth surfaces, such as Hermitian cubic elements (Watson, 1982). If the exact boundary geometry is retained in the analysis, the need for elements in the conventional sense is greatly reduced, and other more convenient ways of interpolating between discrete points chosen on the boundary are likely to be developed.

The boundary element approach eliminates the need to solve for unknowns within the region of interest. In many practical problems, however, it is not necessary to determine even all the unknowns on the boundary, but only those close to particular geometric features. For example, in stress concentration and crack problems, only the peak stress or stress intensity factor is usually required. This selection of local information can be done very efficiently using a local boundary integral equation technique based on series expansions of the unknowns, and results in very small sets of equations to be solved. So far, details of this technique have only been published for two-dimensional potential problems (Fenner & Watson, 1988), although some work on elastostatics is now complete (Askin, 1989). It also provides a means of locally refining approximate solutions obtained by boundary element or other methods.

An important practical application of the boundary element method is to design optimisation. Given a particular design criterion, such as minimum weight or limiting value of maximum stress, an iterative procedure can be set up to progressively improve the shape of a component until the optimum design is obtained. There are two reasons why boundary elements are especially suitable for this purpose. Firstly, because only the boundary is discretised, quite large changes in the geometry can be accommodated without having to redesign the mesh of elements. Secondly, a differentiated form of the BIE can be used to determine directly the required rates of change of the variables with changes in position of the boundary nodes (Moghaddasi Tafreshi, 1989).

REFERENCES

Askin, S. (1989). Local Boundary Integral Equation Analysis of Stress and Potential Problems. PhD Thesis, University of London.

Brebbia, C.A. and Dominguez, J. (1989). Boundary Elements: An Introductory Course. McGraw-Hill, London.

Cruse, T.A. (1988). Boundary Element Analysis in Computational Fracture Mechanics. Kluwer, Boston.

Fenner, R.T. (1983). The Boundary Integral Equation (Boundary Element) Method in Engineering Stress Analysis. *J. Strain Analysis*, 18, 199-205.

Fenner, R.T. and E.M. Remzi (1983). Boundary Integral Equation Evaluation of Compression Moduli of Bonded Rubber Blocks. *J. Strain Analysis*, 18, 217-223.

Fenner, R.T. and Watson, J.O. (1988). A Local Boundary Integral Equation Method for Potential Problems. *Int. J. Num. Meth. Engng.*, 26, 2517-2529.

Karami, G. and R.T. Fenner (1986). Analysis of Mixed Mode Fracture and Crack Closure Using the Boundary Integral Equation Method. *Int. J. Fract.*, 30, 13-29.

Moghaddasi Tafreshi, A.M. (1989). Design Optimisation using the Boundary Integral Equation Method. PhD Thesis, University of London (in preparation).

Watson, J.O. (1982). Hermitian Cubic Boundary Elements for Plane Problems of Fracture Mechanics. *Res Mechanica*, 4, 23-42.

Algorithms for the Rapid Computation of Response of an Oscillator with Bounded Truncation Error Estimates

V. G. COLVIN and T. R. MORRIS

SDRC Engineering Services Ltd., The Genesis Centre,
Science Park South, Birchwood, Warrington,
Cheshire WA3 7BH, UK

ABSTRACT

In this paper an algorithm is developed for the accurate and efficient computation of the response of a linear SDOF, subjected to an arbitrary excitation, known deterministically as a set of discrete ordinates sampled uniformly in time. Bounded local truncation error estimates have been established, and indicate that the algorithm has the same order of relative error as currently published methods, whilst requiring far fewer multiplications per time step.

Furthermore, two recursive algorithms have been developed for the rapid computation of the response spectrum defined uniquely by the given exciting time history. Both algorithms provide accurate estimates of response ordinates in sensitive regions of the spectra, while avoiding the calculation of large numbers of spectral values in areas showing only minor irregularities. Estimates of error bounds have been derived for each method.

Finally, results from the algorithms presented in this paper have been compared to closed solutions for SDOF having homogeneous initial conditions. The results indicate a good agreement within the expected tolerance.

KEYWORDS

SDOF; time history; response spectrum; oscillator.

INTRODUCTION

One of the problems most central to the field of engineering dynamics, is the efficient calculation of the response of a linear single-degree-of-freedom system, subjected to an arbitrary forcing function. The modal decomposition of large sets of linear differential equations, coupled with the solution of the SDOF problem, provides one of the most powerful tools available to the analytical engineer.

39

The methods for step by step integration most commonly used over the past few years (Nigam & Jennings, 1969. Newmark, 1959. Bathe & Wilson, 1973 etc.) require that more than one response is evaluated at each time step, chosen from displacement, velocity or acceleration. For example, Nigam and Jennings technique evaluates both displacement and velocity at each time step. More recently, a method of solution obviating the need to compute unrequired responses has been developed (Beck & Dowling, 1988), which allows any desired response to be calculated independently of the other two. There are numerous advantages to this method, not least that it requires only five multiplications per time step and would appear to possess favourable stability and accuracy characteristics.

This paper presents an integration algorithm which extends the Beck & Dowling method, greatly reducing the computational effort whilst maintaining excellent stability and accuracy properties. Algorithms, which then utilise the integration scheme for the calculation of the response spectra extensively used in seismic engineering, are then developed. These use recursive binary search and cubic spline interpolation techniques to reduce greatly the number of spectral ordinates required for calculation whilst maintaining a specified accuracy. The spectra show a high degree of accuracy near peaks and troughs and may be altered to suit requirements of tolerance and speed of solution, thereby allowing the engineer freedom to adjust sensitivity parameters.

DEVELOPMENT OF ALGORITHM FOR OSCILLATOR RESPONSE

We shall state the equation of motion for a SDOF linear oscillator as:

$$\ddot{x}(t) + 2h\omega_o\dot{x}(t) + \omega_o^2 x(t) = r(t) \qquad \text{-- (1)}$$

; where $\ddot{x}(t)$, $\dot{x}(t)$, $x(t)$ are the required response, ω_o is the natural frequency of oscillation, h is the damping ratio and $r(t)$ is the forcing function.

With arbitrary initial conditions $x(o) = x_o$, $\dot{x}(o) = \dot{x}_o$, the solution is in the form of the standard convolution integral:

$$x(t) = e^{-h\omega_o t}\left[x_o\cos\omega_d t + \left(\frac{\dot{x}_o + x_o h\omega_o}{\omega_d}\right)\sin\omega_d t \right]$$

$$+ \frac{1}{\omega_d}\int_o^t e^{-h\omega_o[t-\tau]}\sin\omega_d[t-\tau]r(\tau)d\tau \qquad \text{-- (2)}$$

; where $\qquad \omega_d = \omega_o\sqrt{(1-h^2)} \qquad \text{-- (2a)}$

41

Adopting the difference equation notation of (Beck & Dowling, 1988) whereby
$i = 0,1 \ldots N$ and $x(t_i) = x_i$ for $t_i = i\Delta t$, we may express firstly x_{i+1} and
then x_{i-1} in terms of x_i and $\dot{x}_i$ by integrating over judiciously selected
limits. By eliminating $\dot{x}_i$, we then have

$$x_{i+1} = 2e^{-h\omega_o \Delta t}\cos(\omega_d \Delta t)x_i - e^{-2h\omega_o \Delta t}x_{i-1}$$

$$- \frac{1}{\omega_d} \int_{t_i}^{t_{i+1}} e^{h\omega_o[\tau-t_{i+1}]}\sin\omega_d[\tau-t_{i+1}]r(\tau)d\tau$$

$$+ \frac{1}{\omega_d} e^{-2h\omega_o \Delta t} \int_{t_{i-1}}^{t_i} e^{h\omega_o[\tau-t_{i-1}]}\sin\omega_d[\tau-t_{i-1}]r(\tau)d\tau \qquad -- (3)$$

Now, if $r(t)$ is known deterministically only at a set of discrete points
evenly sampled at $t_i = i\Delta t$, then the integrands of equation (3) are known only
at the same sampling points. Hence equation (3) may be integrated using the
trapezoidal rule and expressed in the form:

$$x_{i+1} = C_1 x_i + C_2 x_{i-1} + C_3 r_i \qquad -- (4)$$

where
$$C_1 = 2e^{-h\omega_o \Delta t}\cos(\omega_d \Delta t) \qquad -- (4a)$$

$$C_2 = -e^{-2h\omega_o \Delta t} \qquad -- (4b)$$

$$C_3 = \frac{\Delta t}{\omega_d} e^{-h\omega_o \Delta t}\sin(\omega_d \Delta t) \qquad -- (4c)$$

The first ordinate x_1, may be obtained by integrating (2) between $t=0$ and
$t=\Delta t$ to give

$$x_1 = C_4 x_o + C_5 v_o + C_6 r_o \qquad -- (5)$$

where
$$C_4 = e^{-h\omega_o \Delta t}\left[\frac{h\omega_o}{\omega_d}\sin(\omega_d \Delta t) + \cos(\omega_d \Delta t)\right] \qquad -- (5a)$$

$$C_5 = \frac{1}{\omega_d} e^{-h\omega_o \Delta t}\sin(\omega_d \Delta t) \qquad -- (5b)$$

$$C_6 = \frac{1}{\omega_d \Delta t} e^{-h\omega_o \Delta t}\sin(\omega_d \Delta t) \qquad -- (5c)$$

We may note at this stage that only three multiplications per time step are necessary and that the errors inherent in the algorithm are those of the first order Newton-Cotes formula (Trapezoidal Rule) having local truncation error:

$$\varepsilon = \frac{(\Delta t)^3}{12} f''(\xi) \qquad\qquad -- (6)$$

;where $\qquad t_i < \xi < t_{i+1}$

and $f(\tau) = |\frac{1}{\omega_d} e^{h\omega_o[\tau-t_{i+1}]} \sin\omega_d[\tau-t_{i+1}]r(\tau)|$

$$+|\frac{1}{\omega_d} e^{-2h\omega_o\Delta t} . e^{h\omega_o[\tau-t_{i-1}]} . \sin\omega_d[\tau-t_{i-1}]r(\tau)|$$

On the assumption that r(t) is known only at discrete time steps, this is the smallest order error that can be achieved. Should r(t) be known algebraically then equation (3) may be integrated exactly.

Furthermore, using the notation of stability and accuracy given in (Bathe & Wilson, 1973) and (Bathe, 1982), it can be shown that the spectral radius $\rho(A) \leqslant 1$ for all $h \geqslant o$ for the integration approximation matrix $\underline{A}$.

; where $\qquad \begin{pmatrix} x_{i+1} \\ x_i \end{pmatrix} = \underline{A} \begin{pmatrix} x_i \\ x_{i-1} \end{pmatrix} + \underline{L}\ r_i$

and $\qquad \underline{A} = \begin{bmatrix} C_1 & C_2 \\ 1 & 0 \end{bmatrix} , \underline{L} = \begin{bmatrix} C_2 \\ 0 \end{bmatrix}$ $\qquad\qquad -- (7)$

Coupling this with the fact that the algorithm defined by equations (4) and (5) calculate exactly the transient part of the oscillator response due to non-zero initial conditions, we may state that the method is unconditionally stable without offering any amplitude damping or period elongation.

By differentiating equation (2) to give $\dot{x}(t)$ and $\ddot{x}(t)$, and applying similar techniques to those described above, equivalent difference equations, for velocity and acceleration response may be easily derived.

In order to assess the accuracy of the techniques so far described, response spectra were generated and compared to the exact solution for a sine wave input. Figures 1 and 2 show displacement spectra relating to a sine wave excitation $r(t) = \sin\omega t$, normalised to the Nyquist frequency $1/2\Delta t$

In Figure 1 the time step Δt is chosen to achieve 5 time steps per natural period $T = 2\pi/\omega$, while Figure 2 displays the same curve using 10 time steps per period. A comparison of computing effort required to achieve this accuracy is given in Table 1.

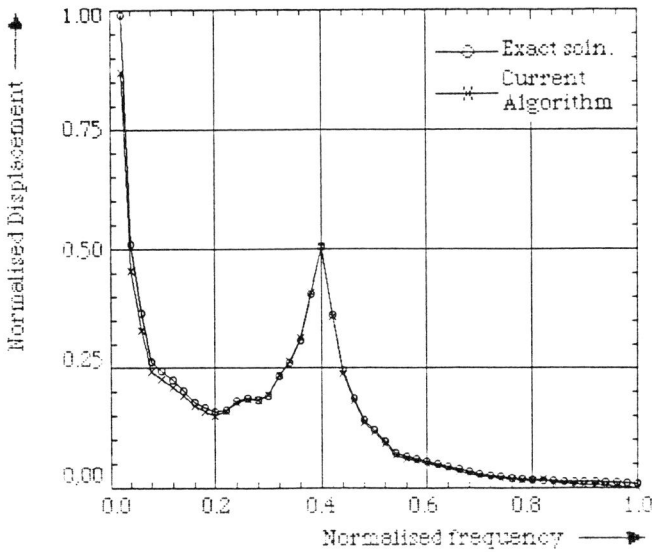

Fig. 1. Displacement Spectra Comparisons for Current
Algorithm using T/Δt =5

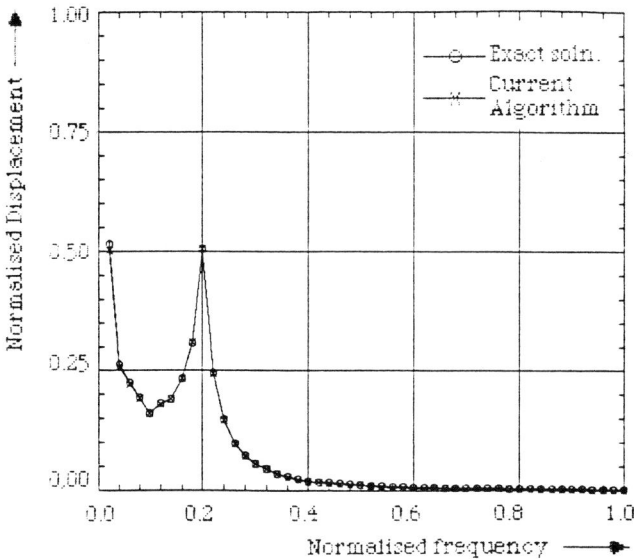

Fig. 2. Displacement Spectra Comparisons for Current
Algorithm using T/Δt =10

RECURSIVE BINARY CHOP SEARCH ALGORITHM FOR SPECTRA GENERATION

A current requirement within the UK nuclear power industry is that equipment essential to the safe operation of nuclear related facilities must maintain a defined degree of integrity under specified earthquake loading. The most common method of ensuring that such equipment is able to withstand the ground accelerations due to a worst case UK earthquake (return period 10000 years) is to use joint testing and analysis techniques. Due to the very large numbers of essential pieces of equipment within a nuclear power plant and the wide variety of on-site configurations of such equipment, the analysis phase can involve the generation of very large numbers of response spectra. The following algorithm allows a reduction in computing effort for the generation of such spectra.

Binary chop search algorithms are a simple and widely used class of curve fitting methods. When coupled with an understanding of the form of the curves to be fitted they can be modified easily to provide quick and accurate solutions.

The nature of single degree of freedom response spectra due to UK earthquake motions has been assessed and the following algorithm has been designed. It provides a sensible balance between accuracy and computational effort, which is applicable to most civil structures and the equipment mounted within them. The advantage of this algorithm is that for a specified number of spectral ordinates which are to be generated, their distribution is more efficiently controlled in the search for maxima and minima. Tolerance values and depth of recursive searching for these maxima and minima may be set by the user to modify the system to his own purposes.

The logic of the algorithm may be delineated as follows, with reference to Figure 3:

♦ Select a fairly large frequency step (Δf).
♦ Select depth of recursive search (maximum depth = 4, say).
♦ Select a tolerance on the minimum variation in adjacent spectral ordinates which may be detected (ΔS_{min}).
♦ Generate spectral ordinates S_1, S_2,.....S_i, S_{i+1}, at intervals of Δf.
♦ At each generation of an ordinate S_{i+1}, check to see if S_i is a point of interest (POI), defined as any ordinate which is a local minimum or maximum as described by S_{i-1}, S_i, S_{i+1}, alone.
♦ On detection of a POI, set depth of search (depth = 1).
♦ If $\left(\dfrac{d^2 S}{df^2} \right) > 0$ then call procedure POIMIN else call procedure POIMAX.
♦ Defining procedure POIMAX called with parameters ($f_{(poi-1)}$, $f_{(poi)}$, $f_{(poi+1)}$, depth).
　(i) Generate two additional ordinates at bisecting frequencies f_{b1} and f_{b2}
　(ii) depth = depth + 1
　(iii) Confirm which is the maximum of [S_{fb1}, S_{fb2}, S_{fpoi}] and set to f_{bmax}

(iv) On $f_{bmax} = S_{fb1}$, S_{fb2}, S_{fpoi} : then $[f_a = f_{(poi-1)}, f_b = f_{poi}]$, $[f_a = f_{poi}, f_b = f_{poi-1}]$, $[f_a = f_{b1}, f_b = f_{b2}]$

(v) If depth $\leqslant$ maxdepth and detected change in maximum value $f_{bmax} > \Delta S_{min}$ then call procedure POIMAX with parameters $(f_a, f_{bmax}, f_b, depth)$.

♦ The definition of procedure POIMIN is similar in nature to POIMAX.

♦ On leaving POIMIN or POIMAX continue generation of ordinates S_{i+2}, S_{i+3}

The maximum possible error in estimation of a spectrum from any given time history depends upon many factors, which are controllable by the user. By way of example, for a sine wave excitation time history $r(t) = \sin\omega t$, $h = 0.05$, $\Delta f = 1/100\Delta t$ and depthmax $= 3$, then the generation of an acceleration spectrum will have a maximum error of less than 0.0773%.

This error is calculated by allowing a linear segment of the fitted curve between two adjacent ordinates to take the value $l(\omega_o)$, whilst the exact response is given by

$$a(\omega_o) = \frac{\omega^2}{\sqrt{((\omega_o^2 - \omega^2)^2 - (2h\omega_o\omega)^2)}} \qquad -- (8)$$

where ω_o is the natural frequency of an oscillator and, ω is the driving frequency (Clough & Penzien, 1982). Then the maximum error occurs when

$$\frac{da(\omega_o)}{d\omega_o} = \frac{dl(\omega_o)}{d\omega_o} \qquad -- (9)$$

(which always exists by the mean value theorem), say when $\omega_o = \omega_m$

Hence the maximum error is given by:

$$\varepsilon_{max} = \left[a(\omega_o) \bigg|_{\omega_o = \omega_m} - l(\omega_o) \bigg|_{\omega_o = \omega_m} \right] \qquad -- (10)$$

Since equation (9) is highly non-linear, ε_{max} is then calculated numerically.

The obvious advantage of this algorithm over the more conventional sweep of frequencies may be seen from Table 1.

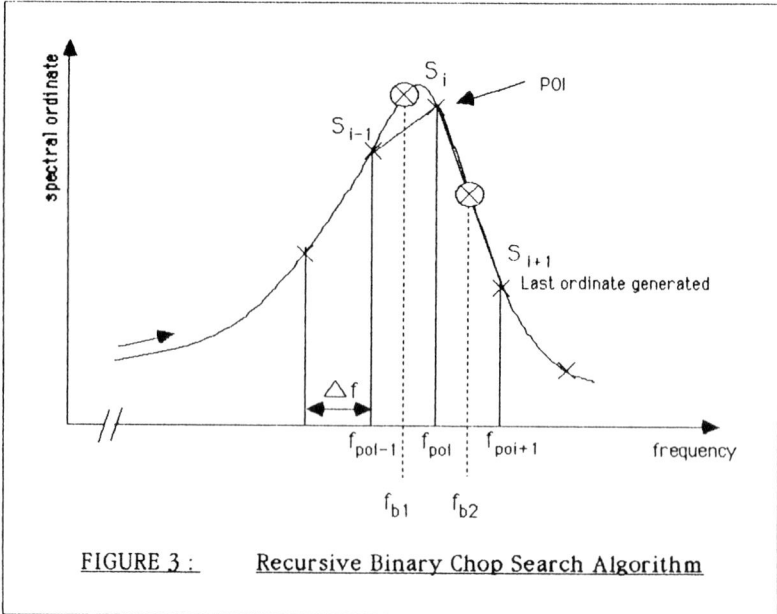

FIGURE 3 :   Recursive Binary Chop Search Algorithm

Table 1.   Comparative Computational Effort for the Calculation
of Time History Response and Spectra Generation

Representative Test Case	Time to Completion (Normalised Units of Time)		
	Current Algorithm	Beck & Dowling	Newmark Constant Average Acceleration
Displacement response of an oscillator; 200 time steps; single frequency forcing function.	1.00	1.36	2.66
Displacement spectra calculation; standard sweep of 100 ordinates; 200 time steps per history, single frequency forcing function. *	100	136	266
Displacement spectrum calculation; recursive binary chop search algorithm; 200 time steps; single frequency forcing function; $\Delta f = 1/60\Delta t$; max depth = 3, tolerance = 10^{-4}*	42	57	112

* Typically similar accuracy.

CUBIC SPLINE INTERPOLANT ALGORITHM FOR SPECTRA GENERATION

The recursive binary chop search algorithm described above is suitable for most engineering applications, but should an unusually high degree of accuracy be required, it may be augmented by the application of a cubic spline interpolant technique.

The principle of the algorithm is that since we have generated a number of spectral ordinates at a location of interest, we may further enhance the accuracy by fitting a clamped or natural cubic spline to that set of local ordinates, using a standard textbook algorithm (Burden & Faires, 1985).

The error bounds for this technique are given in (Burden & Faries, 1985) and (Birkhoff & de Boor, 1946) for clamped and natural splines respectively, although, in general, clamped splines offer a more accurate fit to any given function, in this case since the nodes are unequally spaced near end points the problem of calculating the required derivatives with the necessary accuracy becomes rather difficult. However, it is instructive to note that for a function $f \in C^4[a,b]$ with $\max_{a \leqslant x \leqslant b} |f^{(4)}(x)| \leqslant M$, then

$$\max_{a \leqslant x \leqslant b} |f(x) - S(x)| \leqslant \frac{5M}{384} \max_{o \leqslant j \leqslant n-1} (x_{j+1}-x_j)^4 \qquad -- (11)$$

; where S is the cubic spline interpolant of f with respect to nodes $a = x_o < x_1 < \ldots < x_n = b$, satisfying the clamped boundary conditions.

Finally, substitution of equation (8) into equation (11) provides an equation which will require a numerical solution.

REFERENCES

Nigam, N. C. and P. C. Jennings (1969). Calculation of Response Spectra from Strong-Motion Earthquake Records. Bull. Seismological Soc. Amer., 59/2, 909-922

Beck, J. L. and M. J. Dowling (1988). Quick Algorithms for Computing either Displacement, Velocity or Acceleration of an Oscillator. Earthquake Eng. Struct. Dynamics, 16, 245-253

Birkhoff, G. and C. de Boor (1964). Error Bounds for Spline Interpolation. Jour. Math. & Mech., 13/5, 827-835

Burden, R. L. and J. D. Faires (1985). Numerical Analysis. PWS Publishers, Boston.

Clough, R. W. and J. Penzien (1982). Dynamics of Structures. McGraw-Hill.

Bathe, K. J. (1982). Finite Element Procedures in Engineering Analysis. Prentice-Hall Inc., New Jersey

Newmark, N. M. (1959). A Method of Computation for Structural Dynamics. Jou. Amer. Soc. Civil. Eng., Proceedings, 67-94

Bathe, K. J. and E. L. Wilson (1973). Stability and Accuracy Analysis of Direct Integration Methods. Earthquake Eng. Struct. Dynamics, 1, 283-291

The Non-contacting Measurement of Full-field Modal Stresses under Random Loading Conditions

N. HARWOOD

*National Engineering Laboratory, East Kilbride,
Glasgow G75 0QU, UK*

ABSTRACT

Recent technological developments at the National Engineering Laboratory
(NEL) have permitted full-field surface stress distributions to be measured
on structures subjected to stationary wide-band random loading conditions.
A computer-aided test system has been programmed to acquire and analyse
thermoelastic response signals from a SPATE 8000 infra-red radiometer during
the scanning of structures and components which are undergoing white-noise
excitation. The following paper discusses the selection of optimal
frequency response function estimators (ie H_1, H_2, H_v or H^c) for an accurate
representation of the quasi-static or modal behaviour of randomly loaded
structures. Experimental data are presented which show the feasibility of
generating a modal database in terms of stress rather than displacement mode
shapes. The use of the system in conjunction with finite element modelling
for design assessment of engineering structures under service-loading conditions is discussed.

KEYWORDS

Full field stress measurement, random loading, modal conditions, frequency
response function, thermoelastic effect, SPATE.

INTRODUCTION

A physical phenomenon known as 'the thermoelastic effect' produces revers-
ible temperature changes in compressible materials when they are subjected
to rapid changes of stress; compression generally produces an increase in
temperature and tension a decrease. Although the direct relationship
between stress changes and the corresponding adiabatic temperature changes
has been known since the middle of the 19th century (Thomson, 1853), the
phenomenon has generally been considered to be of negligible significance
for most cases of elastic stress analysis, since the temperature changes in
solid materials are very small. However, technological developments within
the past 10 years in the field of highly sensitive infra-red scanning

detectors have permitted the practical exploitation of this theoretical knowledge in order to estimate surface stress distributions on dynamically loaded structures. Adiabatic temperature changes are proportional to changes in volume, which are in turn proportional to the changes in the sum of the principal stresses (ie the first stress invariant). Therefore the parameter measured by thermoelastic equipment is a scalar value proportional to the effect of an applied hydrostatic pressure (Harwood and Cummings, 1986); maximum and minimum principal stress vectors are not generated.

The SPATE 8000 instrumentation, which was the first commercial thermoelastic stress measurement system (Mountain and Webber, 1978), was developed specifically for uniform cyclic loading applications, and this equipment has been used successfully at NEL over the past five years to measure quasi-static stress distributions over a wide range of offshore, transport, and other engineering structures (Cummings and Harwood, 1987). The SPATE system produces surface stress patterns as 16-colour contour maps in the form of an array of square pixels which indicate the amplitude of the stress and its phase polarity relative to a reference signal (eg a load cell). The pattern is built up gradually, pixel-by-pixel, in real time as the infra-red radiometer scans the required area on the surface of the test structure. The equipment has a temperature sensitivity of 1 mK which equates approximately to a stress resolution of 1 MPa in steel and 0.4 MPa in aluminium.

However, despite the success of the SPATE 8000 system, it has generally been recognised that the use of thermoelastic equipment simply for quasi-static applications does not exploit the full potential of a technique which is essentially dynamic in nature. Moreover, the fact that the infra-red radiometer is non-contacting has great advantages for measuring stress in lightly damped structures which are behaving modally, since the act of measurement itself cannot alter the mass distribution of a test structure which may be very sensitive to such changes. Although the non-contacting nature of thermoelastic equipment does lend itself to modal applications, experimental modal analysis is, in many cases, rather limited and laborious to use if sinusoidal excitation must be applied at each natural frequency in turn, as in the SPATE 8000 system.

The major scientific effort of a small team at NEL has been aimed at extending the capability of the SPATE equipment to the types of complex, non-periodic signals which are commonly encountered in experimental modal analysis applications and under service-loading conditions, and thus extend considerably the potential engineering applications of thermoelastic stress analysis.

CALCULATION OF THE FREQUENCY RESPONSE FUNCTION

The main body of the investigations at NEL of structures subjected to non-periodic excitation has concentrated on wide-band random load waveforms. The measurement of frequency response functions between the thermoelastic response and a load or strain reference signal forms the basis of the stress estimation technique. Several different ways of calculating frequency response functions have been developed over recent years in order to improve their accuracy in the presence of noise. The results of all the methods coincide in a noise-free environment, but when noise is present, the various estimators will give different values.

The traditional form of the frequency response function (H_1) is computed by dividing the cross-power spectrum by the auto-power spectrum of the reference, ie

$$H_1 = \frac{G_{yx}}{G_{xx}}. \tag{1}$$

H_1 is likely to be the most accurate estimator when the reference signal has very low noise at the frequencies of interest. However, H_1 tends to give a low estimate at resonance on lightly damped structures, since there may be a poor signal-to-noise ratio on the force input signal as it falls to very low amplitudes at natural frequencies with high structural gain, due to the displacement limitation of the exciter.

Provided that the noise level in the response is relatively low, the most accurate estimator for lightly damped resonances is likely to be the inverse estimator, known as H_2, which is computed by dividing the auto-power spectrum of the response by the complex conjugate of the cross-power spectrum, ie

$$H_2 = \frac{G_{yy}}{G_{xy}}. \tag{2}$$

H_1 and H_2 are related via the coherence function. Provided that leakage errors are not significant and that noise on the reference and response signals is uncorrelated, H_1 and H_2 must form the lower and upper bounds respectively for the range within which the actual frequency response function must lie.

Considerable efforts have recently been made to find a frequency response function estimator with optimal characteristics. An estimator known as H_v has been developed for multi-point excitation conditions (Rocklin et al, 1984). The computation of this estimator involves noise-vector matrix manipulations which require considerable user effort and relatively long computational times. Under single-input conditions, however, the H_v estimator reduces to the geometric mean of H_1 and H_2 and it may therefore be used as a general estimator which will give acceptable accuracy over a wide range of modal conditions and which, unlike H_1 and H_2, has uniform properties at resonances and anti-resonances.

Note that H_1, H_2 and H_v all have identical phase values preserved in the cross-spectrum term; only the gains may vary. It is particularly advantageous to use H_v when there is significant noise present on both the input and output signals.

The three estimators mentioned previously are each calculated from a combination of cross-power and auto-power spectra. Since a cross spectrum is effectively a correlated function, it is insensitive to incoherent noise. However, measured auto-power spectra tend to have a greater amplitude than the actual value due to the addition of noise, which is why H and H_2 are low and high biased estimators respectively.

The fact that auto-spectra are prone to noise errors, even when averaged considerably, has led to the development of an 'unbiased' estimator, known as H^c (Mitchell et al, 1987), which is computed entirely from cross-spectra. This estimator does, however, have the disadvantage that it requires an extra data acquisition channel for every excitation position. The extra signal which must be connected is the function generator or power amplifier output which is the source of the excitation. H^c is computed by dividing the cross-spectrum between the response and the drive signal by that between the reference and drive signal.

For many modal conditions the H^c estimator tends to be more accurate than H_1 or H_2 when the coherence is low. H^c is liable to vary in both gain and phase from the other three estimators discussed earlier.

TEST RESULTS FOR PRE-RESONANT RANDOM LOADING

In order to demonstrate the effectiveness of the random signal analysis software developed at NEL, a simple cruciform component was analysed under 1–40 Hz white noise loading in a servo-hydraulic test machine, using a load cell output as a reference signal. A universal joint was attached to each end of the component in order to minimise non-axiality. Since this component has two axes of symmetry the scan area was restricted to one quadrant. The stress distribution measured by the NEL software is shown in Fig. 1. The stress concentration at the corner of the joint can be seen clearly when displayed in a colour format.

At NEL stress patterns have been measured on a welded cruciform joint using a strain gauge as a reference. Under service conditions a strain gauge would normally provide the most appropriate method of obtaining a reliable reference signal. The welded joint, which was tested as part of a wider programme investigating the fatigue behaviour of offshore structures, was subjected to 4–40 Hz white noise loading. The structure and the measured stress distribution is shown in Fig. 2; the quality of the pattern is similar to that which would be expected for sinusoidal conditions, despite the more difficult noise-rejection problems encountered in random signal analysis applications.

TEST RESULTS FOR MODAL CONDITIONS

A theoretical modal analysis was performed on a cantilever using the ANSYS software. The dimensions of the cantilever were adjusted until three structural modes were predicted to lie below 40 Hz. The 40 Hz frequency limit was chosen so that attenuation due to conduction to and through a paint coating would not significantly affect the quality of measured thermoelastic responses. The cantilever was then manufactured and an experimental modal analysis performed using an electromagnetic shaker to determine experimentally the natural frequencies and shapes for the first three modes. These mode shapes are displayed in Fig. 3. SPATE was then used to determine stress patterns at each of the natural frequencies in turn. The scan area was restricted to the base of the cantilever around two bolt-heads since the stresses remote from these fixtures were found to be very low. A frequency response function was then determined between the thermoelastic response adjacent to a bolt-head and the input force reference. This thermoelastic frequency response function (Fig. 4) showed the resonant frequencies clearly. A SPATE scan was then performed under 40 Hz baseband white-noise excitation, using the NEL software as described previously. Stress patterns were extracted from the stored frequency response functions at each of the three natural frequencies. The simultaneously excited stress patterns (Fig. 5) produced by random loading are comparable with the data measured by the standard SPATE system using sinusoidal excitation at each frequency in turn. As expected, the random loading data are noisier than the sinusoidal, but the quality of these modal stress distributions is easily sufficient to make engineering judgements from the data produced under random loading. The advantages of random loading would be most apparent in situations where there is a high modal density within the bandwidth of interest and for cases in which consistent excitation of lightly damped modes is difficult to achieve using a sinusoidal waveform.

Thermoelastic modal data acquired at NEL has confirmed that H_1 tends to produce a low estimate at lightly damped resonances due to the poor signal-to-noise ratio in the reference, whereas H_2 usually produces a high estimate due to the high noise level which is invariably present in the response. H_2 also has more sensitivity to the presence of mains harmonics in the SPATE output. The H_v estimator has generally appeared to be significantly smoother than H_1 or H_2 for the same number of averages, although it tends to be more sensitive than H_1 to the presence of mains harmonics in the SPATE signal. Experimental data acquired at NEL indicate that H^c can be used if very few ensemble averages are being taken, but although ensemble averaging may remove bias, performing a large number of ensembled averages does not appear to be as effective at smoothing this estimator as it is for H_v. The suitability of the H_v estimator for conditions in which there is noise on both the input and output signals makes it appropriate for use in the estimation of modal stresses at lightly damped resonances.

CORRELATION OF SPATE DATA WITH IDEALISED MODELS

Thermoelastic stress analysis can be very useful in the design evaluation of structures and components, although the use of the current SPATE instrument is somewhat limited due to a lack of the capability to separate principal stress vectors from the thermoelastic stress patterns. However, the use of SPATE in conjunction with the finite element (FE) technique can provide the experimental data to confirm theoretical solutions obtained from idealisations of the structure. Such experimental confirmation gives confidence in the accuracy of the FE model. Thermoelastic stress patterns may be used to optimise the model by indicating high and low stress gradients where a finer or coarser mesh size would be more appropriate. The data from the FE analysis and the experimental stress patterns can be matched in the form of the sum of the principal stresses to provide a method of validating the model. The distributions of the principal stress vectors may then be obtained from the verified model which of course can also then be used to calculate deformations or effective stresses for multi-axial conditions such as the von Mises yield criterion used for failure prediction in ductile materials.

Thus, although the direct value of SPATE stress data is limited because it is in the form of the sum of the principal stresses, the thermoelastic technique does extend stress measurement into areas which are not presently available to stress analysts and is well suited to integration with theoretical analysis in order to extract parameters more relevant to engineers involved in design for strength, durability and structural efficiency. This methodology is applicable to both static and modal data and offers considerable potential for savings in cost and time if incorporated in the design procedures for engineering components and structures. Examples of the use of thermoelastic stress analysis in conjunction with FE modelling have been described by Harwood and Cummings (1989).

DISCUSSION

The development of the random-signal processing software at NEL has allowed an advanced facility for full-field experimental stress analysis under non-periodic loading to be established. The random-loading investigations have indicated that the thermoelastic technique has great potential as a unique method of measuring surface stress distributions on structures sub-jected to complex loading conditions. A hardware/software system has been developed which can be combined with the standard SPATE 8000 equipment to form a transportable system which is capable of analysing thermoelastic sig-nals typical of those produced by many service-loading or field conditions.

The greatest potential of the technique is realised in its application to modal conditions, when stored data from a single scan can subsequently be analysed to extract a range of stress patterns at all resonant frequencies which lie within the excitation bandwidth, thus producing a modal database in terms of stress rather than the displacement data generated by conven-tional experimental modal analysis techniques. Although it is theoretically possible to calculate stress/strain distributions from experimentally measured displacement data, such data would normally contain insufficient degrees of freedom to enable stresses to be estimated reliably. SPATE, therefore, has great potential for experimental modal analysis applications, since a database may be generated in terms of quantitative stress values, rather than the generally less useful normalised displacement displays produced by traditional techniques. An additional important benefit is that a thermoelastic system which is capable of generating frequency response functions would be readily able to detect changes in dynamic stiffness, such as are produced by crack growth, delamination or a loosening of joints.

The estimation of modal stresses under random loading is a more difficult measurement situation than the quasi-static condition, since bandwidth averaging cannot be performed to reduce noise, and at frequencies above 40 Hz there is likely to be a progressive substrate response signal attenu-ation due to the thermal properties of the paint coatings which must normally be applied to metallic components to enhance emissivity and minimise any correlated reflections from surrounding objects. Experimental investigations at NEL indicate that H_v is the most suitable frequency response function estimator for modal conditions in which the signal-to-noise ratio of the reference signal is poor at the frequencies of interest.

Although the thermoelastic data acquired under random loading is likely to be somewhat noisier than can be achieved under sinusoidal conditions, the pattern-smoothing software which has been developed at NEL allows the random data to be presented in a form which is virtually indistinguishable in quality from sinusoidal patterns measured with the standard SPATE equipment. Stress patterns may be displayed in monochrome on the portable signal analy-sis system and subsequently plotted in a conventional colour-coded format at a fixed workstation.

Calibration software has been written to determine quantitative sum-of-the-principal stresses values from the random data, although the calibration procedure is inevitably somewhat more complicated than that required under sinusoidal conditions. The use of a frictional strain gauge probe and associated software may allow principal stress vectors to be separated at key points on the surface of the structure.

ACKNOWLEDGEMENTS

This paper is published by permission of the Director, National Engineering Laboratory, Department of Trade and Industry. It is Crown copyright.

The SPATE 8000 equipment was developed by SIRA Ltd and manufactured by Ometron Ltd.

REFERENCES

Cummings, W.M. and N. Harwood. (1987). Review of SPATE applications at the National Engineering Laboratory. SPIE, 817, 96-108.
Harwood, N. (1988). An assessment of a frictional strain gauge probe for dynamic strain measurement. Strain, 24(1), 67-70.
Harwood, N. and W.M. Cummings (1986). Calibration and qualitative assessment of the SPATE stress measurement system. National Engineering Laboratory Report No 705.
Harwood, N. and W.M. Cummings (1989). Frequency-domain analysis techniques developed for SPATE response signals. SPIE, 1084.
Mitchell, L.D. et al (1987). An unbiased frequency response function estimator. Proc. 5th Int. Conf. on Modal Analysis, 1, 364-373.
Mountain, D.S. and J.M.B. Webber (1978). Stress pattern analysis by thermal emission (SPATE). Proc. Soc. Photo-Opt. Instrm. Engr., 164, 189-196.
Rocklin, T.G., J. Crowley and H. Vold (1984). A comparison of H_1, H_2 and H_v frequency response functions. Proc. 3rd Int. Conf. on Modal Analysis, 272-278.
Thomson, W. (1853). On the dynamical theory of heat. Trans. Royal Soc. of Edinburgh, 20, 261-283.

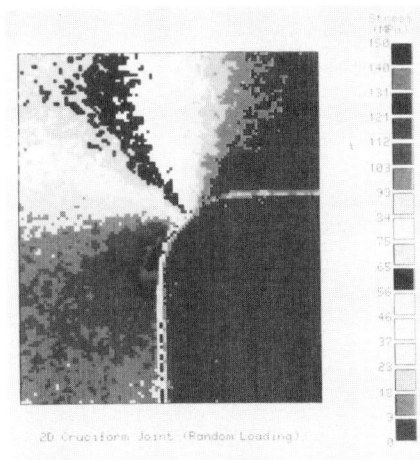

Fig. 1   Random Loading Stress Pattern
from Flat Cruciform Testpiece

Fig. 2   Random Loading Stress Pattern
from Welded Cruciform Component

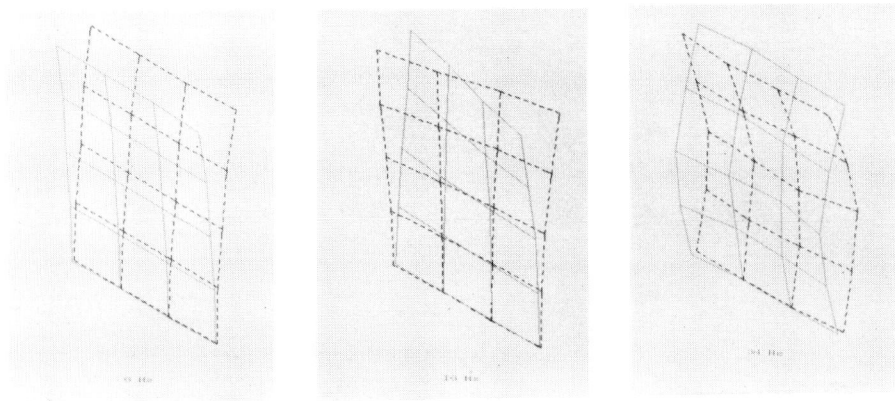

Fig. 3   Mode Shapes from Cantilever

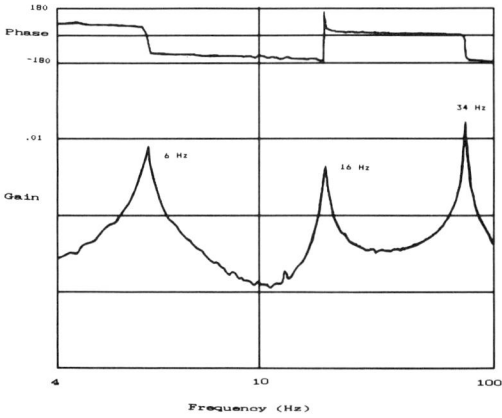

Fig. 4 Thermoelastic Frequency Response
Function from Cantilever

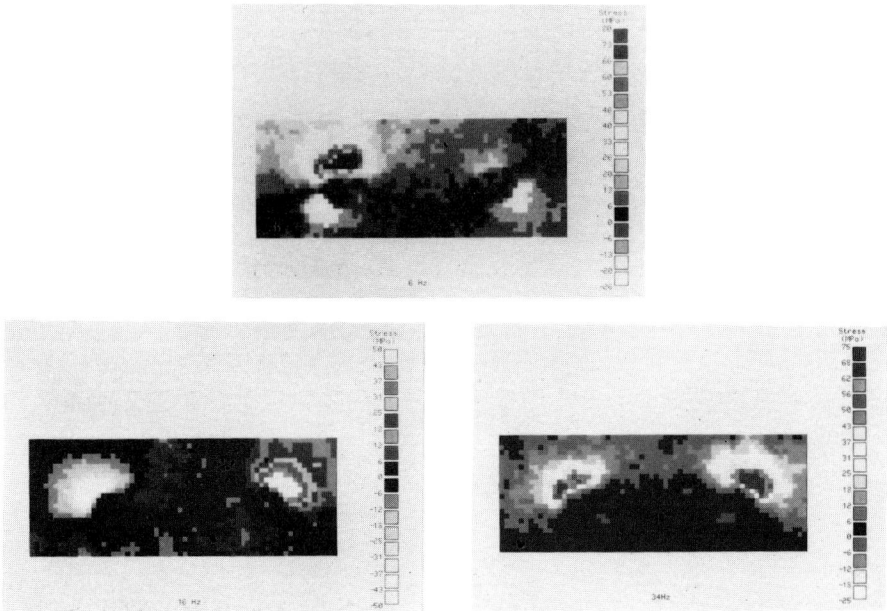

Fig. 5 Stress Patterns from Cantilever
under Random Loading

A Strategy for Quality Assurance in Theoretical and Experimental Modal Analysis

J. A. BRANDON* and H. G. D. GOYDER**

*School of Engineering, University of Wales, College of Cardiff,
Cardiff CF2 1XH, UK
**Harwell Laboratories, Oxfordshire OX11 0RA, UK

ABSTRACT

The paper considers the legal and ethical obligations of the engineer. The current conditions which affect the introduction of standards are reviewed. An example is given which demonstrates that a "systems approach" to experimental and theoretical analysis is desirable.

KEYWORDS

Quality Assurance; Structural Dynamics; Modal Analysis.

INTRODUCTION

The "State of the Art in Mobility Measurement" project (SAMM) gave a disturbing insight into the prevailing state of experimental determination of structural properties. (This was reported in a series of papers by Ewins (Ewins (1981a), Ewins (1981b), Ewins and Griffin (1981))).

Similar difficulties in the theoretical literature have been encountered which bring into question basic levels of competence in numerical analysis. For example the equivalence of modal identifiability and numerical rank is widely overlooked (see Brandon (1987,1989)).

Although no new technical results will be presented the paper will use published algorithms and results to demonstrate the need for the establishment of objective independent standards for quality assurance. The authors have both recently emphasised the importance of developing procedures for Quality Assurance in Structural Dynamics (Brandon (1988), Goyder (1988)). The paper discusses the needs of Modal Analysis, with its interdependent problems of experimental procedure, signal processing and model definition and validation. The roles of researchers, software and equipment suppliers, standards organisations, customers and government agencies are discussed,

59

using as a reference the aims and organisation of the National Agency for Finite Element Methods and Standards (Mair (1985)). In the current paper an experimental example is presented which suggests that a collaborative approach, between equipment and software suppliers, customers and regulatory authorities, is necessary for effective quality assurance. An example of the failure of an identification algorithm is given by Brandon and Cremona (1989).

PROFESSIONAL OBLIGATIONS OF THE ENGINEER

Engineers have both a legal and ethical duty of care, not only to their clients but also to the public at large. This has been described by Jensen and Land (quoted in Holland (1986)), who suggest that the law assumes that the engineer:

> (1) Possesses the required degree of learning, skill and experience possessed by others in the profession and which is regarded by the community as necessary to qualify him or her to practice the profession;
> (2) will use reasonable and ordinary care and diligence in the exercise of his or her skill and in the application of knowledge to accomplish the objectives for which he or she is employed;
> (3) will exert skill with diligence and care; and
> (4) will use his or her best judgement on the employer's behalf.

The first (of seven) 'fundamental canons' of the ethical code of the American Society of Mechanical Engineers states:

> "Engineers shall hold paramount the safety, health and welfare of the public in the performance of their professional duties."

The Hyatt Regency Hotel disaster, in July 1981, led to a further clarification of the duty of the engineer, at least in the American context. Becker (1986) interpreted the legal judgement with respect to the obligations of the engineer, defined in the State Registration Law (Missouri in this case) in the following terms:

> "He emphasized the contractor's profit motive and stated that the engineer is obliged to look out for the State's (public) interest and must ensure that the structural integrity of the design is maintained."

Thus it was recognised that the engineer's professional duty may conflict with the interests of his/her employer and that public duty must take precedence over loyalty to the employer.

THE NEED FOR STANDARDISATION

The legal and ethical requirements, outlined above, may appear initially to place an overwhelming burden on the engineer in general practice. A necessary, though perhaps not always sufficient, part of a defence against litigation on the grounds of negligence is that the engineer has applied up to date theoretical and experimental validation procedures.

Standards in structural dynamics will perform some or all of the following functions (a non exhaustive list):

> disseminate currently available knowledge in a form accessible to engineers in general practice (particularly interpreting solutions to complex mathematical relationships as nomograms or other graphical representations);
> define and describe good design practices;
> specify documentation procedures;
> specify representative test conditions;
> specify test configurations.

It should be realised that, in devising a design and test procedure, the engineer will be (at least partially) protected against both ethical and legal complaints if he/she shows that the procedure takes account of the relevant published standards. Thus the general duty of care may be attenuated in this case.

SCOPE OF EXPERIMENTAL AND THEORETICAL MODAL ANALYSIS

The ISO Draft Standards on Mobility Measurement identify the following areas of concern in experimental structural dynamics:

1 Predicting Dynamic Response of Structures;
2 Determining the Modal Properties of a Structure;
3 Predicting Dynamic Interaction of Interconnected Structures;
4 Checking the Validity and Improving Accuracy of Mathematical Models;
5 Determining Dynamic Properties of Materials.

Although the ISO standards are primarily concerned with experimental procedures, the applicability of the list can be extended readily to encompass the theoretical aspects of modal analysis. Indeed, a theoretical model, albeit occasionally implicit, is a prerequisite for the processes described.

CURRENT ISSUES AFFECTING QUALITY ASSURANCE AND STANDARDISATION IN MODAL ANALYSIS

There are potentially three parties to the quality assurance of modal analysis, suppliers, customers and independent agencies. The willingness of the parties to introduce procedures for quality assurance depends on a variety of factors.

Successful introductions of recent standards have been due to one of the parties taking a strong (and self interested) initiative, often exploiting a monopolistic market position (for example the supplier standardised IBM PC (see Chposky and Leonsis (1989)) and the customer led Manufacturing Automation Protocol, championed by General Motors (see Turff (1988))).

Suppliers

The suppliers of goods and services in modal analysis are

working in a limited market with specialised products. Development costs must be recovered over short production runs, with correspondingly arduous payback criteria. A substantial proportion of the value added, in both goods and services, is "intellectual property". Despite the fact that there must be a strong common element in the underlying processes, the companies exploit novel features of both devices and software to give fast processing, robust algorithms etc.

In these circumstances, reinforcement of customer perceptions as to the power and versatility of proprietary products may take precedence over comparability and verifiability of test results. There is little advantage for the supplier to take an initiative to encourage standardisation, as this may imply the sharing of the intellectual property which constitutes the majority of the added value.

Notwithstanding the above comments, the suppliers of modal analysis hardware and software have, in the main, contributed generously to the literature. apparently taking the view that an informed customer is likely to be a satisfied customer.

Customers

There is a considerable variation of requirements of customers, depending on the equipment to be tested, the type of test required, the degree of expertise available in-house and the resources available for test facilities. There is consequently a corresponding variability in the perception of the need for standards. The people who would have most to gain from, and are perhaps most in need of, objective external standards are small companies with only an occasional need for structural testing. Large companies, with regular need and sufficient resources, are capable of developing quality strategies, including "make or buy" decisions on provision of structural testing.

Independent Standards Agencies

As has already been mentioned, the International Standards Organisation has already taken some interest in the subject. The material produced to date has been largely of relevance at the operational level.

Multi function agencies

There is a significant number of organisations which combine more than one role. For example large aerospace companies, nuclear and energy agencies, and general defence contractors may sell equipment or expertise, developed in-house, through spun-off manufacturing or consultancy subsiduaries. It is not unusual for such organisations to develop reference standards.

EXPERIMENTAL ASPECTS

The measurement of frequency response functions (frfs) lies at the heart of experimental analysis. These functions completely characterise a linear structure. However surprisingly little is known about the accuracy with which these functions may be determined. It is the thesis of this paper that serious practitioners of experimental vibration analysis should be quantifying the errors associated with these measurements. To illustrate the difficulties associated with determining the accuracy of a frf two serious sources of error will be examined here.

There are several methods for measuring frfs. Each method involves applying a force to a structure and measuring the induced motion (see for example Ewins (1984)). Associated with each of the methods are a mixture of systematic and random errors. (Systematic errors distort data by applying some form of weighting function; random errors, on the other hand, lead to each measured value being combined with a random variable.) In the current paper two causes of systematic error are considered. Those selected are particularly troublesome since they lead to frfs which are incorrect, but completely credible, and could belong to the structure under inmvestigation.

The two sources of error are associated with the manner in which a shaker is attached to the structure. This subject is considered in part 2 of the ISO standard. The figure shows a structure connected to a shaker via a flexible link and force transducer.

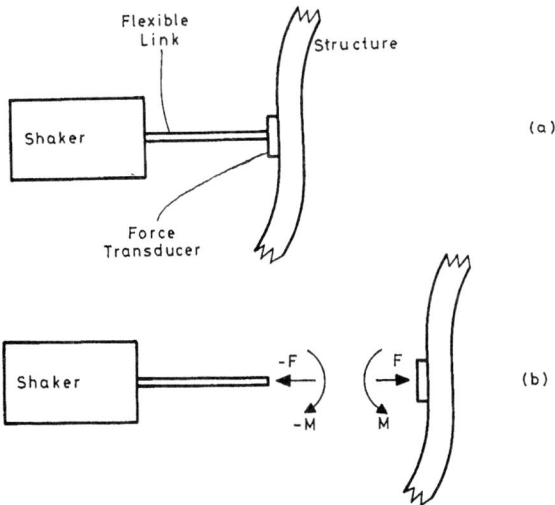

Figure 1(a) Experimental configuration for measuring frequency response functions (b) Forces and moments transmitted by flexible link.

The errors may be introduced by the flexible link which can apply an unwanted bending moment to the structure, in addition to the desired normal force. This bending moment constrains the rotation of the structure and modifies its dynamic response. Thus one source of error is because the dynamics of the structure are coupled with the dynamics of the shaker via the bending moment in the force link.

The second source of error is the force transducer. Most force transducers are sensitive to bending moments and their application produces an output signal. Consequently the correctly measured normal force is contaminated with an additional signal due to the bending moment.

The effect of these two errors may be investigated by using linear models. (Nonlinear models might be more precise but lack of knowledge means that they are not yet applicable.)

The motion of the structure due to the normal force and bending moment may be written

$$\emptyset_1 = H_{11} F_1 + H_{12} M_2 \qquad (1)$$

$$\theta_2 = H_{21} F_1 + H_{22} M_2 \qquad (2)$$

where harmonic motion is assumed and F_1 and M_2 are the (complex) amplitudes of the normal force and bending moments, $\emptyset_1$ and θ_2 are the normal displacement and rotation, while H_{11}, H_{12} and H_{21} are frequency response functions for the structure. These equations represent the response of the structure as a linear superposition of the excitation forces. An additional relationship between θ_2 and M_2 may be obtained by considering the dynamics of the flexible link and shaker.

THe flexible link will exert a bending moment that is proportional to the rotation of the structure:

$$\theta_2 = - H_{33} M_2 \qquad (3)$$

where H_{33} is a frf and the negative sign arises because the bending moment in the link is a reaction force (ie equal but opposite in sense to the moment on the structure). Using the above equations to eliminate θ and M leads to the relationship between $\emptyset$ and F

$$\emptyset_1 = \left[H_{11} - \frac{H_{12}^2}{H_{22} + H_{33}} \right] F_1 \qquad (4)$$

In the absence of errors due to the bending moment then $\emptyset$ and F would be given by

$$\emptyset_1 = H_{11} F_1 \qquad (5)$$

Hence equation 4 represents a considerable modification of the desired relationship.

The second source of error is due to the erroneous measurement of bending moment. The output from the force transducer will be given by

$$F_1^* = F_1 + \alpha M_2$$

where $F_1{}^*$ is the corrupted output signal and α is the frequency response function for the bending moment response. (Bruel and Kjaer quote a value of 100pc/Nm for the 8200 transducer, leading to an effective α of 25 N/Nm). Combining this error with the previous equation leads to the expression

$$\emptyset_1 = \frac{\left[H_{11} - \dfrac{H_{12}{}^2}{H_{22} + H_{33}} \right]}{\left[1 - \dfrac{\alpha\, H_{12}}{H_{22} + H_{33}} \right]} F_1{}^*$$

The difficulties that this equation causes can be seen immediately. If $\emptyset_1$ and $F_1{}^*$ are assumed to be perfect measurements of the displacement and force then the frequency response function they produce will appear to be completely reasonable. It is, after all, built up from the frequency response functions that are completely authentic in their own context.

The only control that the experimenter has over the frf relating $\emptyset_1$ and $F_1{}^*$ is the selection of H_{33}. It can be seen that, in order to eliminate the unwanted terms, H_{33} must be relatively (very) large. In particular it is important that

$$H_{22} \ll H_{33}$$

This condition is particularly difficult to achieve. Under resonant conditions H_{22} will be large (unless the damping is large). Thus at resonance, the very conditions which are most carefully examined, the effects of the bending moment will be most severe.

Lessons for Quality Assurance

In the example described, the overall measured frequency response function includes contributions from structure, force link, transducer and exciter. It is by no means uncommon for each of these to be procured from a separate supplier, each of whom may have no conception of the interaction of the devices and, even if this interaction is appreciated, would be unlikely to be in possession of performance data on the coupling of the devices. As has already been mentioned, it cannot be assured that the user has the requisite knowledge to calibrate the equipment and hence quantify the errors. Whilst the ISO standards give considerable guidance at the detail level (and would be of use in the example cited) they are unhelpful in determining a company's overall quality assurance strategy.

THE NATIONAL AGENCY FOR FINITE ELEMENT METHODS AND STANDARDS, A MODEL FOR QUALITY ASSURANCE

Similar problems prevail(ed?) in the finite element community, as decribed by Mair (1985). The National Agency for Finite Element Methods and Standards was established, by the U K National Engineering Laboratory, to form a focus for these concerns. Their remit includes, for example, Benchmarking,

66

Education and Specialist meetings.

Their quoted areas of concern include

"(a) poor understanding of the physical problem.
(b) inadequate understanding of the assumptions and
limitations...
(c) poor representation of the physical problem.....
(d) inadequate safeguards...
(h) a badly conditioned problem....
(i) poor selection or assessment of output data..

It can be seen that many of these concerns are of equal interest
to the modal analysis community. The group has published a
useful finite element primer and a code of practice. Perhaps
their most important role however is as a focus for discussion
and activity.

Whilst there are a number of formal (eg AGARD, ISO) and informal
groups (UK Modal Analysis Users Group) making valuable
contributions at the operational level, the authors perception
is that no group is currently taking an initiative in
development of Quality Assurance Standards at the strategic
level. Unfortunately the history of Engineering Practice is
that it takes a disaster for Quality Assurance procedures and
practices to be reviewed (see Wearne (1979) and Bignell and
Fortune (1984)).

CONCLUDING REMARKS

The large variety of equipment used, often procured from a
number of different specialist suppliers, couples experimentally
in a manner that is almost impossible to predict analytically
and almost as difficult to measure experimentally (always
assuming that the experimentalist is aware of the coupling
problems). This problem can only be tackled by a strategic
viewpoint rather than the (admittedly useful) operational
approach currently used.

There is no identifiable single agency taking a lead in Quality
Assurance in Modal Analysis.

ACKNOWLEDGEMENT

The authors are grateful for the advice and assistance of
Professor D J Ewins during the preparation of this paper. The
work described in this paper was undertaken as part of the
underlying research programme of the United Kingdom Atomic
Energy Authority.

REFERENCES

Becker E.P. (1986). Who should be responsible for structural
steel design, American Society of Civil Engineers, Journal of
Professional Issues in Engineering, 112(2), pp134-140.
Bignell, V and Fortune, J. (1984). Understanding Systems
Failures, Manchester University Press
Brandon, J.A. (1987). The significance and practice of rank

estimation in structural dynamics identification algorithms, in: Numerical Techniques for Engineering Analysis and Design, (G N Pande and J Middleton Eds.) #S5 pp1-8.

Brandon, J.A. (1988). On the use of oversized models for modal identification in structural dynamics, 3rd International Conference : Recent Advances in Structural Dynamics, Institute for Sound and Vibration Research, Southampton, pp319-328

Brandon, J.A. (1989). On Numerical Analysis Needs for Modal Analysis, Seventh International Modal Analysis Conference, Las Vegas.

Brandon, J.A. and Cremona, C.F. (1989). On the limitations of spatial parameter estimation methods in structural dynamics, International Conference: Modern Practice in Stress and Vibration Analysis, Institute of Physics, Liverpool University.

Chposky, J. and Leonsis, T. (1989). Blue magic. the people, the power and the politics behind the IBM Personal Computer, Grafton.

Ewins, D.J. (1981a). State-of-the-Art Assessment of Mobility Measurement Techniques (SAMM)- Summary of Results, Journal of the Society of Environmental Engineers, March, pp1-11.

Ewins, D.J. (1981b). State-of-the-Art Assessment of Mobility Measurement Techniques- A Summary of European Results, Shock and Vibration Bulletin, 51(1), pp15-36.

Ewins, D.J. and Griffin, J (1981). A State-of-the-Art Assessment of Mobility Measurement Techniques Results for the Mid-range Structures (30-3000 Hz), Journal of Sound and Vibration, 78(2), pp197-222.

Ewins, D.J. (1984). Modal Testing Theory and Practice, Research Studies Press

H G D Goyder, (1988). Three new methods for determining natural frequencies and damping ratios from vibration spectra, 3rd International Conference : Recent Advances in Structural Dynamics, Institute for Sound and Vibration Research, Southampton, pp279-288

Holland, J.P. (1985). Professional liability of the Architect and Engineer, American Society of Civil Engineers, Journal of Professional Issues in Engineering, 111(2), pp57-65.

International Standards Organisation, Vibration and Shock-Experimental Determination of Mechanical Mobility
    Part 1: Basic Definitions and Transducers ISO/DIS 7626/1
            (Draft International Standard 1984)
    Part 2: Measurements Using Single-Point Translation
            Excitation with an Attached Vibration Exciter
            ISO/DIS 7276/2
            (Draft International Standard 1984)
    Part 5: Measurements Using Impact Excitation with an
            Exciter Which is Not Attached to the Structure
            ISO/DP 7276/Part 5 (Draft Proposal 1986)

Mair, W.M. (1985). The objectives of the National Agency for Finite Element Methods and Standards, Computers and Structures, 21(5), pp875-9.

Turff, J. (1988). MAP- Practical Aspects of Life after ENE, International Conference: Factory 2000: Integrating Information and Material Flow, Institute of Electronic and Radio Engineers, Cambridge, pp89-95

Wearne, S. H. (1979). A review of reports of failures, Proceedings of the Institution of Mechanical Engineers, 193, pp125-136.

Frequency Analysis of an Ultrasonically Excited Thick Cylinder

G. M. CHAPMAN and M. LUCAS

Department of Mechanical Engineering,
Loughborough University of Technology,
Loughborough, Leics. LE11 3TU, UK

ABSTRACT

When dealing with vibration analysis at ultrasonic frequencies it is necessary to consider carefully the techniques to be used and the parameters to be monitored. Conventional sensors and their associated electronics invariably are designed with linear characteristics in the audible range only. Ultrasonic machines cannot readily be adapted to monitor the transmission line forces as interruption of the systems geometry will detune the system.

This paper presents the approach taken to analyse an ultrasonically excited thick cylinder using finite element analysis, electronic speckle pattern interferometric analysis and modal analysis supported by software for structural modification.

KEYWORDS

Ultrasonic, vibration, cylinder, modal analysis, ESPI, structural modification, finite element.

INTRODUCTION

A novel method of current interest for the forming of metal uses the ability of high frequency vibrations to reduce the apparent friction between the forming element and the material being formed. Commercial exploitation of such a process applied to wire drawing has been reported previously (Jones 1968; Sansome, 1970; Langenecker 1978). Most processes make use of conventional ultrasonic technology at the lower end of the ultrasonic frequency range by arranging for specific resonant conditions in the mechanical components. In the case of wire drawing the forming element is

69

basically a thick cylinder which vibrates in its fundamental radial mode.

At these comparatively high mechanical frequency levels most structures are subject to a large number of close natural frequencies representing a variety of different families of vibration modes. It is desirable therefore to be able to identify the mode shapes near to the desired operating frequency and to redesign the system to move unwanted modes away from the region of concern. As with most vibration analysis problems it is unwise to rely upon a single method of frequency determination and so this investigation used modal analysis to identify first the location of frequencies, then electronic speckle pattern interferometry to identify mode shapes and finally modal analysis supported by a commercially available software package to effect modifications to the structure. Finite element analysis was undertaken alongside the experimental investigation so that a mathematical model could be developed which could then be used to verify the effect upon frequency shifting for the modifications predicted by the modal analysis.

LABORATORY CONFIGURATION

This investigation was based upon a thorough analysis of the performance of an ultrasonically excited thick cylinder as shown in Fig.1.

Fig. 1.   Thick cylinder with ultrasonic exciter

Vibration of the thick cylinder was stimulated by a magnetostrictive acoustic horn rigidly attached to the circumference of the cylinder. A magnetostrictive exciter was preferred to the more efficient piezo-electric horn as it offered the opportunity to excite the system over a fairly wide range of frequencies, whereas a piezo-electric horn being highly tuned with a high Q factor, offers little opportunity for vibration excitation away from its tuned resonance.

The initial vibration analysis concentrated on using accelerometers to detect motion. To reduce the effect upon the modal characteristics caused by the added mass of the accelerometers to the surface of the cylinder, light weight accelerometers were attached at single discrete locations and were moved for each sample point of investigation. The sub-miniature accelerometers were selected with linear characteristics beyond 30 kHz and the associated output signals were monitored through modified high gain amplifiers capable of a linear output characteristic up to 25 kHz. A suitable adhesive had to be identified to withstand the high g values generated by small amplitude ultrasonic motion.

Initial investigations into the modal characteristics of the system were attempted using two accelerometers. One accelerometer was attached close to the excitation point alongside the joint between horn and cylinder to act as a reference whilst the second accelerometer was moved around the circumference of the cylinder to monitor the response relative to the reference. This approach effectively identified the natural frequencies which produced modal deflection radially and assisted in identifying nodal lines on the surface of the cylinder. However the data was considerably limited in its value as no significance could be placed upon its relevance for further modal analysis.

For progress to be made it was essential to be able to either monitor the force at the injection point or to identify a signal which could be considered to be both in-phase with and proportional to the force input. It was impractical to insert a force transducer between the horn and the cylinder as this upset the tuned characteristics of the system. Attempts to insert a thin piezo-electric film at the horn/cylinder interface also failed because the requirement for a positive connection between horn and cylinder conflicted with the flexibility required in the force sensitive film.

Attention was therefore turned to monitoring the voltage/current relationship supplied to the coils in the horn. Frequency response functions (FRFs) of the acceleration/current and acceleration/voltage were measured.

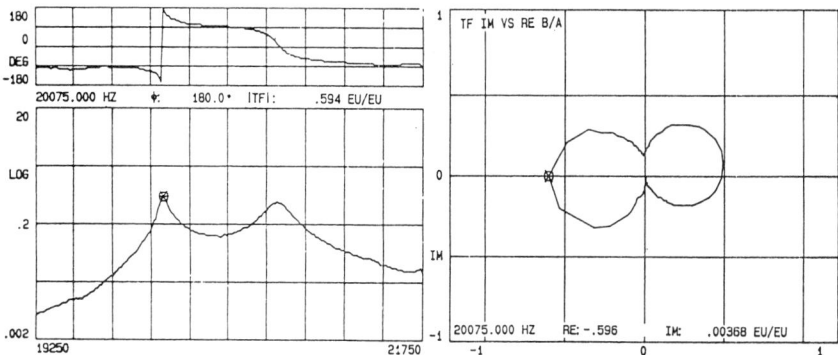

Fig. 2.   FRF from acceleration/voltage measurement

It was noticed that the phase information from the frequency response functions in both cases exhibited a phase shift at each resonance frequency as expected but the phase angles were 90° shifted from those obtained by a conventional acceleration/force FRF. The result was that plots of the frequency response function generated using the coil voltage in place of the conventional force signal when displayed in Nyquist form, produced circular plots displaced by approximately one quadrant and exhibited an equal phase shift for all natural frequencies. This is illustrated in Fig.2.

A signal proportional to the voltage across the windings could therefore be used as a signal which represented the force being transmitted at the interface.

INITIAL MODAL ANALYSIS INVESTIGATION

The configuration used for modal analysis followed conventional practice. An isolated signal generator was used to drive the amplifier of the acoustic horn, whilst a dual channel fast fourier analyser with an extended frequency range was used to monitor both the reference signal and the monitoring sensor. This is illustrated in the line diagram of Fig.3 where additional equipment is shown linked to the analyser for later mathematical processing for the structural modification to be described later in the paper.

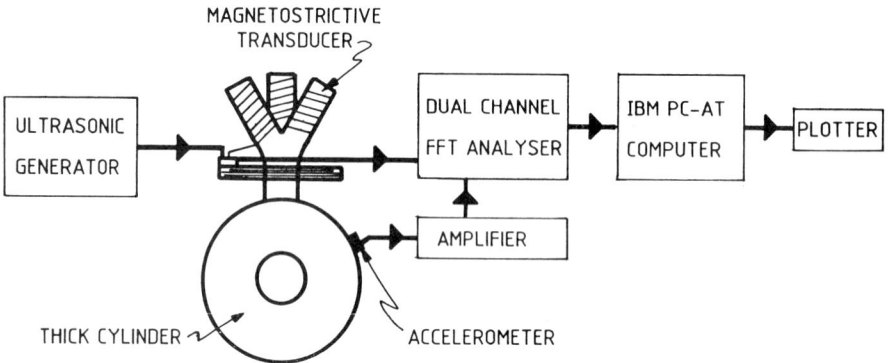

Fig. 3.   Schematic layout of modal analysis

The initial investigation using two accelerometers identified several natural frequencies around the chosen fundamental frequency. A typical frequency response spectrum (FRS) for the monitoring accelerometer is shown in Fig.4

The FRS can be seen to identify several frequencies of interest and highlights the likelihood of coupled natural frequencies caused by different mode families with natural frequencies close to each other. Those which exhibited strong resonance were investigated in an attempt to identify the mode shapes and also to give guidance to the electronic speckle pattern interferometry analysis.

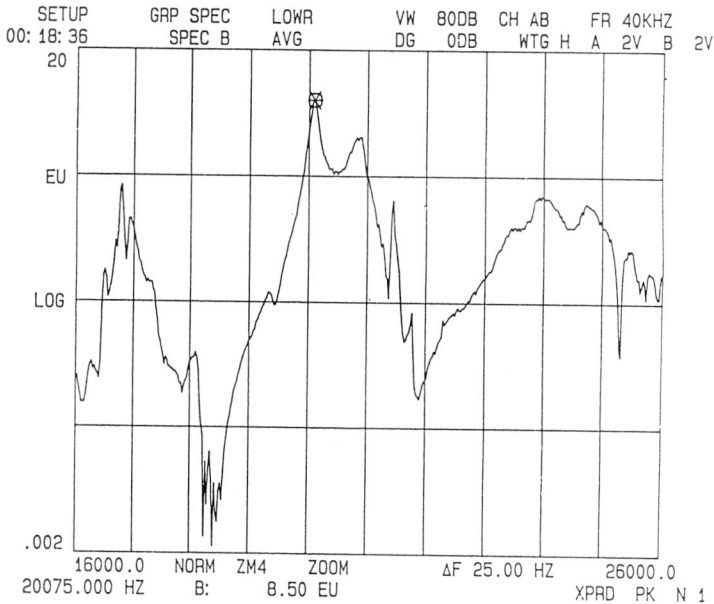

Fig. 4.   Frequency response spectrum for an
accelerometer attached to the outer
surface of the thick cylinder.

ELECTRONIC SPECKLE PATTERN INTERFEROMETRY

Electronic speckle pattern interferometry (ESPI) was selected as this
technique had the capability of visually displaying a representation of the
amplitude of motion of all points on the vibrating surface. No difficulties
were experienced with the use of this technique at ultrasonic frequencies
under consideration, although care had to be exercised with the
interpretation of the images produced. Similar findings were noted during
vibration studies of turbocharger blading (Chapman and Wang, 1989). The
technique depends basically on the interference of two laser beams which
originate from a single source. One beam is caused to travel a fixed
distance representative of the distance between the source and the object
under investigation, whilst the other beam travels directly between the
source and the object under investigation before recombination of both beams
is effected in a video monitor. Small alterations in the path length of the
second beam cause interference with the original beam and consequently
either reinforce or cancel the beam intensity depending upon whether the
difference in path length is a whole wavelength or a half wavelength. The
image is then displayed on a monitor and the fringe patterns produced give
an indication of the objects vibratory motion.

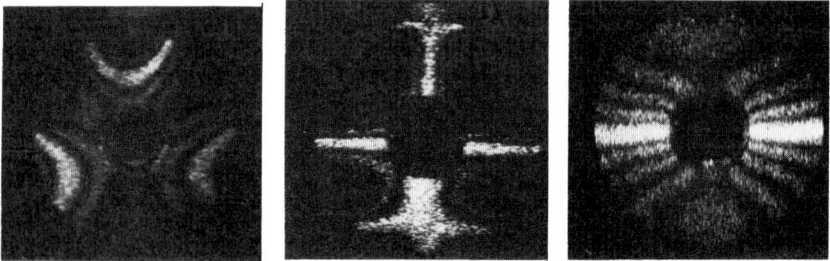

Fig. 5. ESPI images obtained for R3 mode
         a) out-of-plane image
         b) in-plane image measured in horizontal axis
         c) in-plane image measured in vertical axis

This technique however has to be applied in one axis at a time (Shellabear and Tyrer, 1988) and by adjusting the optical configuration it is possible to measure either out-of-plane motion or in-plane motion.

This technique was applied to the ultrasonically excited thick cylinder to identify the true three dimensional mode shapes. Figure 5 displays the recorded ESPI images for the thick cylinder vibrating in R3 mode, where the cylinder is deforming with three radial waves.

FINITE ELEMENT MODELLING

To support the analysis a mathematical model of the thick cylinder was developed so that future modifications could be simulated as a check on any predicted alterations from the experimental analysis. An unrestrained boundary constraint was used as this most closely represented the configuration for the experimental set-up, where the thick cylinder was suspended at the free anti-node of the acoustic horn.

Initial investigations concentrated on modelling the cylinder with two dimensional elements as these offered an economic solution for radial mode identification. By modelling the cylinder along the cylinder axis and using axi-symmetric analysis it was also possible to determine estimates of the cylinders torsional vibration modes. A more detailed understanding of the thick cylinder vibration characteristics was determined using three dimensional elements to model the complete thick cylinder. Increase in computational time for analysis of all modes was limited by adopting a segmentation analysis of the structure using an increased density of elements to achieve the desired accuracy of solution. Visualisation of mode shapes at the determined natural frequencies was enhanced by using an inter-active graphics package to rotate the three dimensional model. The mesh used and the results obtained for the R3 mode are illustrated in Fig.6.

The three dimensional approach proved to be beneficial for later use as a comparison model for the structural modification analysis, where the ability to adjust mass at discrete points was required. It also offered the potential for future modelling of the geometries used for ultrasonically assisted metal forming tools.

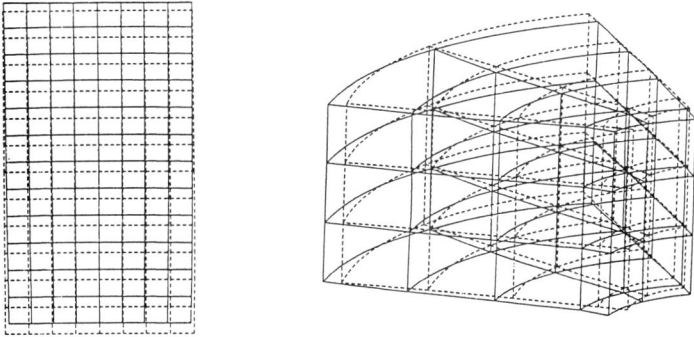

Fig. 6.   Finite element representation of thick cylinder.
           a) deflected shape for R3 mode by 2D analysis
           b) deflected shape for R3 mode by 3D segment analysis

FULL MODAL ANALYSIS OF THICK CYLINDER

The predicted modal patterns obtained by finite element analysis were verified by the ESPI analysis results. As a further check a more detailed modal analysis was carried out based upon the voltage across the coils of the ultrasonic horn as an indicator of the transmitted force.

Monitoring of the response at discrete points around the vibrating cylinder was carried out using both a proximity probe and sub-miniature accelerometers. The displacement probe was a modified inductive proximity device which used a carrier frequency well in excess of the ultrasonic range of concern. Electronic filters were fitted to reject all low frequency signals and therefore reduced the need for rigid mountings to ensure absolute amplitude readings. However as a consequence of the low amplitudes associated with high frequency vibrations, detection of all but the most dominant frequencies was difficult.

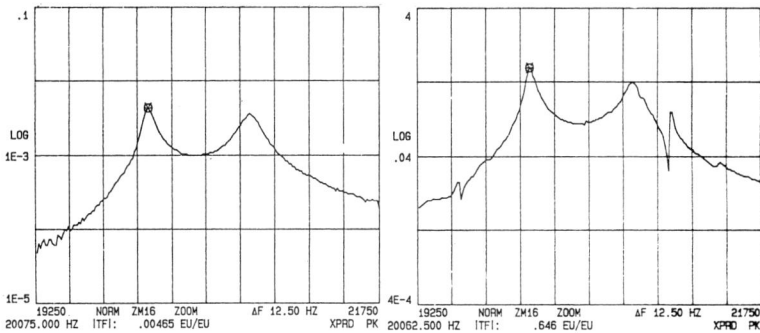

Fig. 7.   Frequency response function
           a) obtained by displacement probe
           b) obtained by accelerometer

Figure 7a shows the frequency response spectrum obtained using the displacement probe. The accelerometers adequately detected natural frequencies but introduced added mass on the circumference of the cylinder. Attachment of the accelerometers to the surface for each location under test required considerable care to ensure a strong bond capable of withstanding the ultrasonic surface vibrations. However, the geometric size of the sub-miniature accelerometers offered the advantage of being able to monitor at a wider range of locations than was possible with the inductance probe.

The frequency response function shown in Fig.7b illustrates the benefits of accelerometers for detecting the less dominant natural frequencies. Fig.7 also illustrates the problem of close modes. Particular attention must be paid to mode coupling, due to damping at each natural frequency, especially around the desired operating frequency, since mode switching can seriously affect the reliability of the metal forming process. Modal analysis proceeded by computer-aided structural modification can be used to attempt to separate close modes.

STRUCTURAL MODIFICATION

Data collected on the analyser was down-loaded through an IEEE interface to a personal computer on which was mounted a state-of-the-art structural modification package. This package was able to simulate the vibratory motion of the measured modal data after first curve fitting. Both single degree of freedom (SDOF) and multidegree of freedom (MDOF) curve fit procedures were available to deal with the problem of close modes. The software was capable of animating the vibrating object and this facility coupled with the ability to rotate the viewing angle considerably assisted with identification of modal behaviour. Typical deflected mode shapes obtained from the animation process for two identified natural frequency mode shapes are shown in Fig.8.

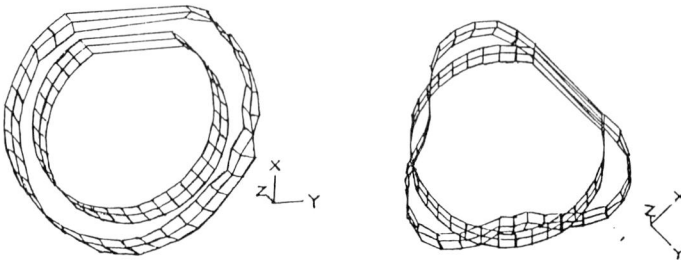

Fig. 8. Deflected mode shapes obtained by modal analysis.
a) fundamental R0 mode
b) R3 mode

Investigation of the effect of changes in mass addition or stiffness addition to discrete points of the structure under test were possible. An option existed to either specify the mass at a point for the software to predict frequency alterations or for the user to specify the desired frequency change for a specific modification location with the software predicting the mass required.

An attempt was made to separate the fundamental (R0) mode and the third harmonic (R3) mode, maintaining R0 at its original frequency, by two methods:

1. Mass was added at R3 antinodal positions on the cylinder circumference and subtracted at nodal positions
2. Mass was subtracted at R3 antinodes and added at nodes.

The results of the modifications are presented in Table 1. This approach was successfully used to shift unwanted natural frequencies away from the operating frequency and thus demonstrated the ability to be able to design ultrasonic metal forming dies.

Table 1. Effect of mass modification on cylinder natural frequencies.

MODE	CYLINDER NATURAL FREQUENCIES (kHz)		
	ORIGINAL CYLINDER	CYLINDER (Case 1)	CYLINDER (Case 2)
RO	20.08	20.38	19.98
R3	20.81	17.60	21.42

CONCLUSIONS

By making use of modal analysis, electronic speckle pattern interferometry and finite element analysis it has been possible to adjust the modal characteristics of an ultrasonically excited thick cylinder by separating natural frequencies and shifting them away from around the operating fundamental mode. All three methods are applicable to vibration analysis at the lower end of the ultrasonic frequency range. Selection of lightweight accelerometers to ensure minimum alteration to the mass of the structure and suitable choice of adhesive to withstand the high acceleration values are areas that require careful consideration.

ACKNOWLEDGEMENTS

The authors wish to thank the Science and Engineering Research Council for its financial support of this project through its Specially Promoted Programme for High Speed Machinery, the Metalbox Company for collaboration with the research and the Department of Mechanical Engineering at Loughborough University of Technology for the research facilities provided.

REFERENCES

Chapman G.M. and Wang X., 1988, Interpretation of experimental and theoretical data for prediction of mode shapes of vibrating turbocharger blades, Journal of Vibration, Acoustics, Stress, and Reliability in Design, ASME, vol 110, No 1, pp 53-58.
Jones J.B., 1968, Tube drawing, draw ironing, flare and flange forming with an ultrasonic assist, Metal Progress.
Langenecker B., 1978, Ultrasonically activated wire drawing, Wire Journal.
Lucas M. and Chapman G.M., 1989, Vibration analysis at ultrasonic frequencies, 12th Biennial Conf. on Mechanical Vibration and Noise, ASME, Montreal, Canada.
Young M., Winsper C. and Sansome D., 1970, Radial mode vibrators for oscillatory metal forming, Applied Acoustics, vol 3.

Prediction of Vibrational Power Transmission through Jointed Beams

J. L. HORNER and R. G. WHITE

*Institute of Sound and Vibration Research, The University,
Southampton SO9 5NH, UK*

ABSTRACT

When attempting to control the vibration transmitted from a machine into and through the structure upon which it is mounted, it is desirable to be able to identify and quantify the vibration transmission paths in the structure. Knowledge of transmission path characteristics enables procedures to be carried out, for example, to reduce vibration levels at points remote from the source, perhaps with the objective of reducing unwanted radiation of sound. One method for obtaining transmission path information is to use the concept of vibrational power transmission.

This paper is concerned with the prediction of vibrational power transmission through a system of jointed beams carrying longitudinal and flexural waves. Models have been developed which determine the wave type which carries most power in each section of the system. By establishing the wave type which predominantly causes power transmission it is then possible to apply the most suitable vibration control technique.

KEYWORDS

Vibration control; vibrational power transmission; flexural waves; longitudinal waves; joints in beam.

INTRODUCTION

When attempting to control the vibration levels transmitted from a machine through the various connections to the mounting structure it is mounted on, it is desirable to be able to identify and quantify the vibration paths in the structure. By absorbing the mechanical energy along the propagation paths in some convenient manner, it is possible to reduce the sound radiated from the structure. One method of obtaining path information is to use the concept of vibrational power transmission. It allows the direction of propagation to be determined and magnitude to be assigned to each path.

By using the concepts developed by Pavic (1976) and implemented by Redman-White (1984), it is possible to measure time-averaged power flow in a beam to within half a wavelength of a discontinuity. The work presented in this paper is concerned with the effect of joints on the power transmission in beams. Methods have been developed for predicting the effect of T-joints on the power transmission in uniform beams. Before it is possible to predict power transmission, it is necessary to know the effect of a joint on an incident wave. Work on beams with bends or branched joints has been carried out by Lee and Kolsky (1972), Desmond (1981), Rosenhouse et al (1981), Doyle and Kamle (1987) and Gibbs and Tattersall (1987). By using the reflection and transmission coefficients for joints, it is possible to predict the vibrational power associated with flexural and longitudinal waves in each arm of the joint.

POWER TRANSMISSION IN A JOINTED BEAM

From a knowledge of the amplitude of a travelling wave in a beam, it is possible to predict the power transmission in the beam. Expressions may be derived from fundamental theories (Bishop and Johnson, 1960) of flexural and longitudinal vibration which relate time averaged power to travelling wave amplitude. These expressions are:

For flexural power

$$<P>_f = EI\omega k^3_f |A_f|^2 \tag{1}$$

and for longitudinal power

$$<P>_l = \frac{1}{2} EAk_l\omega |A_l|^2 \tag{2}$$

Before it is possible to calculate the power transmission through a discontinuity in a beam, it is necessary to predict the proportion of flexural and longitudinal waves reflected and transmitted. The amplitudes of the reflected and transmitted waves are found by solving sets of simultaneous equations derived from the continuity and equilibrium conditions at the joint.

Conditions of Continuity and Equilibrium at a Joint

Figure 1 shows a branched beam. The wave motion in each arm is as follows. Expressions in **bold** script only apply when the joint is impinged on by a flexural wave and expressions in *italic* script apply when the joint is impinged on by a longitudinal wave. These expressions describe the impinging wave.

Arm 1
$$W_1(x,t) = (A_1 e^{k_f x} + A_3 e^{ik_f x} + \mathbf{A_4 e^{-ik_f x}}) e^{i\omega t} \tag{3}$$
$$U_1(x,t) = (A_1 e^{ik_l x} + A_l e^{-ik_l x}) e^{i\omega t} \tag{4}$$

Arm 2
$$W_2(\psi,t) = (B_2 e^{-k_f \psi} + B_4 e^{-ik_f \psi}) e^{i\omega t} \tag{5}$$
$$U_2(\psi,t) = B_1 e^{-ik_l \psi} e^{i\omega t} \tag{6}$$

Arm 3
$$W_3(\Theta,t) = (C_2 e^{ik_f \Theta} + C_4 e^{-ik_f \Theta}) e^{i\omega t} \tag{7}$$
$$U_3(\Theta,t) = C_l e^{-ik_l \Theta} e^{i\omega t} \tag{8}$$

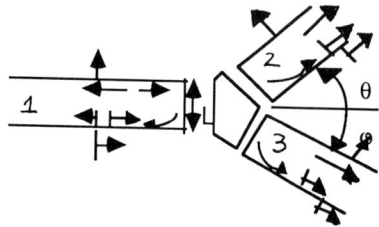

FIG1 : WAVE MOTION IN A BRANCHED BEAM

where A_4, A_3, B_4 and C_4 are travelling flexural waves. A_1, B_2 and C_2 are near field flexural waves and A_l, B_l and C_l are longitudinal travelling waves. Doyle and Kamle (1987) assumed that the joint between the three beams was a rigid mass.

Joint mass, $M_j = \dfrac{\rho_j \pi L^2 W_j}{4}$ Moment of inertia of joint, $I_j = \dfrac{M_j L^2}{8}$

By considering the conditions of continuity and equilibrium at the joints, the following nine statements can be written.

$$U_1 = U_2 \cos\theta - W_2 \sin\theta + \frac{L}{2} \sin\theta \frac{\partial W_2}{\partial \psi} \qquad (9)$$

$$U_1 = U_3 \cos\Phi - W_3 \sin\Phi + \frac{L}{2} \sin\Phi \frac{\partial W_3}{\partial \Theta} \qquad (10)$$

$$W_1 = U_2 \sin\theta - W_2 \cos\theta - \frac{L}{2}(1 + \cos\theta) \frac{\partial W_2}{\partial \psi} \qquad (11)$$

$$W_1 = U_3 \sin\Phi + W_3 \cos\Phi - \frac{L}{2}(1 + \cos\Phi) \frac{\partial W_3}{\partial \Theta} \qquad (12)$$

$$\frac{\partial W_1}{\partial x} = \frac{\partial W_2}{\partial \psi} \qquad (13)$$

$$\frac{\partial W_1}{\partial x} = \frac{\partial W_3}{\partial \Theta} \qquad (14)$$

$$E_1 I_1 \frac{\partial^2 W_1}{\partial x^2} + \frac{1}{2} E_1 I_1 \frac{\partial^3 W_1}{\partial x^3} = E_2 I_2 \frac{\partial^2 W_2}{\partial \psi^2} + E_3 I_3 \frac{\partial^2 W_3}{\partial \Theta^2} -$$

$$- \frac{L}{2} E_2 I_2 \frac{\partial^3 W_2}{\partial \psi^3} - \frac{L}{2} E_3 I_3 \frac{\partial^3 W_3}{\partial \Theta^3} - I_j \frac{\partial W_1}{\partial x} \qquad (15)$$

$$E_1 A_1 \frac{\partial U_1}{\partial x} = E_2 A_2 \frac{\partial U_2}{\partial \psi} \cos\theta + E_3 A_3 \frac{\partial U_3}{\partial \Theta} \cos\Phi + E_2 I_2 \frac{\partial^3 W_2}{\partial \psi^3} \sin\theta +$$

$$+ E_2 I_2 \frac{\partial^3 W_3}{\partial \Theta^3} \sin\Phi - mj \frac{\partial^2 U_1}{\partial t^2} \qquad (16)$$

$$- E_1 I_1 \frac{\partial^3 W_1}{\partial x^3} = E_2 A_2 \frac{\partial U_2}{\partial \psi} \sin\theta + E_3 A_3 \frac{\partial U_3}{\partial \Theta} \sin\Phi - E_2 I_2 \frac{\partial^3 W_2}{\partial \psi^3} \cos\theta -$$

$$- E_3 I_3 \frac{\partial^3 W_3}{\partial \Theta^3} \cos\Phi - mj \frac{\partial}{\partial t^2} \left(W_1 - \frac{L}{2} \frac{\partial W_1}{\partial x} \right) \qquad (17)$$

Solution for Power Transmission

By substituting the wave equations (3 - 8) into the above, it is possible to form nine simultaneous equations. The nine unknown wave coefficients can be found by solving the set of simultaneous equations. The matrix of the equations is given in Appendix 2. It was inverted using Choleski's method (A.J. Spencer et al, 1977) of lower and upper decomposition. By substituting the correct wave amplitudes into equations (1) and (2), it is possible to determine the transmitted longitudinal and flexural power in each arm. The following figures show the power transmitted in each arm for a system with arm 2 at various angles between zero and 180° and arm 3 set at zero angle. For each arm of the system, the percentage of impinging power is plotted against the angle of arm 2. The notation used on each curve is described in Appendix 1.

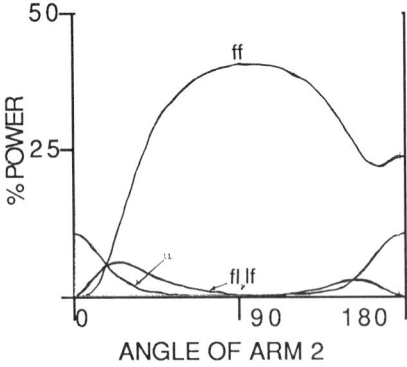

FIG 2 : REFLECTED POWER IN ARM 1

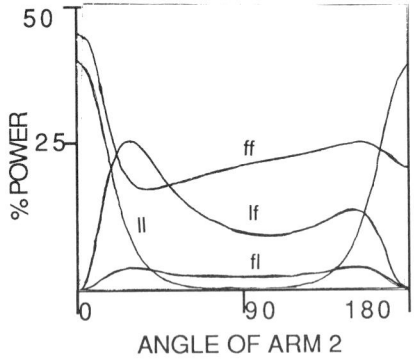

FIG 3: TRANSMITTED POWER IN ARM 2

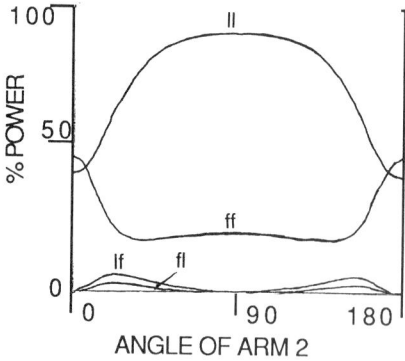

FIG 4: TRANSMITTED POWER ARM 3

Figure 2 shows the reflected power in arm 1 of the joint system. As the angle of arm 2 varies from zero to 180°, the proportions of reflected power in the arm change. It can be seen that different impinging wave types produce different power reflection characteristics. For example, the flexural power produced by a flexural wave impinging is a maximum at 90°, whilst the longitudinal power produced by a longitudinal wave impinging is a maximum at both 24° and 156°. Similar trends can be seen in figs. 3 and 4, the power transmission in arms 2 and 3 respectively.

Power transmission in a system with the angle of arm 2 varying between zero and 180° and arm 3 set at 90° is shown in figs. 5 to 7.

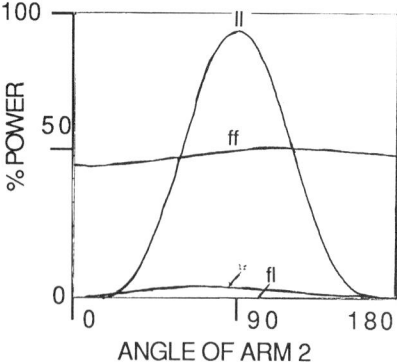

FIG 5: REFLECTED POWER IN ARM 1

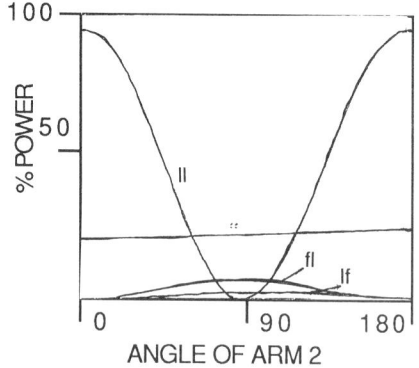

FIG 6: TRANSMITTED POWER IN ARM 2

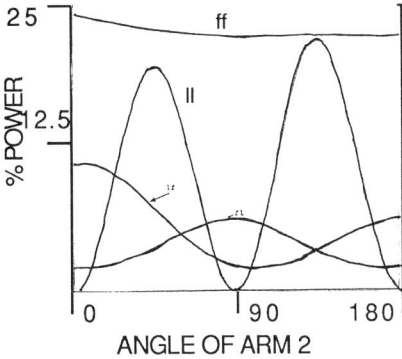

ANGLE OF ARM 2

FIG 7 : TRANSMITTED POWER IN ARM 3

Again, it can be seen that different impinging wave types produce different reflection and transmission characteristics. From both sets of figures (i.e., 2 - 4 and 5 - 7) it can be determined that:-

(i) in the arm carrying the impinging wave, the dominant power component is of the same wave type as the impinging wave;

(ii) when a flexural wave impinges on a joint, the transmitted flexural power is approximately constant and almost independent of arm angle, except at extreme angles;

(iii) when a longitudinal wave impinges on a joint, minimum transmitted longitudinal power is induced in an arm when it is 90° off axis;

(iv) when a flexural wave impinges on a joint, minimum transmitted longitudinal power is induced when the angle of arm 2 is zero;

(v) when a longitudinal wave impinges on a joint, the majority of nett power is induced in the arm with the least off axis angle whilst when a flexural wave impinges, the majority of nett power is carried in the arm with the greatest off axis angle.

Thus, it can be concluded that when designing a beam system for either minimum of maximum power transmission, care must be taken in the choice of arm angle. It is possible to greatly change the power transmission in an arm by only changing the arm angle by a few degrees. It should also be noted that different impinging wave types produce different power transmission characteristics. It is important to determine the wave type impinging on a joint before the angles of the joint are chosen. An arm angle which minimises power when a flexural wave impinges may maximise it when a longitudinal wave impinges on the joint.

CLOSED FORM SOLUTION FOR MULTIPLE REFLECTIONS

In a finite system of beams and joints, multiple reflections occur between two discontinuities. Figure 8 shows such a system at four different time intervals.

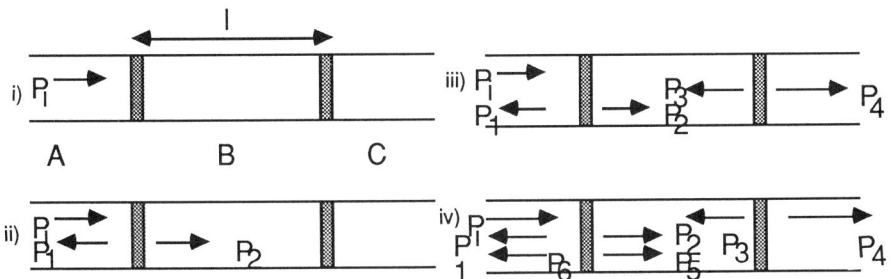

FIG 8 : MULTIPLE POWER REFLECTIONS IN A FINITE SYSTEM

Obviously, it would be of great use to obtain one expression to describe the sum of all the reflections within a section of beam. The following theory assumes that there is no power dissipation in the system. If Pi is the power impinging on discontinuity 1 with a reflection coefficient of r_1 and a transmission coefficient t_1, then

Reflected power $P_1 = r_1$ Pi and Transmitted power $P_2 = t_1$ Pi .

Then, if P_2 impinges on discontinuity 2 with a reflection coefficient of r_2 and a transmission coefficient of t_2, then

Reflected power $P_3 = r_2 P_2 = r_2 t_1$ Pi and Transmitted power $P_4 = t_2 P_2 = t_1 t_2$ Pi

Now P_3 impinges on discontinuity 1 from the -ve direction, its reflection coefficient is now r_3 and transmission coefficient now t_3.

Reflected power $P_5 = r_3 P_3 = r_2 r_3 t_1$ Pi and Transmitted power $P_6 = t_3 P_3 = r_2 t_1 t_3$ Pi

P_5 impinges on discontinuity 2 again.

Reflected power $P_7 = r_2 P_5 = r_2^2 r_3 t_1$ Pi and Transmitted power $P_8 = t_2 P_5 = r_2 r_3 t_1 t_2$ Pi

P_7 impinges on discontinuity 1.

Reflected power $P_9 = r_3 P_7 = r_2^2 r_3^2 t_1$ Pi and Transmitted power $P_{10} = t_3 P_7 = r_2^2 r_3 t_1 t_3$ Pi

P_9 impinges on discontinuity 2.

Reflected power $P_{11} = r_2 P_9 = r_2^3 r_3^2 t_1$ Pi and Transmitted power $P_{12} = t_2 P_9 = r_2^2 r_3^2 t_1 t_2$ Pi

Thus nett power in section A,

$$Nett_A = Pi - P_1 - P_6 - P_{10} - ...$$

$$= Pi \ (1 - r_1 - r_2 t_1 t_3 (1 + r_2 r_3 + r_2^2 \ r_3^2 + ...)$$

The bracketed term is an infinite series but if r_2 and r_3 are less than 1, the series is convergent and therefore has a sum which yields:

$$Nett_A = Pi \left(1 - r_1 - \frac{r_2 t_1 t_3}{1 - r_2 r_3} \right) \tag{18}$$

Similarly, the nett power in section B,

$$Nett_B = Pi \left(t_1 \left(\frac{1 - r_2}{1 - r_2 r_3} \right) \right) \tag{19}$$

and the nett power in section C,

$$Nett_C = \frac{Pi t_1 t_2}{1 - r_2 r_3} \tag{20}$$

where the reflection coefficients are

$r_1 = | \ \alpha \ |^2$ where α = ratio of reflected and impinging wave amplitudes for joint 1

$t_1 = R \mid \beta \mid^2$

where for flexural waves $R = \dfrac{E_B I_B k^3_B}{E_A I_A k^3_A}$

and for longitudinal waves $R = \dfrac{E_B A_B k_B}{E_A A_A k_A}$

β = ratio of transmitted and impinging wave amplitude for joint 1

$r_2 = \mid \gamma \mid^2$

where γ = ratio of amplitudes of wave reflected by and wave impinging on joint 2

$t_2 = Q \mid \varepsilon \mid^2$

where for flexural waves $Q = \dfrac{E_A I_A k^3_A}{E_B I_B k^3_B}$

and for longitudinal waves $Q = \dfrac{E_A I_A k_A}{E_B I_B k_B}$

ε = ratio of amplitudes of wave transmitted and wave impinging on joint 2

r_3 and t_3 are of the same form as r_2 and t_2 but with the correct material properties and wave amplitudes inserted. Thus, it is possible to determine the nett power in a beam between two discontinuities from a knowledge of the reflection and transmission coefficients at the discontinuities.

CONCLUSIONS

It has been shown that it is possible to make predictions for the power transmitted through different arms of a joint. From these predictions it is possible to determine the proportions of nett power which are associated with each wave type. This greatly assists the process of applying vibration control techniques to the structure. Also developed in this paper are closed form solutions for the multiple power transmission within finite sections of structures. The two sections of theory may be jointly applied to assess the power transmission through a multiple joint structure. Thus, it would be possible to design a system for optimal power transmission.

ACKNOWLEDGEMENT

This work has been carried out with the support of the Procurement Executive, Ministry of Defence.

REFERENCES

Bishop, R.E.D. and D.C. Johnson (1960). *The Mechanics of Vibration.* Cambridge University Press.

Desmond, T.P. (1981). Theoretical and experimental investigation of stress waves at a junction of three bars. *J. Appl. Mech.,* 48, 148-154.

Doyle, J.F. and S. Kamle (1987). An experimental study of the reflection and transmission of flexural waves at an arbitrary T-joint. *J. Appl. Mech.,* 54, 136-140.

Gibbs, B.M. and J.D. Tattersall (1987). Vibrational energy transmission and mode conversion at a corner junction of square section rods. *J. Vib. Acoust., Stress, Rel. Design,* 109, 348-355.

Lee, J.P. and H. Kolsky (1972). The generation of stress pulses at the junction of two non-collinear rods. *J. Appl. Mech.,* 39, 809-813.

Pavic, G. (1976). Techniques for the determination of vibration transmission mechanisms in structures. PhD Thesis, University of Southampton.

Redman-White, W. (1984). The measurement of structural wave intensity. PhD Thesis, University of Southampton.

Rosenhouse, G., H. Ertel and F.P. Mechel (1981). Theoretical and experimental investigations of structure-borne sound transmission through a T-joint in a finite system. *JASA,* 70, 492-499.

Spencer, A.J.M. et al (1977). *Engineering Mathematics,* Vol. 1, Van Nostrand Reinhold.

APPENDIX 1

Notation

A	-	area
A_{suffix}	-	wave amplitude
B_{suffix}	-	wave amplitude
C_{suffix}	-	wave amplitude
E	-	Young's modulus of elasticity
I	-	moment of inertia
k_f	-	flexural wave number
k_l	-	longitudinal wave number
L	-	radius of joint
l	-	length
m_j	-	mass of joint
$<P>_f$	-	time averaged flexural power
$<P>_l$	-	time averaged longitudinal power
t	-	time
U(x,t)	-	longitudinal wave displacement
W(x,t)	-	flexural wave displacement
W_j	-	width of joint
x	-	distance
θ	. -	angle of arm 2 of joint
ρ	-	density
ψ	-	distance
ω	-	frequency in radians per sec
Θ	-	distance
Φ	-	angle of arm 3 of joint
ff	-	flexural power produced at a joint by an incident flexural wave
fl	-	flexural power produced at a joint by an incident longitudinal wave
lf	-	longitudinal power produced at a joint by an incident flexural wave
ll	-	longitudinal power produced at a joint by an incident longitudinal wave

APPENDIX 2

A = amplitude of impinging wave of either type

$a = \dfrac{A_1}{A}$ $\qquad b = \dfrac{B_2}{A}$ $\qquad c = \dfrac{A_3}{A}$ $\qquad d = \dfrac{B_4}{A}$ $\qquad e = \dfrac{C_2}{A}$

$f = \dfrac{C_4}{A}$ $\qquad g = \dfrac{C_1}{A}$ $\qquad h = \dfrac{B_1}{A}$ $\qquad i = \dfrac{A_1}{A}$

The matrix to be solved is of the form

$$[X]\,[Y] = [Z]$$

where

$X_{11} = k_{f1}$ $\quad X_{12} = 0$ $\quad X_{13} = ik_{f1}$ $\quad X_{14} = 0$ $\quad X_{15} = k_{f3}$ $\quad X_{16} = ik_{f3}$
$X_{17} = 0$ $\quad X_{18} = 0$ $\quad X_{19} = 0$ $\quad X_{21} = k_{f1}$ $\quad X_{22} = k_{f2}$ $\quad X_{23} = ik_{f1}$
$X_{24} = ik_{f2}$ $\quad X_{25} = 0$ $\quad X_{26} = 0$ $\quad X_{27} = 0$ $\quad X_{28} = 0$ $\quad X_{29} = 0$

$X_{31} = E_1 I_1 k^3_{f1} + mj\omega^2(1 - \frac{1}{2}k_{f1})$ $\qquad X_{32} = E_2 I_2 k^3_{f2}\cos\theta$ $\qquad X_{33} = mj\omega^2(1 - i\frac{L}{2}k_{f1}) - E_1 I_1 ik^3_{f1}$

$X_{34} = - E_2 I_2 ik^3_{f2}\cos\theta$ $\qquad X_{35} = E_3 I_3 k^3_{f3}\cos\Phi$ $\quad X_{36} = - E_3 I_3 k^3_{f3}\cos\Phi$

$X_{37} = - E_3 A_3 ik_{l3}\sin\Phi$ $\qquad X_{38} = - E_2 A_2 ik_{l2}\sin\theta$ $\quad X_{39} = 0$

$X_{41} = 0$ $\qquad X_{42} = \sin\theta(1 + \frac{L}{2}k_{f2})$ $\quad X_{43} = 0$ $\qquad X_{44} = \sin\theta(1 + \frac{L}{2}k_{f2})$

$X_{45} = 0$ $\qquad X_{46} = 0$ $\qquad X_{47} = 0$ $\qquad X_{48} = - \cos\theta$ $\quad X_{49} = 1$ $\quad X_{51} = 1$

$X_{52} = 0$ $\qquad X_{53} = 1$ $\qquad X_{54} = 0$ $\qquad X_{55} = - \cos\Phi - \frac{L}{2}(1 + \cos\Phi)\,k_{f3}$

$X_{56} = - \cos\Phi - \frac{L}{2}(1 + \cos\Phi)\,ik_{f3}$ $\quad X_{57} = - \sin\Phi$ $\quad X_{58} = 0$ $\qquad X_{59} = 0$

$X_{61} = - E_1 I_1 k^2_{f1} - \frac{L}{2}E_1 I_1 k^3_{f1} + Ij\omega^2 k_{f1}$ $\qquad X_{62} = E_2 I_2 k^2_{f2} + \frac{L}{2}E_2 I_2\, k^3_{f2}$

$X_{63} = E_1 I_1 k^2_{f1} + ik^3_{f1}\frac{L}{2}E_1 I_1 + Ij\, ik_{f1}\omega^2$ $\qquad X_{64} = - E_2 I_2\, k^2_{f2} - \frac{L}{2}E_2 I_2\, ik^3_{f2}$

$X_{65} = E_3 I_3 k^2_{f3} + \frac{L}{2}E_3 I_3 k^3_{f3}$ $\qquad X_{66} = - E_3 I_3 k^2_{f3} - \frac{L}{2}E_3 I_3 ik^3_{f3}$

$X_{67} = 0$ $\qquad X_{68} = 0$ $\qquad X_{69} = 0$ $\qquad X_{71} = 0$
$X_{72} = 0$ $\qquad X_{73} = 0$ $\qquad X_{74} = 0$

$X_{75} = \sin\Phi + \frac{L}{2}k_{f3}\sin\Phi$ $\qquad X_{76} = \sin\Phi + i\frac{L}{2}k_{f3}\sin\Phi$

$X_{77} = - \cos\Phi$ $\quad X_{78} = 0$ $\qquad X_{79} = 1$ $\qquad X_{81} = 1$

$X_{82} = - \cos\theta - \frac{L}{2}(1 + \cos\theta)\,k_{f2}$ $\quad X_{83} = 1$ $\qquad X_{84} = - \cos\theta - \frac{L}{2}(1 + \cos\theta)\,ik_{f2}$

$X_{85} = 0$ $\qquad X_{86} = 0$ $\qquad X_{85} = 0$ $\qquad X_{88} = - \sin\theta$ $\quad X_{89} = 0$ $\qquad X_{91} = 0$

$X_{92} = - E_2 I_2 k^3_{f2}\sin\theta$ $\qquad X_{93} = 0$ $\qquad X_{94} = E_2 I_2 ik^3_{f2}\sin\theta$

$X_{95} = - E_3 I_3 k^3_{f3}\sin\Phi$ $\qquad X_{96} = E_3 I_3 ik^3_{f3}\sin\Phi$ $\qquad X_{97} = - E_3 A_3\, ik_{l3}\cos\Phi$

$X_{98} = - E_2 A_2\, ik_{l2}\cos\theta$ $\qquad X_{99} = mj\omega^2 - E_1 A_1 ik_{l1}$

$Y_1 = a$ $\qquad Y_2 = b$ $\qquad Y_3 = c$ $\qquad Y_4 = d$ $\qquad Y_5 = e$
$Y_6 = f$ $\qquad Y_7 = g$ $\qquad Y_8 = h$ $\qquad Y_9 = i$

$Z_1 = ik_{f1}$ $\qquad Z_2 = ik_{f1}$ $\qquad Z_3 = - mj\omega^2(1 - i\frac{L}{2})\,k_{f1} - E_1 I_1 ik^3_{f1}$

$Z_4 = - 1$ $\qquad Z_5 = - 1$ $\qquad Z_6 = - E_1 I_1 k^2_{f1} + \frac{L}{2}E_1 I_1\, ik^3_{f1} + Ij ik_{f1}\omega^2$

$Z_7 = - 1$ $\qquad Z_8 = - 1$ $\qquad Z_9 = - E_1 A_1 ik_{l1} - mj\omega^2$

Experimental Study on Control of Building Structures by Active Cables

J. RODELLAR*, F. LÓPEZ-ALMANSA**, P. WANG***,
A. REINHORN*** and T. T. SOONG***

*Department of Applied Mathematics III, Technical University
of Catalonia, C/J. Girona Salgado, 31 08034 Barcelona, Spain
**Department of Architecture Structures, Technical University
of Catalonia, Avda. Diagonal, 649 08028 Barcelona, Spain
***Department of Civil Engineering, State University of New York
at Buffalo, Buffalo, NY 14260, USA

ABSTRACT

Actuators for reducing the vibration of building structures subjected to seismic excitations based on active cables have been proposed and tested in recent years. Active cables are connected to the structure and to hydraulic servosystems which can modify their tensions as a function of the measured structural response. This paper describes the implementation of an active tendon control system to an experimental six stories metal building structure excited by a shaking table which simulates earthquake base accelerations. The experimental set-up is installed in the State University of New York at Buffalo. The control strategy used to command the tendon actuators is predictive control. Experimental results shows a significant damping of the response when the structure is under control.

KEYWORDS

Active structural control, active tendons, predictive control, building structures, experimental dynamics of structures, seismic design.

INTRODUCTION

Between the familiy of actuators proposed for active control of building structures, those based on active cables have deserved a particular attention in recent years. Active cables consist of tendons attached to the structure whose tension can be quickly modified since they are attached to the hydraulic cylinders whose motion is commanded by servovalves. These servovalves are connected with a feedback control system in such a way that the desired movement of the cylinders is a function of the structural response measured by sensors. This function is formulated based on an automatic control strategy. Optimal control has been one of the control strategies used by some authors combined with active cables (Yang and Samali, 1983; Leipholz and Abdel–Rohman, 1986; Chung et al., 1988) as well as predictive control strategy (Rodellar et al., 1989; López Almansa, 1988; López Almansa and Rodellar, 1989). Although most work has been done using computer simulation for assessing the control performance, experimental results have been reported with optimal control (Chung et al., 1988) and with predictive control (Rodellar et al., 1989) using a small scale of a three story building existing at the State University of New York (SUNY) at Buffalo. This paper describes the implementation and presents the first results of a predictive control algorithm in a new 1:4 scale six story building structure with active cables which has been installed in the SUNY laboratory (Reinhorn et al., 1989).

EXPERIMENTAL SET-UP

Figure 1 shows a block diagram with the basic elements of the experimental set-up. The structure is a 1:4 scale, three bays, six stories metal model. The structure and the additional masses placed at the different floors weights 19.18 tons. Figure 2 shows a front view of the structure. Feedback signals to the control system are the displacements and velocities of each floor relative to the ground. In addition, accelerometers are used to monitor the structure's response. Analog feedback signals are sampled and converted to discrete-time values by analog/digital converters. These values are used by a computer to calculate the desired values of the control sequences which are translated into analog control signals by digital/analog converters. These control signals feed the servovalves of two tendon controllers, placed between the ground and first floor and between the second and third floors respectively, which apply horizontal control forces on the structure at floors 1, 2 and 3. A more detailed description on how the controllers work is given in the next section.

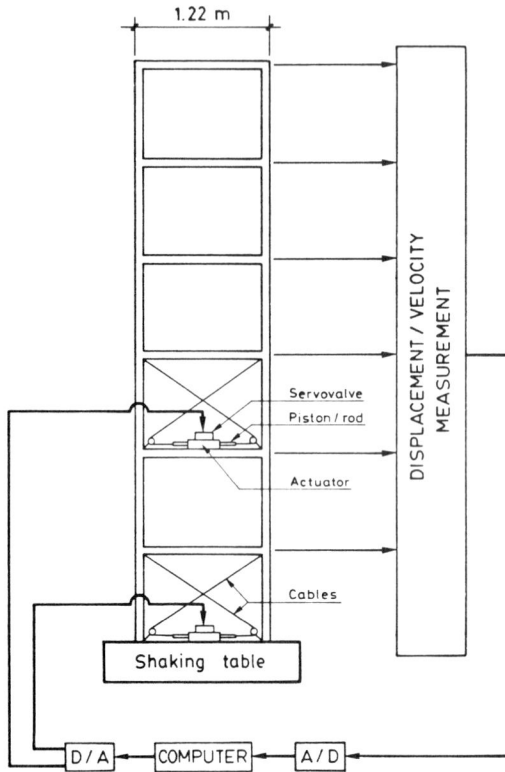

Fig. 1.– Experimental control loop.

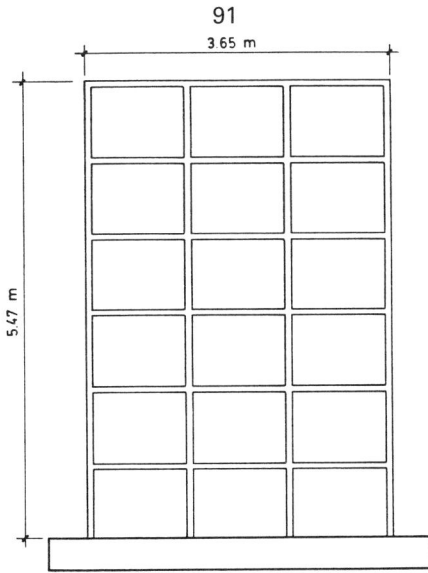

Fig. 2.– Front view of the model structure

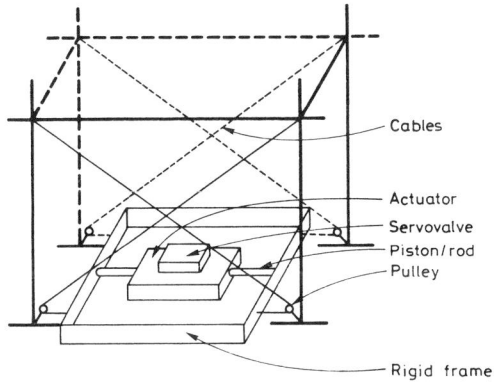

Fig. 3.– Active tendon controller.

CONTROL ACTION

Figure 3 shows a diagram of a tendon controller. It is composed by four cables and an hydraulic actuator. The cables are braced to the upper floor by one of their ends while the other ends are attached to a horizontal rigid frame through four pulleys. The frame is connected to the piston rod of an hydraulic actuator whose motion is commanded by a servovalve proportionally to the difference between the analog signal from the D/A converter and the actual displacement of the piston/rod. In this way the tensions of the cables are actively modified, which results in horizontal control forces on the structure. In fact, consider the actuator 1 at instant t when the first floor relative displacement is $d_1(t)$. Tensions at cables, as represented in Fig. 4, verify

$$T_{1d} = T_0 + K_t \left[d_1(t) \cos \alpha_1 + u_1(t) \right] \tag{1a}$$

$$T_{1q} = T_0 - K_t \left[d_1(t) \cos \alpha_1 + u_1(t) \right] \tag{1b}$$

where T_0 is a pretension in order to prevent the tension release during control application. K_t is the stiffness of the cables and u_1 is the displacement of the actuator. The resulting horizontal control force f_1 at floor 1 is then

$$f_1(t) = 4\, K_t \cos \alpha_1 \left[d_1(t) \cos \alpha_1 + u_1(t) \right] \tag{2}$$

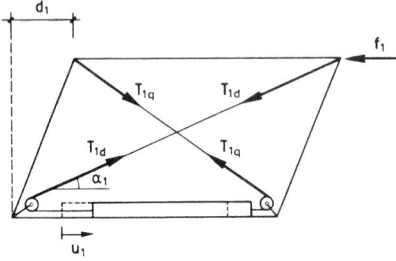

Fig. 4.–Control force applied on the structure by the actuator 1.

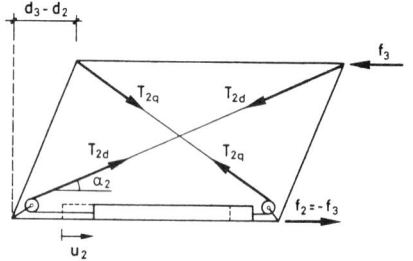

Fig. 5.–Control forces applied on the structure by the actuator 2.

The same analysis for the actuator 2 (see Fig. 5) shows the presence of control forces at floors 2 and 3 verifying:

$$f_2(t) = -4\, K_t \cos \alpha_2 \left[\{ d_3(t) - d_2(t) \} \cos \alpha_2 + u_2(t) \right] \tag{3a}$$

$$f_3(t) = -f_2(t) \tag{3b}$$

The structure can be modelled as a 6 degrees of freedom lumped mass model whose motion is described by

$$M\,\ddot{d} + C\,\dot{d} + K\,d = -f(t) - M\,j\,\ddot{d}_0(t) \tag{4}$$

where M, C and K are the mass, damping and stiffness matrices, d the floor displacement vector, j the unit vector and f the control force vector $f^T = [f_1, f_2, f_3, 0, 0, 0]$ which can be written, according to eqs. (2) and (3), in the form:

$$f = K_p d + L u \tag{5}$$

where

$$K_p = \begin{bmatrix} 4\,K_t \cos^2 \alpha_1 & 0 & 0 & 0 & 0 & 0 \\ 0 & 4\,K_t \cos^2 \alpha_2 & -4\,K_t \cos^2 \alpha_2 & 0 & 0 & 0 \\ 0 & -4\,K_t \cos^2 \alpha_2 & 4\,K_t \cos^2 \alpha_2 & 0 & 0 & 0 \\ 0 & 0 & 0 & 0 & 0 & 0 \\ 0 & 0 & 0 & 0 & 0 & 0 \\ 0 & 0 & 0 & 0 & 0 & 0 \end{bmatrix} \tag{6a}$$

$$L = \begin{bmatrix} 4\,K_t \cos^2 \alpha_1 & 0 \\ 0 & -4\,K_t \cos^2 \alpha_2 \\ 0 & 4\,K_t \cos^2 \alpha_2 \\ 0 & 0 \\ 0 & 0 \\ 0 & 0 \end{bmatrix} \tag{6b}$$

By substituting (5) into (4) one can write

$$\boldsymbol{M}\ddot{\boldsymbol{d}} + \boldsymbol{C}\dot{\boldsymbol{d}} + \boldsymbol{K}'\boldsymbol{d} = -\boldsymbol{L}\boldsymbol{u}(t) - \boldsymbol{M}\boldsymbol{j}\ddot{d}_0(t) \tag{7}$$

where

$$\boldsymbol{K}' = \boldsymbol{K} + \boldsymbol{K}_p \tag{8}$$

Equation (7) describes the motion of the structure under control by active cables. One can see the effect of the cables is double: they supply a passive control action $\boldsymbol{K}_p \boldsymbol{d}$ and an active control $\boldsymbol{L}\boldsymbol{u}$. The first one is equivalent to a change in the stiffness matrix of the structure which is now $\boldsymbol{K}'$. The active control is represented by $\boldsymbol{L}\boldsymbol{u}$ and is produced by the displacement $\boldsymbol{u}$ of the actuators as commanded in closed loop by the control system. Next section describes how this is accomplished in this implementation.

CONTROL ALGORITHM

The control algorithm is formulated through a discrete-time law which calculates the vector $\boldsymbol{u}(k\,T)$ at each instant $k\,T$ (k is an integer and T is the sampling period) as a function of the structural response sampled at instant $k\,T$.

The first step to formulate the control algorithm is to write a discrete-time state model of the control loop (Rodellar et al., 1987). In order to do this, eq. (7) is rewriten in the state form:

$$\dot{\boldsymbol{x}}(t) = \boldsymbol{F}\boldsymbol{x}(t) + \boldsymbol{G}\boldsymbol{u}(t) + \boldsymbol{w}(t) \tag{9}$$

where

$$\boldsymbol{F} = \begin{bmatrix} \boldsymbol{0} & \boldsymbol{I} \\ -\boldsymbol{M}^{-1}\boldsymbol{K}' & -\boldsymbol{M}^{-1}\boldsymbol{C} \end{bmatrix} \qquad \boldsymbol{G} = \begin{bmatrix} \boldsymbol{0} \\ \boldsymbol{M}^{-1}\boldsymbol{L} \end{bmatrix} \qquad \boldsymbol{w} = \begin{bmatrix} \boldsymbol{0} \\ -\boldsymbol{j} \end{bmatrix}\ddot{d}_0 \tag{10}$$

$\boldsymbol{x}$ being defined as $\boldsymbol{x}^T = [\boldsymbol{d}, \dot{\boldsymbol{d}}]$.

From (9) one can derive a discrete-time model of the form:

$$\boldsymbol{x}(k\,T + T) = \boldsymbol{A}\boldsymbol{x}(k\,T) + \boldsymbol{B}\boldsymbol{u}(k\,T - r\,T) \tag{11}$$

where

$$\boldsymbol{A} = \exp(T\,\boldsymbol{F}) \qquad \boldsymbol{B} = \boldsymbol{F}^{-1}(\boldsymbol{A} - \boldsymbol{I}) \tag{12}$$

In model (11) the excitation $\boldsymbol{w}$ is not included since it is assumed to be unknown for the control algorithm. For the control vector $\boldsymbol{u}$, the zero-order interpolation of the D/A converter has been taken into account as well as the presence of a time delay in the control loop. Time delay comes from the time required for the operations involved at each time step along the control loop: measurement by sensors, computation time and, primarily, the physical inertia of the actuators since they cannot move instantaneously to the desired position. Time delay is one of the main problems in a real time implementation which can lead to instability. One way to solve this problem is to include an integer number r in the discrete-time model (11) where $r\,T$ is a measure of the total time lag in the control loop (Rodellar et al., 1989).

The second step to formulate the control algorithm is to use a control strategy to obtain $\boldsymbol{u}(k\,T)$ as a function of $\boldsymbol{x}(k\,T)$, both of them related by model (11). In this implementation we have used predictive control. This is a control strategy which has shown its potential in previous works (Rodellar et al., 1987; López-Almansa, 1988; Rodellar et al., 1989; López-Almansa and Rodellar

1989). It is based on using model (11) to predict the response at each instant k for a future fictitious scenario and calculating the control $\boldsymbol{u}(kT)$ which verifies a performance criterion. The resulting control law has the form:

$$\boldsymbol{u}(kT) = \boldsymbol{D}\boldsymbol{x}(kT) + \sum_{i=1}^{r} \boldsymbol{Z}_i\,\boldsymbol{u}(kT - iT) \qquad (13)$$

where $\boldsymbol{D}$ is the gain matrix and $\boldsymbol{Z}_i$ are matrices contributing to the control action at instant kT proportionally to the control action applied at r previous instants taking into account the existence of time delay along the control loop. The computation of these matrices can be found in the reference Rodellar et al. (1989).

EXPERIMENTAL RESULTS

Identification tests have been performed before running the active control system. The structure has been excited by the shaking table with a white noise base acceleration. By using modal techniques, dynamic characteristics have been obtained from the measured acceleration response of each floor in the frequency domain. Table 1 shows the frequencies and damping factors. This has been done with the cables prestressed, which means the structure is under passive control. Thus the mass, damping and stiffness matrices obtained from identification are those corresponding to eq. (7). These matrices have been used to compute matrices (12) with a sampling period T=7.5 msec which, in turn, have been used to calculate the matrices of the predictive control law (13). A number r=3 has been considered which approximates to the actual time delay along the control loop which has been observed to be 25 msec.

For the active control experiments, the control law (13) has been implemented on-line. A number of test series have been carried out with different excitations simulating earthquakes. Here a sample is presented to illustrate the effectiveness of the control system. Figure 6 shows the displacement of the top floor relative to the base without active control (passive control exists due to the prestressing of the cables) when the shaking table simulates the Myagioki earthquake (40% of its full scale). Figure 7 shows the same response under active control when the first actuator is active commanded by the predictive control law while the second actuator is fixed giving a passive control. Figure 8 corresponds to the case of active control for only the second actuator, the first one being passive. Both cases of active control in Figs. 7 and 8 show a significant reduction of the response as compared with the case with passive control (Fig. 6). The reduction of the maximum peak is greater in the case of active control with the upper actuator but the response has a higher frequency as observed when comparing Figs. 7 and 8.

Table 1. Dynamic characteristics of model structure

Mode No.	Frequency (Hz)	Damping (%)
1	1.560	2.34
2	4.590	0.59
3	7.910	0.61
4	10.450	2.10
5	11.520	0.81
6	13.380	1.11

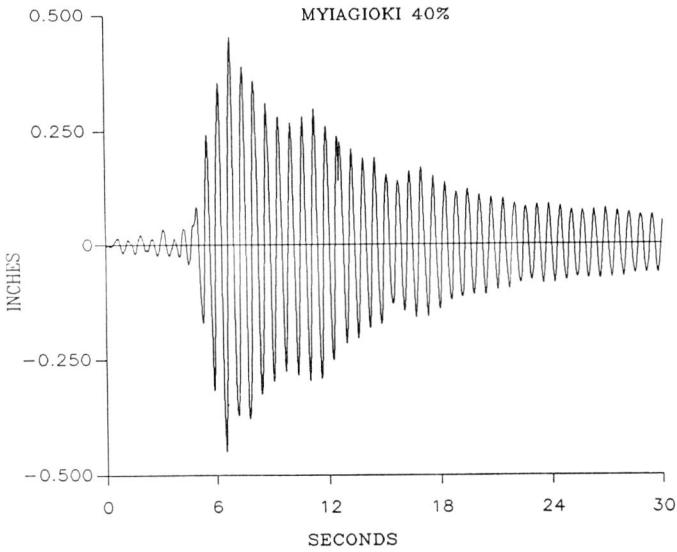

Fig. 6.– Top floor relative displacement without active control.

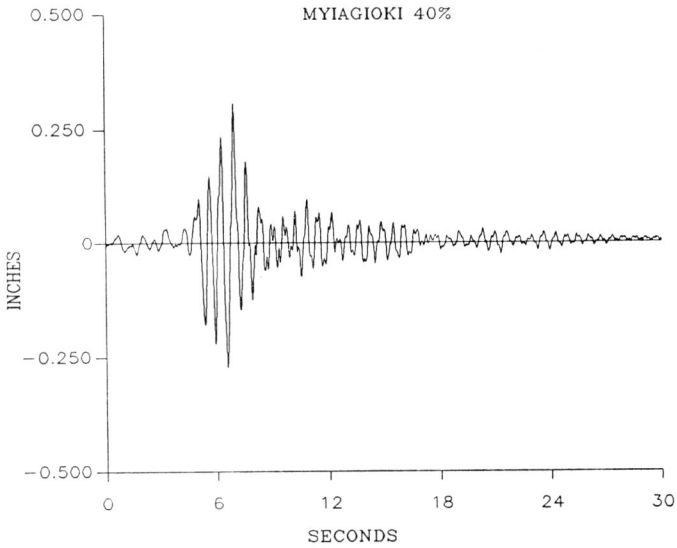

Fig. 7.– Top floor relative displacement with active control using actuator 1.

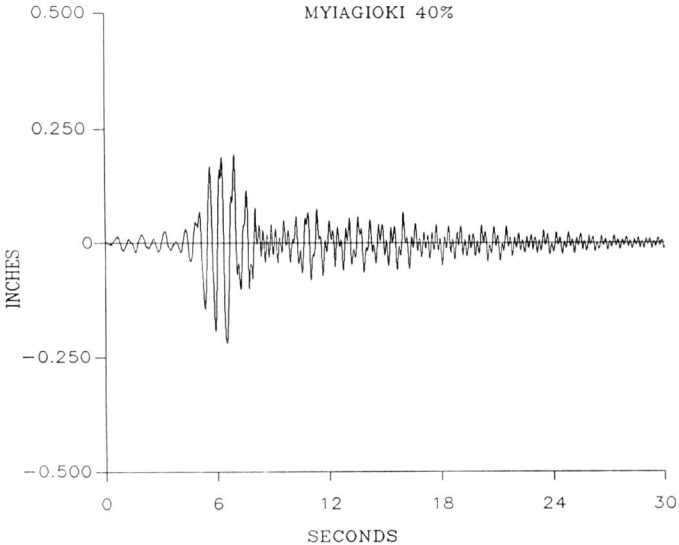

Fig. 8.– Top floor relative displacement with active control using actuator 2.

CONCLUSIONS

An active tendon control system commanded by a predictive control law has been used to control an experimental model of a six stories building structure subjected to earthquake excitation. Results presented here show the control system is effective to reduce the response in a significant factor. More results are being analyzed from an extensive experimental program. They will lead to an assessment of the performance, feasibility and reliability of the control system in more detail.

ACKNOWLEDGEMENTS

This work has been partially supported by the DGYCIT –Spanish Government– (project PB87-0844) and by the CIRIT –Catalan Government– (project EE88/1). Authors are grateful to Dr. R. C. Lin for assisting in the experiments and to Mr. M. Pitman the laboratory operator/test engineer for his assistance.

REFERENCES

Chung L.L., Reinhorn A.M. and Soong T.T. (1988). Experiments on active control of seismic structures. *Journal Engineering Mechanics Division ASCE* 114(2) 241–256.

Leipholz H.H.E. and Abdel-Rohman M. (1986). *Control of structures.* Martinus Nijhoff Publishers.

López-Almansa F. (1988). *Contribution to the development of control systems of building structures by active tendons.* Ph.D. Thesis. Technical University of Catalonia (in Spanish).

López-Almansa F. and Rodellar J. (1989). Control systems of building structures by active cables. *Journal Structural Engineering Division ASCE.* in print.

Reinhorn A.M., Soong T.T., Lin R.C., Fukao Y., Nakai M., Haniuda N. and Clark A. (1989). Experimental study of actively controlled structures subjected to earthquake ground motion. Part I: Set-up. National Center for Earthquake Engineering Research (State University of New York at Buffalo), in print.

Rodellar J., Barbat A.H. and Martín-Sánchez J.M. (1987). Predictive control of structures. *Journal Engineering Mechanics Division ASCE* 113(6) 797–812.

Rodellar J., Chung L.L., Soong T.T. and Reinhorn A.M. (1989). Experimental digital predictive control of structures. *Journal Engineering Mechanics Division ASCE* in print.

Yang J.N. and Samali B. (1983). Control of tall buildings in along-wind motion *Journal Structural Engineering Division ASCE* 109(EM1) 50–68.

The use of Constrained Layer Damping in Vibration Control

G. R. TOMLINSON

*Department of Mechanical Engineering, University of Manchester,
Oxford Road, Manchester M13 9PL, UK*

ABSTRACT

Constrained layer damping is widely used in engineering practise for minimising unwanted vibration. This paper describes an application to an outlet guide vane whereby damping treatments are considered in terms of external constrained layers and internal fillers. The problem of optimising the characteristics of shear damping at room and elevated temperatures is shown to be a difficult problem.

KEYWORDS

Damping, vibration isolation, shear damping, constrained layer damping.

INTRODUCTION

This paper is concerned with an investigation into the vibration of an outlet guide vane (OGV) which necessitated the use of a passive damping mechanism to minimise the levels of vibration induced by air flowing over the vane. The investigation centred on the use of a form of constrained layer damping (CLD) which is often referred to as shear damping.

It would appear that in the mid 17th centrury a form of CLD played an important (but possibly unknowingly) role in the manufacture of violins. The famous Italian violin manufacturer Antonio Stradivari (circa 1644 - 1737) bought the wood to manufacture his violins from Venice. During this period, the varnish employed for sealing the wood was apparently made from a mixture of resin and ground gem stones. This combination of the gem stone particles in the resin matrix possibly created a form of constraining layer/friction mechanism which may have produced sufficient damping to explain whey many of his violins had a "rich full tone"

Today CLD has a wide variety of applications ranging from Nuclear Submarines to jet engines. The operating environment however, plays an important role in the effectiveness of CLD treatments. This is because the constrained or shear layer is usually a viscoelastic material which is both frequency and temperature dependent. In order to induce a high shear strain in this, the relative flexibility of the constraining structure, the mode of vibration (or operating deflection shape) and the location of CLD play significant roles in the optimisation of CLD treatments.

Detailed in this paper is an investigation into the use of shear damping methods on a hollow vane. Attention is given to the use of CLD materials applied to the external surfaces and to the use of damping materials injected into the blade cavity. It is shown that for room temperature applications there is a wide range of CLD materials which can offer significant improvements in reducing vibration levels, both as external (classical CLD) or internal (fillers) applications. Unfortunately, at higher operating temperatures (e.g. $\geq 100^{\circ}C$) few of these materials offer any worthwhile improvements due to their reduced shear moduli at these higher temperatures.

AN OUTLINE OF THE THEORETICAL BASIS OF CONSTRAINED LAYER DAMPING (CLD)

In general, surface damping treatments fall into one of two categories; free or contrained layer damping. Free layer damping is simply a surface layer added (bonded) to the structure of interest in which energy is dissipated as a result of extensional strain energy arising from the flexure of the structure of interest. Since this is not in our area of interest, the reader is referred to (Cremer et al; 1972, Nashif et al; 1985) where further details are available.

CLD, in its simplest form consists of three layers as shown in Figure 1. The analysis of such a configuration was initially carried out by Ross, Kerwin and Ungar (Ross et al; 1959) and is referred to in the literature as the RKU analysis; it can be employed with both shear and extensional damping treatments. Before describing the basic RKU equations, some preliminaries with regard to the dynamic behaviour of viscoelastic materials is necessary.

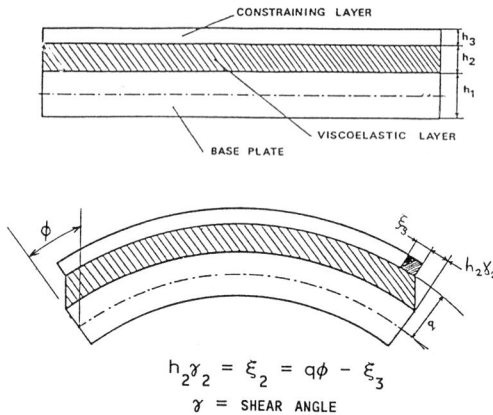

$$h_2\gamma_2 = \xi_2 = q\phi - \xi_3$$
$$\gamma = \text{SHEAR ANGLE}$$

Fig. 1   Constrained layer damping deformation mechanism for flexural vibration.

Materials which exhibit a marked temperature - frequency dependence, (a typical material property master curve is shown in Figure 2(a))exhibit three important regimes as shown in Figure 2(b).

These regions are the glassy or low temperature region, the transition region, and the rubbery or high temperture region. It can be seen that there is a point of inflexion or maximum slope in the modulus curve of Figure 2(a) in the transition region and it is at this point that the damping or loss factor is a maximum. In fact, the two curves, modulus and damping, are uniquely related and it is possible to calculate the damping from the modulus curve (Kennedy and Tomlinson; 1988).

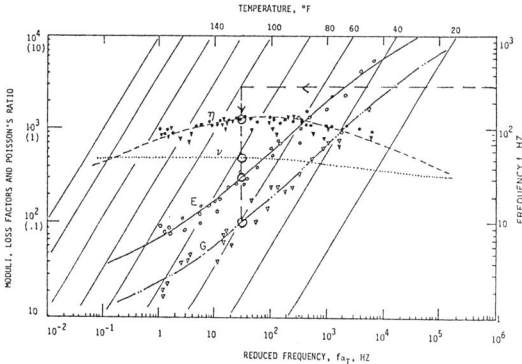

Fig. 2(a) Typical master curve of a viscoelastic polymer showing the complex moduli properties.

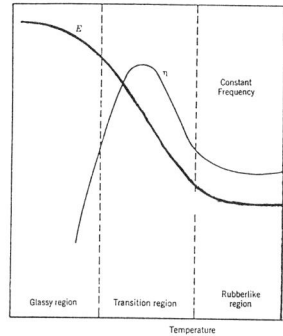

Fig. 2(b) Important material property regimes as a function of temperature for a constant frequency.

The simplest function which describes the energy loss per cycle is the Complex Modulus,

$$E^* = E' + iE'' = E'(1 + i\eta); \quad i = (-1)^{1/2} \tag{1}$$

$$G^* = G' + iG'' = G'(1 + i\beta) \tag{2}$$

$$E' = 2(1 + \nu)G' ; \quad \eta \approx \beta \tag{3}$$

Equation (1) represents the complex flexural modulus and (2) the complex shear modulus. The terms η, β represent the material loss factors which are similar in magnitude as a result of the similar shapes for E and G as shown in Figure 2(a). Figure 3 shows how the loss factor is a function of the slope and not the magnitude of the modulus curve.

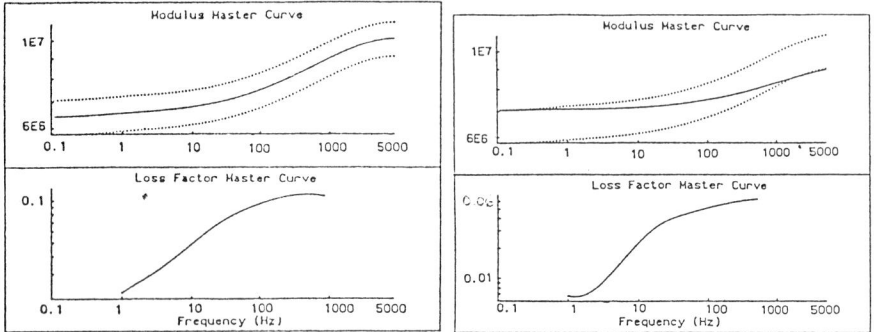

Fig. 3   Dependence of the loss factor on the slope of
the Modulus master curve.

The important parameters in equations (1) and (2) are the imaginary parts
or loss modulus, real parts being termed the storage modulus and the ratio
of the imaginary to the real being the loss factor.   In order to combine
these properties with the geometry of the structure to be damped, the RKU
analysis produces three dimensionless parameters; Geometric (or Stiffness)
parameter Y, Shear parameter g and the composite loss factor η_c.

The stiffness parameter, which defines the overall flexural rigidity for
the three layer panel system shown in Figure 1 can be written as

$$Y = \frac{(E_1 A_1)(E_3 A_3) \ d^2}{(E_1 A_1 + E_3 A_3)(E_1 I_1 + E_3 I_3)} \; ; \; \text{for } E_1 > E_2 < E_3 \qquad (4)$$

and $d = \dfrac{(h_1 + h_3)}{2} + h_2$

A_i = cross sectional area of ith layer

I_i = 2nd moment of area about neutral axis of ith layer

For layers of equal width and the same material, equation (4) reduces to a
purely geometrical property,

$$Y = \frac{12 \ h_1 h_3 d^2}{(h_1 + h_3)(h_1^3 + h_3^3)} \qquad (5)$$

The second property, the Shear parameter, relates to the shear properties
of the constrained viscoelastic layer which is expressed as,

$$g = \frac{G_2 b}{h_2 k^2} \left[\frac{1}{E_1 A_1} + \frac{1}{E_3 A_3} \right] \; ; \; G_2 = \text{shear storage modulus} \qquad (6)$$
$$b = \text{width of layer}$$

where $k^2 = \omega\sqrt{\dfrac{m}{B}}$; m = mass/unit length of composite system

B = bending stiffness of the composite system

ω = circular frequency

It is usual to find that $\dfrac{1}{E_3 A_3} > \dfrac{1}{E_1 A_1}$ and equation (6) reduces to the more common form,

$$g = \frac{G_2}{E_3 h_2 h_3 k^2} \qquad (7)$$

g is dependent upon the frequency of operation and the shear modulus which in turn is also dependent upon frequency and temperature.

Equations (4) and (6) may be combined to give the composite loss factor (Cremer et al; 1973) as,

$$\eta_c = \frac{\beta_2 g Y}{1 + g(2 + Y) + g^2(1 + Y)(1 + \beta_2^2)} \qquad (8)$$

where β_2 = loss factor of the viscoelastic layer.

Equation (8) is a maximum at an optimum value of g given when $\dfrac{\partial \eta_c}{\partial g} = 0$, e.g.

$$g_{optimum} = \left[(1 + Y)(1 + \beta_2^2)\right]^{-1/2} \text{ and, } \quad \eta_{max} = \frac{\beta_2 Y}{2 + Y + 2/g_{opt}} \qquad (9), (10)$$

3. Application to an Outlet Guide Vane Study

There are a wide range of materials commercially available for external CLD applications. Many of these materials are very easy to apply as they use a self-adhesive procedure. The viscoelastic damping material basically acts as an adhesive, this being chemically bonded to an aluminium constraining layer; thus all one has to do is basically attach this to the structure of interest. Since many of these CLD materials are only 0.5mm thick there are relatively few problems in attaching them to flat or curved surfaces. In order to assess their usefulness a simulated vane was designed using FE methods which had similar dynamic properties (mode shapes and frequencies) as the actual OGV. The design was based upon a modal analysis of an actual OGV. Typical mode shapes at the frequency of interest indicated a fundamental bending mode coupled with a rotation, the leading edge suffering the largest amplitudes. The test rig used for evaluating the CLD materials consisted of two angle plates bolted to a large cast iron base. The simulated (or actual) vane with the CLD attached was secured between these plates and the vane was enclosed in a thermal chamber. Hot air guns were used to heat the assembly up to a known temperature which was then allowed to "soak" until steady state conditions were achieved.   An accelerometer and force transducer was attached to the vane via fixing studs. The vane was excited with an electrodynamic exciter. During each temperature increment one end of the fixture (the non-exciter end) was free to slide and then re-clamped before a vibration test was carred out. The measured frequency response functions were recorded and curve-fitted to extract the modal damping values.

The materials giving optimum damping obtained from the simulated vane tests were applied to an acutal OGV and the best results achieved for the maximum temperture of 100°C are shown in Figure 4. This indicates an improvement of 100% in the level of the original virgin vane damping. When an additional constrained layer was added this increased the damping level by a further 50%.

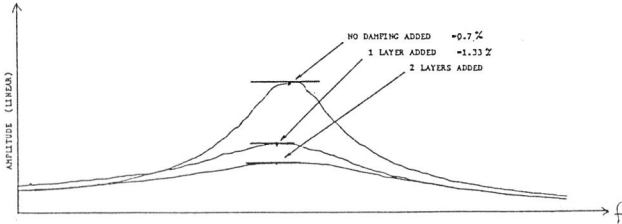

Fig. 4   The effect of adding CLD material to the actual vane.

It is interesting to compute the expected damping or loss factor for the OGV for a given damping material. Let us assume the following conditions:

Damping material: Soundcoat DYAD 609 (pg 402, Nashif et al)
Viscoelastic layer 0.5mm thick,
Constraining layer 0.6mm thick, Aluminium
Operating temperature, 100°C
Operating frequency, 300 Hz
Cross sectional Area of CLD :   50mm^2
Cross sectional Area of vane: 150mm^2
Youngs Modulus of constraining layer: $70 \times 10^9 N/m^2$
Youngs Modulus of vane material: $200 \times 10^9 N/m^2$

Thus from equation (4) we have,

$$Y = \frac{(E_1 A_1)(E_3 A_3) \ d^2}{(E_1 A_1 + E_3 A_3)(E_1 I_1 + E_3 I_3)}$$

$$= \frac{(200 \times 10^3 \times 150)(70 \times 10^3 \times 60) \times (2.55)^2}{(200 \times 10^3 \times 150 + 70 \times 10^3 \times 60)(200 \times 10^3 \times 100 \times \frac{1.5^3}{12} + 70 \times 10^3 \times \frac{100 \times 0.6^3}{12})}$$

$$\therefore \ Y = \underline{1.55}$$

The shear parameter, g, is given as,

$$g = \frac{G_2 b}{h_2 k^2} \left[\frac{1}{E_1 A_1} + \frac{1}{E_3 A_3} \right] \tag{11}$$

$$\approx \frac{G_2}{E_3 h_3 h_2 k^2} = \frac{G_2}{E_3 h_2 h_3 \omega} \sqrt{\frac{B}{m}} \tag{12}$$

Since the bending stiffness of the aluminium constraining layer is much lower than that of the vane, we can approximate B as follows:

$$B \approx \frac{E_1 h_1^3}{12} = \frac{200 \times 10^3 \times 1.5^3}{12} = 56.25 \times 10^3 Nmm$$

Taking the value of G for DYAD 609 at $100°C$ and 300Hz as approximately $5 \times 10^6 N/m^2$, and ρ_{DYAD} = 970 kg/m^3 gives a value of the shear parameter from equation (10) as, g = 0.07.

Substituting g = 0.07 and Y = 1.55 in equation (8) gives the composite loss factor as,

$$\eta_{Cmax} = \frac{0.4 \times 0.07 \times 1.55}{1 + 0.007 (2 + 1.55) + .07^2 (1 + 1.55) (40.4^2)}$$

$$\eta_c = \underline{0.035}$$

Actual tests with DYAD 609 applied to the vane produced loss factors which were in the range $0.01 \leq \eta_c \leq 0.025$.

One factor which is overlooked in the above analysis is that the damping of the original structure is assumed to be zero. Obviously this is not true and if the damping in the original structure is significant compared with the added damping then no worthwhile gains can be achieved. In practise the initial structure will exhibit a certain level of inherent damping. To take this into account a procedure was developed (Vandeurzen 1982) which utilises an energy loss expression.

Application of internal fillers as an equivalent CLD

Having established the viability of external CLD treatments as a possible means of reducing forced vibration levels of an OGV, it was decided to investigate the use of "fillers". By filling the cavity of the vane with a viscoelastic material it was thought that due to the different geometries of the upper and lower faces of the vane (convex and concave) sufficient shear strain could be induced in the damping material to reduce the forced vibration levels. Again a procedure was developed whereby a "simulated hollow vane" was designed and employed to establish the most suitable materials for this application. This simulated vane had a facility whereby the top cover-plate could be removed and the internal cavity cleaned. With the top cover replaced the filler material was injected via a high pressure gun into the cavity. Several materials were shown to be very effective with this procedure at room temperature but few were effective at $100°C$ (this follows the same trends found with the external CLD treatment).

However, at an elevated temperature most materials exhibited low damping behaviour due to the reduction in their shear modulus. The best material at the elevated temperature was found to be a special compound developed by Soundcoat, which gave the results shown in Figure 5. This returned an increase in the damping levels of a factor greater than two at $100°C$.

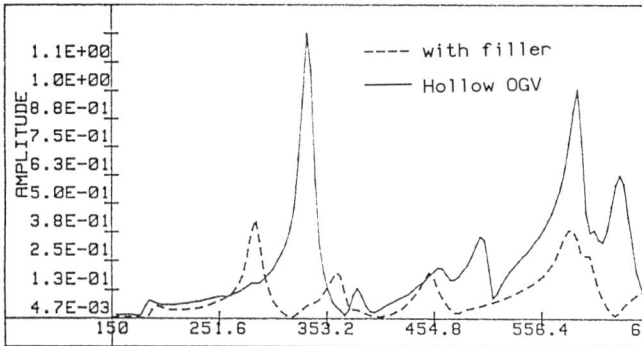

Fig. 5   The effect of using an internal filler
(no external damping material) to suppress
vibration at a temperature of 100°C.

CONCLUSIONS

The use of CLD treatment can produce worthwhile reductions in vibration levels if the optimum conditions are met.   These conditions are basically that the constraining layer should exhibit the maximum loss factor or damping level at the frequency and temperature of the operating environment.   If the environment is at a relatively high temperature (e.g. > 100°C) then there are few viscoelastic materials available which can be used.   It should also be recognised that the composite loss factor may differ significantly from the loss factor of the actual constraining materials. This is often due to high levels of inherent damping in the original structure and the inability to accurately define the areas of maximum strain energy.

It has been demonstrated that the use of appropriate internal fillers in hollow vanes can return improved levels of damping.   However, no theory covering such an application is apparently available and the results shown in this paper cannnot be generalised.

REFERENCES

Cremer L, M. Hecth and E. E. Ungar (1972).
        In : Structure Borne Sound. Springer Verlag

Kennedy, I. and G. R. Tomlinson (1988).   Torsional vibration
        transmissibility characteristics of reinforced viscoelastic
        flexible pipes.   Jn. Sound and Vibration, 122(1), 149-169

Nashif A. D., D. I. G. Jones, J. P. Henderson (1985).
        In : Vibration Damping.   Wiley Intersciences

Reid M. C., S. O. Oyadiji and G. R. Tomlinson (1988)
        Relating material properties and wave effects in vibration
        isolators.   59th Shock and Vibration Symposium, Alberquerque,
        NM.   139-156

Ross D., E. E. Ungar, E. M. Kerwin Jnr. (1959). Damping of plate flexural vibration by means of viscoelastic laminate Structural Damping, ASME, New York, 49-88.

Vandeurzen U. (1982). Identification of damping in materials and Structures. Ph.D. Thesis, K. U. Leuven, Department of Mechanical Engineering

ACKNOWLEDGEMENTS

The author wishes to thank Rolls Royce PLC (Derby) for their kind permission to publish the results of this paper.

APPENDIX 1

MANUFACTURERS OF VISCOELASTIC MATERIALS AND ADHESIVES USED IN TESTS

MATERIAL OR ADHESIVE	MANUFACTURER
SJ-2052X	3M
DYAD 606	SOUNDCOAT
DYAD 609	SOUNDCOAT
SOUNDFOIL ION2	SOUNDCOAT
SOUNDFOIL 15D2	SOUNDCOAT
B-FLEX	
V55	SWEDAC
V515	
V5MVM	
LW520	DU-PONT
LOCTITE 499	LOCTITE
ARALDITE HY932	CIBA-GEIGY

MANUFACTURERS OF VISCOELASTIC FILLERS AND PRIMERS

DG 20	"SWEDAC" supplied by Industrial and Marine Acoustics LTD
RTV630 + RTB11	General Electics
93 - 118	Dow Corning
VUP 9157	Resinous Chemicals LTD
LW 520	Uniroyal Chemicals LTD
CILCAST 100	
CIL MONOTHANE	Compounding Ingredients LTD
SVDC1	Sound Service (Oxford) LTD
3MXA8283	3M

Parameter Estimation Techniques for Monitoring Machines and Structures

C. D. FOSTER and J. E. MOTTERSHEAD

Department of Mechanical Engineering, University of Liverpool, UK

ABSTRACT

Techniques for monitoring the condition of plant and machinery based on measured vibration data are in widespread use in the manufacturing and process industries. The development of such techniques has taken place in the wake of advances in the capability of modern electronic instrumentation for signal processing. In recent years methods have emerged which offer the promise of estimating physical (structural) parameters from vibration measurements. It is thus timely to consider how they might be applied to attribute changes in the observed motion of a machine to incipient failure of a machine component. The onset of malfunctioning of a component might typically be indicated by an identified stiffness or mass element which differs from a previously identified value. An example showing the application of the parameter estimation technique is provided using experimental data. It is demonstrated that the identified parameters are physically meaningful in the sense that finite element parameters are meaningful.

KEYWORDS

Parameter estimation; condition monitoring; mass, stiffness and damping parameters.

INTRODUCTION

Condition monitoring is performed by extracting information from machines to indicate their condition so that they may be operated and maintained efficiently. The broad range of condition monitoring techniques have been outlined by Michael Neale and Associates (1979) in a guide prepared for the Department of Industry. Vibration data has been used mainly in the monitoring of rotating machinery, and to a lesser degree in reciprocating machinery. In particular by the application of spectral analysis, it has become straightforward to relate changes in observed spectra to specific mechanisms in simple machinery (see for example Table 8.1 in Tustin and

Mercado (1985)). In complex machinery there may be many mechanisms at work and in that case simple inspection of measured spectra may be insufficient in itself, to allow the discrimination of the particular machine element (which is responsible for the observed spectral changes) from many potentially faulty components.

A class of structures of current interest to the authors are the steel foundations used in modern power stations to support turbo-alternators. In the power generation industries it is of great importance to be able to predict future service problems by interpreting vibration records from turbomachinery. In this respect, measurements of lateral vibration at the bearing pedestals, taken during the run-down regime, are especially valuable: during this regime the forces generated by rotor unbalance sweep through the entire frequency range and are liable to excite the most significant modes of vibration of the rotor, oil-film bearings and foundations. Assuming reliable mathematical models to be available for the bearings and rotor (with known unbalance) and using measured vibrations at the pedestals, Lees (1988) has provided a method for the determination of the bearing forces which are transmitted to the foundations. The measured pedestal vibrations recorded during run-down carry information on the condition of the foundations which may be extracted with the aid of the computed bearing forces. Clearly for a complex structure such as a foundation it would be totally inappropriate to attempt to identify precise locations on the structure without the availability of powerful signal processing techniques, which are capable of providing a depth of insight which is not available by visual inspection of vibration spectra alone.

Advances have been made recently in the system identification of vibrating structures which now offer the promise of estimating structural parameters (elements of mass, stiffness and damping matrices) directly from measured data. Mottershead (1988) developed a family of least-squares, frequency domain filters for structural identification and this work has since been extended (Mottershead, submitted) to allow the correction of reduced-order finite element models using incomplete vibration measurements. It has been demonstrated that the method is capable of identifying a model of lower order than the system under test even when the number of modes present in the measured data is fewer than the number of measurement stations (i.e. the order of the identified model). Such an identified model has been shown to hold good when the loading conditions are changed from the loads applied in the test.

Mass, stiffness and damping parameters can be important in the monitoring of machines and structures since they relate to precise locations on a physical structure. Thus for example, if changes are identified in the structural parameters of a foundation then such information can perhaps lead immediately to the part of the structure which is responsible for the observed changes. In this paper the authors present some recent results on the estimation of structural parameters using experimental data.

BRIEF REVIEW OF THE THEORY

A full account of the theory can be found in (Mottershead, 1988) and (Mottershead, submitted) but the essential details will now be reiterated to ensure completeness of the account presented here. Consider the following model of a vibrating structure,

$$q(\omega) = \underline{B}(\omega) \, \{\underline{z}(\omega) - \underline{\xi}\} \qquad (1)$$

where,

$\underline{z}(\omega)$ is a vector of measured displacements
$\underline{\xi}$ is a vector of measurement noise
$\underline{q}(\omega)$ is a vector of known input forces

and,

$$\underline{B}(\omega) = -\omega^2 \underline{M} + j\omega \, \underline{C} + \underline{K} \qquad (2)$$

such that $\underline{B}^{-1}$ is a matrix of displacement/force frequency response functions. The model is allowed to be of lower order than the system under test which may (in theory) contain an infinite number of vibration modes in the case of a continuous structure. The number of modes present in the range of frequencies used in the test may be smaller than the order of the model.

Let the unknown mass, stiffness and damping parameters be assembled in a vector $\underline{x}$ and corresponding finite element values in $\underline{x}_{fe}$. Then denoting a vector of correcting terms by,

$$\underline{e} = \underline{x} - \underline{x}_{fe} \qquad (3)$$

and considering discrete frequency data at N intervals a least-squares estimate $\underline{\hat{e}}$ of $\underline{e}$ can be computed from,

$$\left(\sum_{i=1}^{N} \underline{H}_i \, \underline{Q} \, \underline{H}_i^{H} \right) \underline{\hat{e}} = \sum_{i=1}^{N} \underline{H}_i \underline{Q} \, (\underline{r}_i - \underline{H}_i^{H} \underline{x}_{fe}) \qquad (4)$$

In the above,

$$\underline{H}_i = \frac{d(\underline{B}_i \, \underline{z}_i)^{H}}{d\underline{x}} \qquad (5)$$

and

$$\underline{r}_i = \underline{q}_i - \frac{d(\underline{B}_i \, \underline{z}_i)^{H}}{d\underline{y}}$$

where $\underline{y}$ is a vector of known structural parameters and $\underline{Q}$ is a positive-definite weighting matrix.

In general the array $\sum_{i=1}^{N} \underline{H}_i \, \underline{Q} \, \underline{H}_i^{H}$ will be rank deficient and then singular value decomposition can be applied to obtain the unique solution of smallest norm $\|\underline{\hat{e}}\|$ to the degenerate least-squares problem defined in (4).

Equation (4) can be posed as a continuous frequency domain filter such that $\underline{\hat{e}}$ is obtained by integration of,

$$\frac{d\underline{\hat{e}}}{d\Omega} = (\underline{P}^{-1})^{+} \, \underline{H}(\Omega) \, \underline{Q} \, (\underline{r}(\Omega) - \underline{H}^{H}(\Omega) \, (\underline{x}_{fe} + \underline{\hat{e}})) \qquad (7)$$

where

$$\frac{dP^{-1}}{d\Omega} = \underline{H}(\Omega) \ \underline{Q} \ \underline{H}^H(\Omega) \tag{8}$$

$(\ )^+$ denotes the pseudo inverse and Ω represents the current frequency step. The physical significance of performing the pseudo-inversion by singular value decomposition is that it forces the solution $\hat{\underline{x}} = x_{fe} + \hat{\underline{e}}$ which deviates least from the initial finite element model. Thus the estimated parameters $\hat{\underline{x}}$ can be said to be physically meaningful in the sense that a finite element model is physically meaningful. Further discussion of the singular value decomposition can be found in (Stewart, 1973; Golub and Van Loan, 1983) and with special regard to its application here, the reader is referred to the Appendix.

EXPERIMENTATION AND SIGNAL PROCESSING

The objective of the experiment was to construct an accurate four-degree-of-freedom model of the portal frame shown in Figure 1 using an initial finite element representation which was to be corrected using measured data. In-plane, transverse vibration measurements were taken using a roving accelerometer at each of the four measurement stations. Pseudo-random excitation was applied transversely at station 1 (in the range 0-200 Hz) where it ws measured using a piezo-electric force transducer. Frequency response functions were computed using a dual-channel, FFT signal analyser for subsequent processing by the least-squares, filter algorithm given by equations (7) and (8).

The first three modes of the portal frame are illustrated in Figure 2. The fourth mode was outside the range of excited frequencies and stations 2 and 3 were nodes of the second vibration mode. The gathered data could therefore be considered to be incomplete.

The initial, finite element model was formed by static condensation of a model containing twenty, two-noded beam elements. The model thus formed was finally massaged by applying a scaling factor to the stiffness matrix so as to obtain a finite element frequency response which offered a reasonable fit to the measured data. Data was processed at forty frequency steps of 0.25 Hz around each of the three peaks of the measured frequency responses.

After obtaining estimates of the complete set of mass, stiffness and damping parameters an additional mass of 9 kg was attached to the portal frame at station 4 and the experiment was repeated. The purpose of the second experiment was to discover whether or not the filter algorithm was capable of identifying the known structural modification.

In both experiments the mass, stiffness and damping matrices were assumed to be symmetric. The estimates were obtained separately for the mass, stiffness and damping matrices such that three separate P^{-1} matrices were computed. Mass, stiffness and damping estimates were calculated in turn at each frequency step so as not to put any undue emphasis on any of the estimated matrices.

RESULTS

Portal Frame Without Added Mass

The singular values corresponding to mass, stiffness and damping estimates are given in Table 1. Two singular values were set to zero in the computation of each $(\underline{P}^{-1})^{+}$ matrix. This was found to provide sufficient flexibility in the available (non-unique) solutions to allow a particular (minimum norm) solution which was very close to the initial finite element model.

The estimated parameters are given in Table 2 and a full set of reconstructed frequency responses are shown in Fig. 3. The estimated parameters are compared with the corresponding finite element values and the reconstructed frequency responses are plotted alongside the measured, and finite element, frequency responses.

Portal Frame With 9 kg Mass Added at Station 4

The computed singular values are given in Table 3. Two singular values were again set to zero in the computation of each pseudo-inverse matrix.

The estimated parameters are compared with the previous estimates in Table 4. The previous estimates (with 9 kg added at station 4) were used in place of a finite element model, as initial estimates in this second experiment. Reconstructed frequency responses are compared with measured data in Figure 4.

DISCUSSION AND CONCLUSIONS

It can be seen from inspection of Tables 2 and 4 that, in the main, the natural symmetry of the portal frame structure has been maintained in the parameter estimates. For example, in the case without added mass, stiffnesses and masses on the leading diagonal show almost exact symmetry ($k_{11} = k_{44}$, $k_{22} = k_{33}$, $m_{11} = m_{44}$, $m_{22} = m_{33}$). The estimated damping matrix does not display this form of symmetry.

The trace of the estimated mass matrix gives a value of 19.7 kg which may be compared with the calculated mass of the portal frame at about 34 kg. Taking into account that portion of the total mass which may be considered to be lumped at the feet, it would appear that the estimated masses are perhaps slightly too small.

By comparing the results of the two experiments it is clear that a considerable change in the estimated mass at station 4 has been identified. The estimated additional mass of 7.8 kg is remarkably close to the known value of 9 kg. The stiffness estimates vary insignificantly between the two experiments. No trend can be detected in the shifts of the damping parameters. This is perhaps likely since finite elements do not provide a consistent representation of damping in the way that they do for stiffness and mass. Thus the initial damping estimates were significantly more uncertain than the initial stiffness and mass estimates.

In the case of both experiments the reconstructed frequency responses show excellent agreement with the measured data in both amplitude and phase.

It can be concluded that the least-squares filtering method offers considerable promise in the monitoring of machines and structures. In the particular experiment described here an additional mass was identified at the correct location. It can further be concluded that the method is capable of providing estimates which are physically meaningful. They are physically meaningful in the sense that a finite element model is meaningful, since they are found by correcting an initial finite element mesh.

ACKNOWLEDGEMENT

This work is supported by the Science and Engineering Research Council under grant GR/E 20592.

REFERENCES

Golub, G. H. and Van Loan, C. F. (1983), Matrix Computations, John Hopkins University Press, Maryland.

Lees, A. W. (1988), The least squares method applied to identify rotor/ foundation parameters, Proc. I. Mech. E. Int. Conf. on Vibrations in Rotating Machinery, 209-216.

Mottershead, J. E. (1988), A unified theory of recursive, frequency domain filters with application to system identification in structural dynamics, Trans. ASME, J. Vibrations, Acoustics, Stress and Reliability in Design, 110, 3, 360-365.

Mottershead, J. E. (submitted), Theory for the estimation of structural vibration parameters from incomplete data, AIAA Journal.

Neale, Michael and Associates (1979), A Guide to the Condition Monitoring of Machinery, Her Majesty's Stationery Office, London.

Stewart, G. W. (1973), Introduction to Matrix Computations, Academic Press, New York.

Tustin, W. and Mercado, R. (1985), Machinery health monitoring - an application of spectrum analysis, J. Soc. Env. Engineers, December, 17-22.

APPENDIX

Application of Singular Value Decomposition

With reference to the filter algorithm given by equations (7) and (8) it is necessary to compute the pseudo-inverse of the matrix $\underline{P}^{-1}$. If the matrix $\int_0^\Omega \underline{H}(\omega) \, \underline{Q} \, \underline{H}^H(\omega) \, d\omega$ is of full rank then a unique least-squares solution $\hat{\underline{e}}$ is available. However, at every frequency step, each element of $\underline{H}$ is a vibration measurement. Thus if the data is incomplete in the sense that some modes are not present or are only weakly represented in the data then $\underline{P}^{-1}$ should be rank deficient or nearly rank deficient. In

this latter case of a degenerate least-squares problem a multiplicity of
least-squares solutions exist and it is then of benefit to apply the
singular value decomposition such that,

$$\underline{P}^{-1} = \underline{U} \ \underline{\Sigma} \ \underline{U}^H \tag{9}$$

where

$$\underline{\Sigma} = \begin{bmatrix} \underline{S} & \underline{0} \\ \underline{0} & \underline{0} \end{bmatrix}. \tag{10}$$

If $\underline{P}^{-1}$ is of order n and rank $(\underline{P}^{-1})$ = r then,

$$\underline{S} = \text{diag} \ (\sigma_1, \sigma_2, \ldots, \sigma_r), \ r \leq n \tag{11}$$

with

$$\sigma_1 \geq \sigma_2 \geq \sigma_3 \ldots \geq \sigma_r > 0$$

The numbers σ_1, $\sigma_2, \ldots, \sigma_r$ represent the non-zero eigenvalues of $\underline{P}^{-1}$ and
the unitary matrix $\underline{U}$ is made up of columns corresponding to the eigen-
vectors of $\underline{P}^{-1}$. The diagonal elements of $\underline{\Sigma}$ are known as the singular
values and the number of non-zero singular values is equal to rank $(\underline{P}^{-1})$.
The pseudo-inverse of $\underline{P}^{-1}$ is then given by,

$$(\underline{P}^{-1})^+ = \underline{U} \ \underline{\Sigma}^+ \ \underline{U}^H \tag{12}$$

where

$$\underline{\Sigma}^+ = \begin{bmatrix} \underline{S}^{-1} & \underline{0} \\ \underline{0} & \underline{0} \end{bmatrix} \tag{13}$$

and
$$\underline{S}^{-1} = \text{diag} \ (^1/\sigma_1, \ ^1/\sigma_2, \ldots, \ ^1/\sigma_r). \tag{14}$$

A least-squares solution can then be found as,

$$\hat{\underline{e}} = \underline{U} \ \underline{\Sigma}^+ \ \underline{U}^H \ \underline{b} \tag{15}$$

where

$$\underline{b} = \int_0^\Omega \underline{H}(\omega) \ \underline{Q} \ (\underline{r}(\omega) - \underline{H}^H(\omega) \ \underline{x}_{fe}) d\omega \tag{16}$$

such that $\hat{\underline{e}}$ is a minimisor of $\|\underline{b} - \underline{P}^{-1} \ \hat{\underline{e}}\|_2^2$.

Let $\underline{z} = \underline{U}^H \ \hat{\underline{e}}$ and $\underline{c} = \underline{U}^H \ \underline{b}$, and partition $\underline{z}$ and $\underline{c}$ in the form,

$$\underline{z} = \left\{ \begin{matrix} \underline{z}_1 \\ \underline{z}_2 \end{matrix} \right\}, \quad \underline{c} = \left\{ \begin{matrix} \underline{c}_1 \\ \underline{c}_2 \end{matrix} \right\} \tag{17,18}$$

then,

$$\|\underline{b} - \underline{P}^{-1} \ \hat{\underline{e}}\|_2^2 = \|\underline{U}^H(\underline{b} - \underline{P}^{-1} \ \underline{U} \ \underline{U}^H \ \hat{\underline{e}}\|_2^2$$

$$= \left\| \left\{ \begin{matrix} \underline{c}_1 \\ \underline{c}_2 \end{matrix} \right\} - \begin{bmatrix} \underline{S} & \underline{0} \\ \underline{0} & \underline{0} \end{bmatrix} \left\{ \begin{matrix} \underline{z}_1 \\ \underline{z}_2 \end{matrix} \right\} \right\|_2^2.$$

$$= \left\| \left\{ \begin{matrix} \underline{c}_1 \\ \underline{c}_2 \end{matrix} - \underline{S}\ \underline{z}_1 \right\} \right\|_2^2 \tag{19}$$

Equation (19) will be a minimum when $\underline{z}_1 = \underline{S}^{-1}\ \underline{c}_1$ and the value of $\underline{z}_2$ is arbitrary. Thus if the particular solution given by,

$$\underline{\hat{e}} = \underline{U} \left\{ \begin{matrix} \underline{S}^{-1}\ \underline{c}_1 \\ \underline{O} \end{matrix} \right\}$$

$$= \underline{U} \left[\begin{matrix} \underline{S}^{-1} & \underline{O} \\ \underline{O} & \underline{O} \end{matrix} \right] \underline{U}^H\ \underline{b}$$

$$= \underline{U}\ \underline{\Sigma}^+\ \underline{U}^H\ \underline{b} \tag{20}$$

is considered, then it can be seen that the singular value decomposition provides the unique solution of smallest norm $\|\underline{\hat{e}}\|_2$.

Table 1. Singular Values – Portal Frame Without Added Mass

Singular Values

Mass	Stiffness	Damping
$.178 \times 10^{15}$	$.161 \times 10^{13}$	$.444 \times 10^{15}$
$.962 \times 10^{14}$	$.903 \times 10^{12}$	$.275 \times 10^{15}$
$.956 \times 10^{14}$	$.842 \times 10^{12}$	$.243 \times 10^{15}$
$.877 \times 10^{14}$	$.581 \times 10^{12}$	$.237 \times 10^{15}$
$.503 \times 10^{14}$	$.112 \times 10^{12}$	$.152 \times 10^{15}$
$.415 \times 10^{14}$	$.575 \times 10^{11}$	$.129 \times 10^{15}$
$.344 \times 10^{14}$	$.527 \times 10^{11}$	$.128 \times 10^{15}$
$.887 \times 10^{12}$	$.480 \times 10^{11}$	$.623 \times 10^{14}$
$.301 \times 10^{12}$	$.394 \times 10^{11}$	$.211 \times 10^{14}$
$.150 \times 10^{10}$	$.801 \times 10^{8}$	$.687 \times 10^{10}$

Table 2. Estimated Parameters – Portal Frame Without Added Mass

		Finite Element	Correcting Term	Estimated Parameters
Mass (kg)	m_{11}	4.314	-0.206	4.107
	m_{21}	0.763	-0.74	0.023
	m_{22}	6.411	-0.755	5.657
	m_{31}	-0.007	0.366	0.359
	m_{32}	2.02	-1.407	0.613
	m_{33}	6.411	-0.65	5.761
	m_{41}	-0.015	0.037	0.022
	m_{42}	-0.007	0.64	0.634
	m_{43}	0.763	-1.014	-0.251
	m_{44}	4.314	-0.207	4.106
Stiffness (MN/m)	k_{11}	1.242	0.091	1.333
	k_{21}	-0.613	0.429	-0.183
	k_{22}	110.66	0.171	110.83
	k_{31}	-0.008	-0.47	-0.487
	k_{32}	-110.27	-0.014	-110.28
	k_{33}	110.66	-0.192	110.47
	k_{41}	0.011	0.013	0.024
	k_{42}	-0.008	0.519	0.511
	k_{43}	-0.613	-0.568	-1.181
	k_{44}	1.242	0.105	1.347
Damping (Ns/m)	c_{11}	17.264	-89.031	-71.777
	c_{21}	3.054	520.64	523.69
	c_{22}	25.644	-174.23	-148.59
	c_{31}	-0.027	-651.62	-651.65
	c_{32}	8.079	76.12	84.199
	c_{33}	25.644	257.52	283.17
	c_{41}	-0.061	-68.56	-68.62
	c_{42}	-0.027	465.26	465.24
	c_{43}	3.054	-550.13	-547.08
	c_{44}	17.254	-70.588	-53.334

Table 3. Singular Values – Portal Frame With Added Mass

Singular Values

Mass	Stiffness	Damping
$.163 \times 10^{19}$	$.171 \times 10^{13}$	$.45 \times 10^{15}$
$.116 \times 10^{19}$	$.969 \times 10^{12}$	$.325 \times 10^{15}$
$.108 \times 10^{19}$	$.858 \times 10^{12}$	$.294 \times 10^{15}$
$.802 \times 10^{18}$	$.617 \times 10^{12}$	$.232 \times 10^{15}$
$.306 \times 10^{17}$	$.109 \times 10^{12}$	$.633 \times 10^{14}$
$.149 \times 10^{17}$	$.764 \times 10^{11}$	$.416 \times 10^{14}$
$.131 \times 10^{17}$	$.758 \times 10^{11}$	$.218 \times 10^{14}$
$.480 \times 10^{16}$	$.444 \times 10^{10}$	$.723 \times 10^{13}$
$.265 \times 10^{16}$	$.367 \times 10^{10}$	$.648 \times 10^{13}$
$.374 \times 10^{13}$	$.184 \times 10^{8}$	$.298 \times 10^{10}$

Table 4. Estimated Parameters for the Two Experiments

	Estimate Without Added Mass	Estimate With Added Mass	Difference
Mass (kg)			
m_{11}	4.107	4.073	-0.035
m_{21}	0.023	-0.596	-0.619
m_{22}	5.657	4.166	-1.491
m_{31}	0.359	0.9	0.541
m_{32}	0.613	0.535	-0.078
m_{33}	5.761	7.155	1.394
m_{41}	0.022	0.142	0.12
m_{42}	0.634	2.485	1.851
m_{43}	-0.251	-2.091	-1.84
m_{44}	4.106	11.873	7.767
Stiffness (MN/m)			
k_{11}	1.333	1.34	0.008
k_{21}	-0.183	0.123	0.307
k_{22}	110.83	111.37	0.543
k_{31}	-0.487	-0.782	-0.304
k_{32}	-110.28	-110.26	0.025
k_{33}	110.47	109.93	-0.534
k_{41}	0.024	0.027	0.004
k_{42}	0.511	0.736	0.225
k_{43}	-1.181	-1.534	-0.353
k_{44}	1.347	1.554	0.206
Damping (Ns/m)			
c_{11}	-71.777	-89.519	-17.74
c_{21}	523.69	700.9	177.21
c_{22}	-148.59	2036.4	2185.
c_{31}	-651.65	-823.36	-171.71
c_{32}	84.199	125.29	41.093
c_{33}	283.174	-1900.	-2183.1
c_{41}	-68.62	-36.918	31.702
c_{42}	465.24	1276.9	811.66
c_{43}	-547.08	-1417.6	-870.53
c_{44}	-53.334	-193.28	-139.94

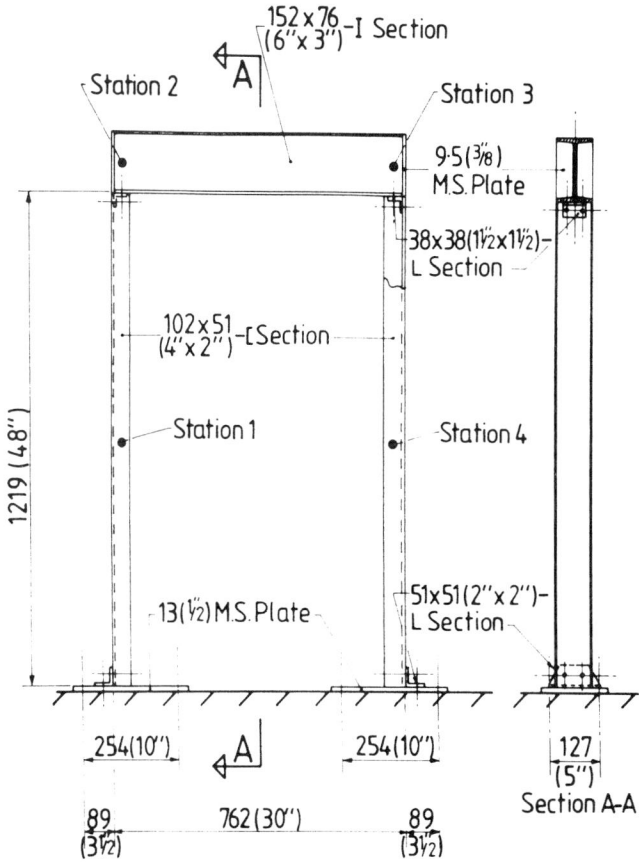

FIG. 1 Portal frame rig.

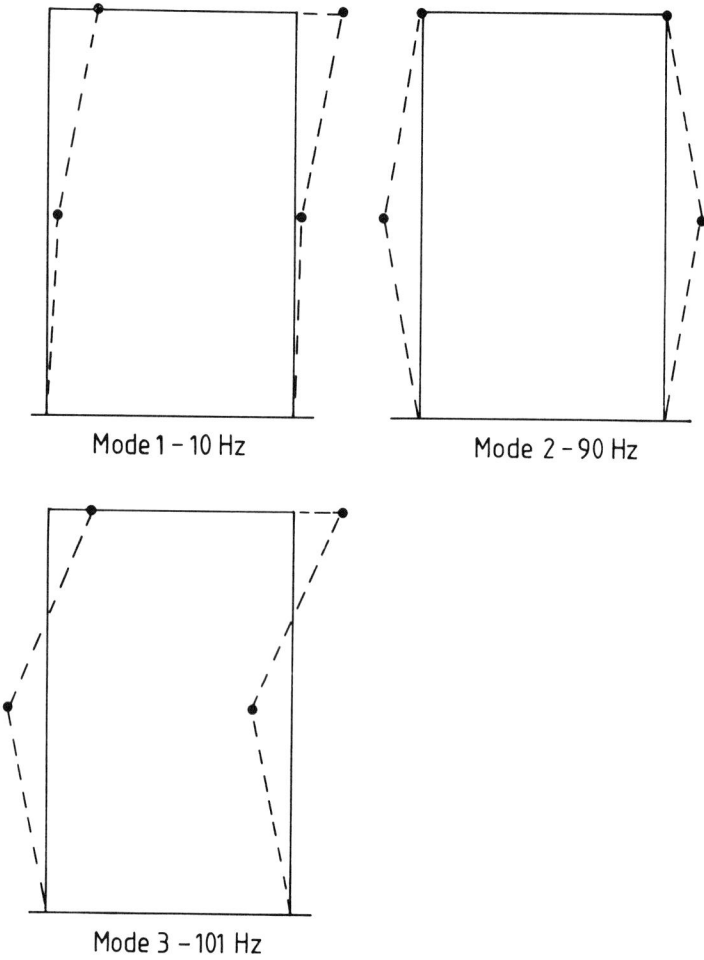

Mode 1 – 10 Hz

Mode 2 – 90 Hz

Mode 3 – 101 Hz

FIG. 2   Mode Shapes of Portal Frame Rig.

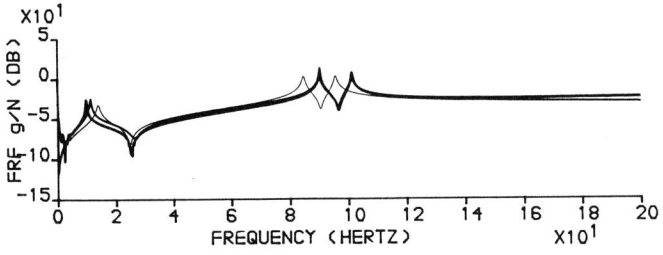

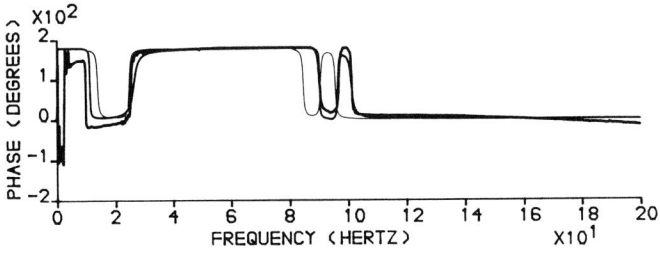

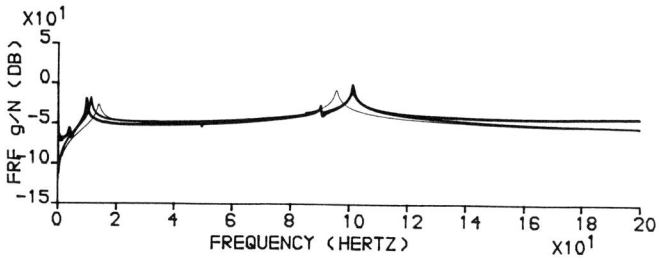

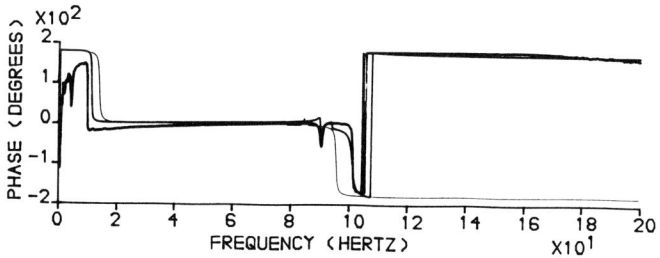

FIG 3 FREQUENCY RESPONSE FUNCTIONS

(a) station 1   (b) station 2

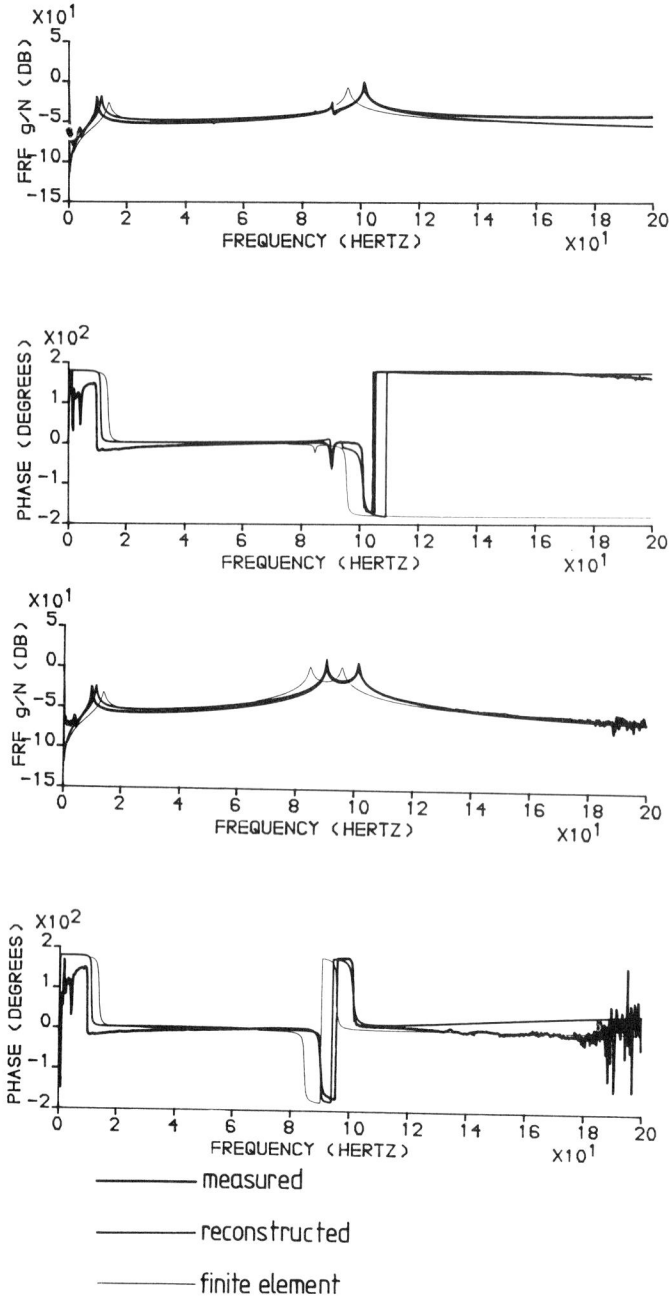

FIG 3   FREQUENCY RESPONSE FUNCTIONS

(c) station 3   (d)station 4

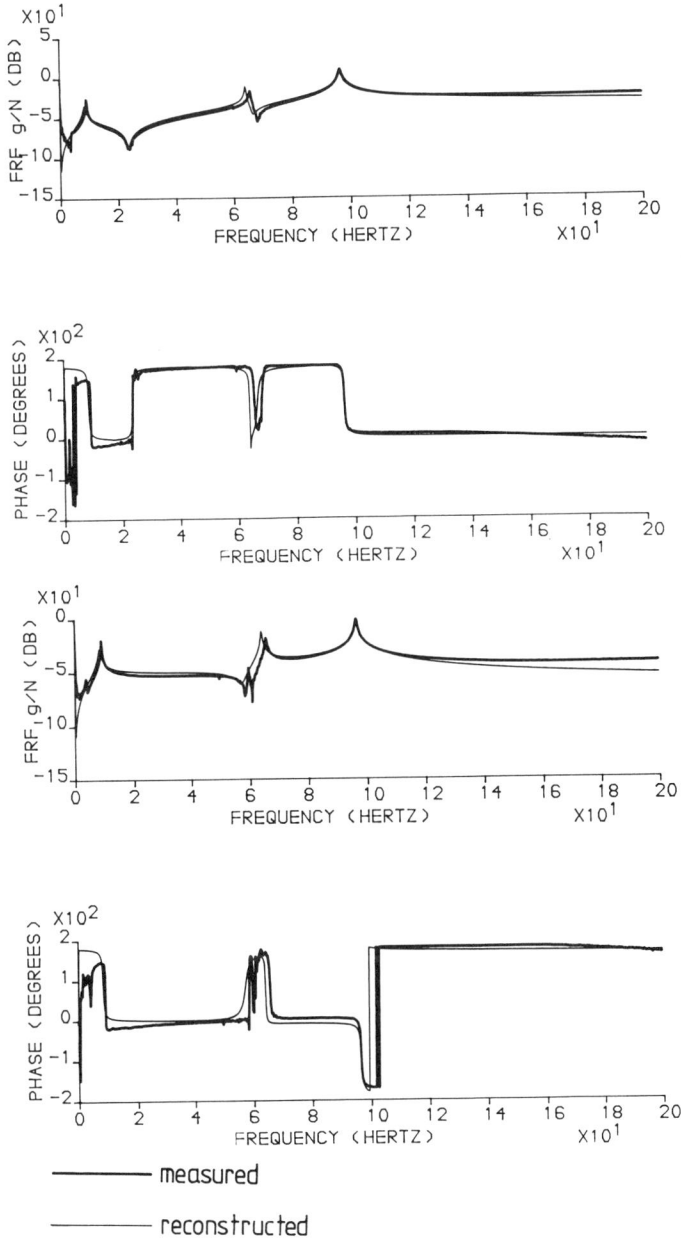

FIG 4 FREQUENCY RESPONSE FUNCTIONS - ADDED MASS

(a)station 1   (b)station 2

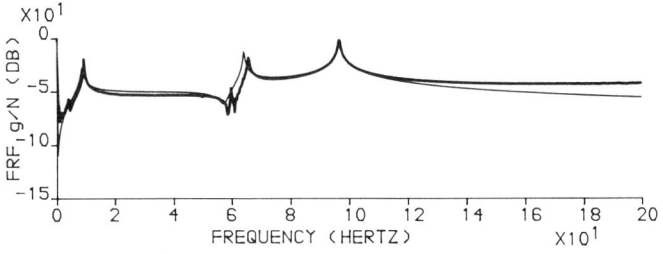

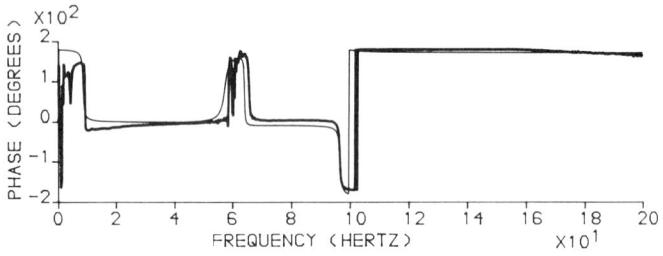

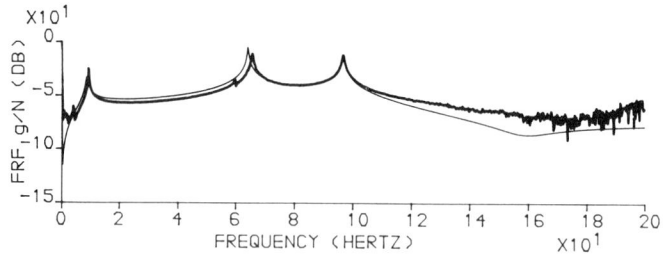

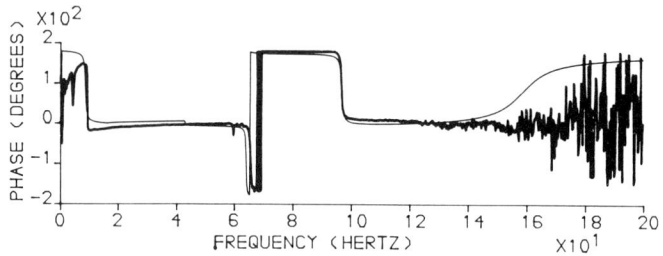

124

FIG 4 FREQUENCY RESPONSE FUNCTIONS – ADDED MASS

(c) station 3   (d) station 4

On the Limitations of Spatial Parameter Estimation Methods in Structural Dynamics

J. A. BRANDON and C. CREMONA

School of Engineering, University of Wales,
College of Cardiff, P.O. Box 917, Cardiff CF2 1XH, UK

ABSTRACT

Relationships between models of different order for the same structure are described. A distinction is made between continuum, discrete full rank, incomplete and low rank models, identifying the modelling discrepancies which can arise. A simple single degree of freedom algorithm is constructed that demonstrates the potential fallibility of the spatial estimation approach.

KEYWORDS

Structural Dynamics; Parameter Estimation; Identification Algorithms.

INTRODUCTION

The matrix analysis of vibration implicitly assumes a topological representation of a structure in terms of a lattice of springs, masses and dampers. Few, if any, practical structures have such a convenient structure.

The work of Fredholm however established that a continuous structure can be represented, to any required degree of accuracy, by a discrete approximation (see for example Lanczos (1957)). Indeed the entire basis of the Finite Element Method, the Rayleigh-Ritz procedure, is founded on this premise (see Strang and Fix (1973)). The continuous structure is considered as a limiting case as the order of the system matrices approaches infinity. The connections between continuum and discrete models are considered in some detail by Meirovitch (1967).

Representations of the same structure, by matrices of different order, and the transformations between these models, will be the central theme of this paper. Of particular importance are the bracketing or inclusion theorems, which provide mutual bounds for the modal properties of different approximations for the

125

same structure. The truncation a model is often described as an incomplete model, a term due to Flannelly and Berman (1972). The analysis of such models, and their analytical consequences, will be of considerable importance in the subsequent discussion.

As has been suggested the discrete model is an approximation to the real structure. The order of a model (corresponding to the dimension of the system matrices) particularly under experimental conditions, often depends on factors beyond the control of the analyst. This may be constrained by the number of channels of data available on the instrumentation (continually increasing) or by access limitations to the structure under test. In contrast finite element models usually have a large number of degrees of freedom but often suffer from inadequacies due to necessary geometrical simplification of the model and the neglect of damping.

BRACKETING PROPERTIES OF EIGENVALUES

The eigenvalues for the generalised eigenvalue problem given by

$$(\sigma^2 \ \underline{M} + \ \underline{K}) \ \underline{\emptyset} \ = \ \underline{0} \qquad (1)$$

are often expressed in terms of the Rayleigh quotient

$$\sigma_i{}^2 \ = \ \frac{\underline{\emptyset}_i{}^T \ \underline{K} \ \underline{\emptyset}_i}{\underline{\emptyset}_i{}^T \ \underline{M} \ \underline{\emptyset}_i} \qquad (2)$$

It can readily be seen that an increase in stiffness will give rise to an increase in the magnitude of the eigenvalue, whereas an increase in mass will cause a decrease in the eigenvalue. The sole exception is the case when the modification is located at a node of the mode, when the eigenvalue will be unchanged.

The change in the value of the Rayleigh quotient is however not unlimited, and is actually asymptotic. The bracketing or inclusion theorems required to understand that theory are also of importance in the comparison of alternative models of the same structure, particularly the reconciliation of experimental and finite element models. The description given here is primarily qualitative since the detail of the theory is well established, see for example Meirovitch (1967, 1980, 1986).

The bracketing theorems relate the eigenvalues of structural approximations when the number of degrees of freedom is varied. Consider the eigenvalues of equation (1) computed according to equation (2). If the number of degrees of freedom is reduced by one this is equivalent to a constraint on the system, making it stiffer and raising the eigenvalues of the system. If these new eigenvalues are denoted by $\sigma_{ic}{}^2$ (for constrained) then the inclusion theorem (Meirovitch (1986) p278) gives

$$\sigma_1{}^2 \ \leq \ \sigma_{1c}{}^2 \ \leq \ \sigma_2{}^2 \ \leq \ \ldots \ \leq \ \sigma_{(n-1)c}{}^2 \ \leq \ \sigma_n{}^2 \qquad (3)$$

Thus each eigenvalue of the reduced system is bounded or "bracketed" by the pair of eigenvalues of the initial system. Conversely increasing the order of a model has the effect of reducing the estimates of the natural frequencies.

The logical extension of this argument is that the continuum model of the structure, viewed as the limit of a sequence of discrete approximations, will give an absolute lower bound to the estimates of any discrete model and the larger the number of degrees of freedom, the better will be the approximation.

Repeated application of the inclusion theorem stated above may lead to an unduly pessimistic view of the quality of low order structural models. For example the introduction of (say) four degrees of freedom into a model may reduce the estimate of the fourth eigenvalue of the augmented model below the first eigenvalue of the initial approximation. In practice however the numerical properties of the Rayleigh quotient usually lead to excellent estimates of the lower eigenvalues, even for rather crude approximations to the eigenvector and coarse spatial discretisation. The quality of the estimation for low order models, both of eigenvalues and eigenvectors, deteriorates progressively as higher values of eigenvalue are taken.

SYNTHESIS OF STRUCTURAL MODELS FROM EXPERIMENTAL DATA: INCOMPLETE MODELS

In an exact analysis, if all of the modes (eigenvectors) and modal masses m_i and stiffnesses k_i are known then the system mass and stiffness matrices may be constructed as follows:

$$\underline{M} = \underline{\Phi}^{-T} \, \underline{m} \, \underline{\Phi}^{-1} \qquad (3)$$

$$\underline{K} = \underline{\Phi}^{-T} \, \underline{k} \, \underline{\Phi}^{-1} \qquad (4)$$

The receptance matrix, which is used to predict the response to external forcing may be defined in spectral form

$$\underline{R} = \underline{\Phi} \, \text{diag} \, \frac{1}{(\sigma^2 m_i + k_i)} \, \underline{\Phi}^T \qquad (5)$$

for undamped systems and

$$\underline{R} = \sum^n \frac{\emptyset_i \, \emptyset_i{}^T}{(\sigma^2 m_i + \sigma c_i + k_i)} \qquad (6)$$

for viscously damped systems.

It can be seen that the response of the system can be constructed as the superposition of the contributions of the responses of the individual modes (the fundamental characteristic of linear systems). Furthermore the response due to an individual mode is a dominant feature of the overall response in the region of its natural frequency. This property is the basis for single mode curve fitting algorithms (see Bishop and Johnson (1963), for the basic theory, and Brandon and Cowley (1983) for a modern implementation). The receptance matrix is often closely approximated by the use of two residual terms to represent out-of-bandwidth response properties ie

$$\underline{R} \approx \frac{1}{\sigma^2} \, \underline{R}_M + \sum \frac{\underline{\Phi} \, \text{diag} \, 1 \, \underline{\Phi}^T}{(\sigma^2 m_i + k_i)} + \underline{R}_K \qquad (7)$$

where the upper and lower residual matrices, $\underline{R}_M$ and $\underline{R}_K$

respectively, are constant and the summation now only applies over the measured modes. The former corresponds to an unconstrained mass effect, or rigid body mode, whilst the latter corresponds to a statically compliant constraint distribution. This type of response model is well proven and widely used (see for example Ewins (1980)).

More controversial, yet as justified on theoretical grounds, has been the construction of approximations to the mass and stiffness matrices of the structure using the measured modal data using rank deficient equivalents to equations (3) and (4):

$$\underline{M}_r = \underline{\Phi}_r{}^{gT} \underline{m}_r \underline{\Phi}_r{}^{g} \qquad (8)$$

$$\underline{K}_r = \underline{\Phi}_r{}^{gT} \underline{k}_r \underline{\Phi}_r{}^{g} \qquad (9)$$

where $\underline{\Phi}_r$ is now nxr, and g denotes the Moore-Penrose generalised inverse (often also called the pseudo-inverse). The definition and application of the pseudo-inverse has been the subject of some misunderstanding and abuse in structural dynamics (see Brandon (1988)). Although the matrices $\underline{M}_r$ and $\underline{K}_r$ are of rank r, ie they have r linearly independent rows/columns, they are of order n. The essential property of such a construction for approximations to the mass and stiffness matrices is that these matrices define an eigenvalue problem which replicates the measured properties of the structure.

The term "incomplete models" was used by Berman and Flannelly (1971) to describe mathematical models depending on this type of representation. Such representations are widely mistrusted within the modal analysis community because of gross discrepancies between the incomplete mass and stiffness matrices, produced from experimental tests, and the corresponding analytically constructed full rank matrices of the same order. Thus the lack of physical significance which can be derived from the models is a bar to their acceptance.

The idea of an incomplete model was originally rationalised by Ross (1971) in the following terms:

"For a forced response analysis, the individual matrix elements need not have physical significance because their primary function is to model accurately the structure's dynamic response. However, when synthesizing high-order models which are to be compared with high-order theoretical models, it is necessary that the individual elements of the synthesized matrices agree with the well-defined meanings of the theoretical elements."

A similar approach to that of Ross was initially made by Berman (1975), although he and his associates have latterly used these methods primarily for adjustments to an existing model (see for example Berman and Nagy (1983)). Latterly both Berman (1984) and Caughey (1982) have explicitly rejected the construction of representative spatial models from response data. More recently Leuridan et al (1986) and Tee et al (1988) have devised new parameter estimation techniques.

CONSTRUCTING ANALYTICAL MODELS FROM EXPERIMENTAL
DATA: LOW ORDER MODELS

The methods described in the previous section involved low rank
(the number of modes, r, used in model construction) but large
order (the number of degrees of freedom considered, n). The
rank of the matrix indicates the number of independent vectors
used in the construction of the rows (or columns) of the matrix.
(The rank of a system estimated from experimental testing can be
an extremely subjective measure- see Brandon (1987)). In
identifying the system in terms of mode shapes and frequency
response functions (corresponding to the information given by
the eigenvalues of the analytical system), it may be sufficient
(depending on the requirements of the analyst), to undertake the
modal survey at only r points (subject to the pedantic
constraint that the r points are not all nodes of at least one
mode of the structure, in which case that mode, or modes will
not be detected and the system will appear rank deficient). In
this case the mass and stiffness matrices defined by equations
(8) and (9) will define an eigenvalue problem which will
reproduce exactly the observed response behaviour of the
structure at the chosen points, although this model may be
extremely sensitive to small perturbations in the structural
parameters. This type of representation is referred to as a low
order model of the structure. The practice of representing
distributed structural properties by local concentrations is
described as "lumping" and the models lumped parameter models.

The analysis of such models has progressed from the initial
ideas of Ross (1971). Successful application of such methods
has been reported by Lincoln (1977) and by Tlusty et al (1978).
Whittaker and Sadek (1980) have proposed the incorporation of
low order models in ".. a complete on-line optimisation package
with regard to the dynamics of machine tools."

The successful application of low order models has been achieved
almost exclusively when applied to machine tool structures.
There is however widespread scepticism about these methods and
anecdotal evidence that the methods break down when applied to
more general structures. (Little has been published since many
researchers are understandably reluctant to report their
failures). That is not to say however that the machine tool
researchers have been misled, since machine tools are
intrinsically amenable to lumped parameter representations,
comprising rigid massive structural members, such as beds,
columns, etc, connected by compliant interfaces at guideways,
slides etc. Further the structural characteristic of interest is
often restricted to the most compliant mode, (measured at the
tool-workpiece interface), which Rayleigh's principle suggests
will be closely approximated with a low order model.

The low order model may be satisfactory for response
calculations but it has serious deficiencies for reanalysis.
This stems from the basic mismatch between the model and the
observed data. The experimental data are discrete observations
of the continuous structure implying, from the inclusion
principle, that the natural frequencies measured will provide
absolute lower bounds to the frequencies predicted from a

discrete representation. Conversely, if estimates to the structure's natural frequencies are provided analytically, through a process of order reduction to match the degrees of freedom of the low order model, then the analytically derived frequencies will be (perhaps substantially) greater than those measured. Thus, for reanalysis purposes, although the measured mode shapes may conform closely to those which would be given by an analytical low order model, the measured frequencies are likely to be substantial underestimates, becoming progressively worse with increasing frequency.

A CASE STUDY

In the literature it is common to demonstrate the power of algorithms by applying them to actual data derived from large complex structures. The approach used here, in demonstrating the fallibility of a parameter estimation algorithm, is to construct a theoretically credible algorithm which fails (comprehensively) on a very simple structure.

Consider the single degree of freedom spring-mass-damper system, subject to forced vibration, described by the equation

$$m \ddot{x} + c \dot{x} + k x = f \tag{10}$$

Taking three successive observations of this system leads to a matrix expression

$$\underline{R} \; \underline{p} \;\; = \;\; \underline{f} \tag{11}$$

where

$$\underline{R} = \begin{pmatrix} \ddot{x}_1 & \dot{x}_1 & x_1 \\ \ddot{x}_2 & \dot{x}_2 & x_2 \\ \ddot{x}_3 & \dot{x}_3 & x_3 \end{pmatrix} \quad \underline{p} = \begin{pmatrix} m \\ c \\ k \end{pmatrix} \quad \underline{f} = \begin{pmatrix} f_1 \\ f_2 \\ f_3 \end{pmatrix}$$

Provided that $\underline{R}$ is nonsingular, the system spatial properties may be derived immediately:

$$\underline{p} = \underline{R}^{-1} \underline{f} \tag{12}$$

Algorithms for multi degree of freedom systems may be devised in a similar manner. The algorithm, as stated, demands no specific signal for excitation. In the subsequent analysis a harmonic input will be used to simplify the problem.

Example

Without loss of generality, the three observations are chosen at t = 0, $p/(2\sigma)$, p/σ where σ is the excitation frequency. In this case, the vector f can be written

$$f = A \cdot (1 \; 0 \; -1)^T \tag{13}$$

where A is the amplitude of excitation.

In order to illustrate the behaviour of the identification it is only necessary to assume an error in $\ddot{x}_3$. All the other measures

131

will be considered as the exact values. Choosing for initial conditions $\dot{x}(0) = x(0) = 0$, Eq.(12) will give a correct mass identification. Thus, the mass can equated to unity. The working hypothesis leads to the study of the system:

$$\delta c\ \dot{x}_2{}^e\ +\ \delta k\ x_2{}^e\ =\ 0$$

$$\delta c\ \dot{x}_3{}^e\ +\ \delta k\ x_3{}^e\ =\ -\ r\ \ddot{x}_3{}^e$$

where r is the ratio of the perturbation and ^{e} denotes the exact value. Taking c = 10 Nsm^{-1} and k = 100000 Nm^{-1}, r = 10%, Fig.1 and Fig.2 give the curves of $\delta c/c$ and $\delta k/k$ as functions of σ.

These curves show that the errors on both coefficients increase with frequency. Moreover, at high frequencies, negative stiffnesses are identified, whereas negative damping is determined at low frequencies. The concavity at low frequencies is in the region of the resonant frequency.

CONCLUSIONS

The identification of real continuous systems, approximated by large discrete system of order n is not possible in practice for a number of reasons

(i) the instrumentation is restricted to a limited frequency range and will be incapable of identifying high frequency modes;
(ii) the necessary inversions will be subject to the problems of numerical ill-conditioning;
(iii) the discrepancy between discrete and continuum models increases with increasing frequency.

In evaluating these methods the viewpoint, and hence the objectives, of the analyst must be considered. Although it is still an active area of research, the attempts to construct physically meaningful mass and stiffness matrices from experimental data will inevitably be defeated by the numerical characteristics of the process.

This type of method has shown considerable promise for adjustment of established analytical models and the authors would have few reservations in applying such techniques to parameter adjustment. There is little doubt that these procedures are less prone to (possibly catastrophic) errors.

REFERENCES

Berman, A. (1975). Determining structural parameters from dynamic testing, Shock and Vibration Bulletin, 7, pp10-17.
Berman, A. (1984). Limitations on the identification of discrete structural dynamic models, 2nd International Conference Recent Advances in Structural Dynamics, ISVR Southampton, pp427-435
Berman A. and Flannelly, W. G. (1971). Theory of incomplete models of dynamic structures, American Institute of Aeronautics and Astronautics Journal, 9, pp1481-7
Berman, A. and Nagy, E. J. (1983), Improvement of a large analytical model using test data, American Institute of

Aeronautics and Astronautics Journal, 21, pp1168-73.

Bishop, R. E. D. and Johnson, D.C. (1963). An investigation into the theory of resonance testing, Proc Royal Society, A255, pp241-280.

Brandon, J. A. and Cowley, A. (1983). A Weighted least squares method for fitting circles to frequency response data, Journal of Sound and Vibration, 89, pp419-424.

Brandon, J. A. (1987). The significance and practice of rank estimation in structural dynamics identification, International Conference Numerical Methods in Engineering: Theory and Applications, U C Swansea, Eds. G Pande and J Middleton, Martinus Nijhoff, section S5, pp1-8.

Brandon, J. A. (1988). On the robustness of algorithms for the computation of the pseudo-inverse for modal analysis, 6th International Modal Analysis Conference, Orlando, Fa

Brandon, J. A., (1989). Numerical Analysis Needs for Modal Analysis, 7th International Modal Analysis Conference, Las Vegas

Caughey, T. K. (1982). Structural Dnamics Ananlyses Testing and Correlation, Report 81-72, Jet Propulsion Laboratory, California Institute of Technology

Ewins, D. J. (1980). On predicting point mobility plots from measurements of other mobility parameters, Journal of Sound and Vibration, 70, pp69-75.

Flannelly W. G. and Berman, A. (1972). The state of the art of system identification of aerospace structures, Proc Symposium on System Identification of Vibrating Structures, 1972 ASME Winter Annual Meeting, pp121-131.

Lanczos, C. (1957). Applied Analysis, Pitman.

Leuridan, J., Brown, D. and Allemang R. J. (1986). Time Domain Parameter Identification Methods for Linear Modal Analysis: A Unifying Approach, Transactions of the American Society of Mechanical Engineers, Journal of Vibration, Acoustics, Stress and Reliability in Design, 108, pp 1-8

Lincoln, A. P. (1977). Modelling of structural behaviour from frequency response data, Technical Report No 83, Institute of Sound and Vibration Research, Southampton University

Meirovitch, L. (1967). Analytical Methods in Vibrations, Macmillan

Meirovitch, L. (1980). Computational Methods in Structural Dynamics, Sijthoff & Noordhoff

Meirovitch, L. (1986). Elements of Vibration Analysis, McGraw-Hill.

Ross, R. G. (1971). Synthesis of stiffness and mass matrices from experimental vibration modes, SAE paper 710787

Strang, G. and Fix, G. J. (1973). An Analysis of the Finite Element Method, Prentice-Hall.

Tee, T. K., Mottershead, J.E., Stanway, R. and Brookfield, D.J. (1988) Application of filtering techniques for system identification of vibrating structures, 3rd International Conference on Recent Advances in Structural Dynamics, Southampton, pp289-298.

Tlusty, J., Ismail F. and Prossler, E. K. (1978). Identification, Modelling and Modification of Structures Defined from Measured Data, Technical Report, Laboratorium fur Werkzeugmaschinnen und Betriebslehre, Technische Hochschule Aachen

Whittaker, A. R. and Sadek, M. M., (1980). Optimisation of machine tool structures using structural modelling techniques,

133

1st International Conference Recent Advances in Structural Dynamics, Southampton, pp211-223.

Fig.1 : variation of error on damping.

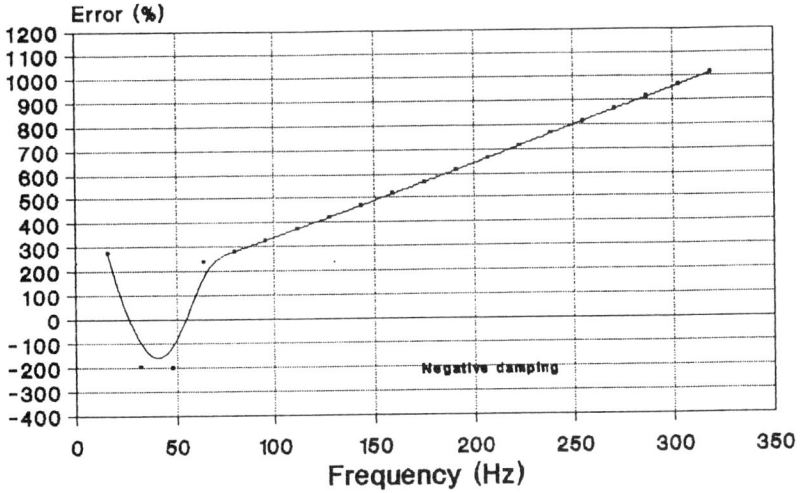

Perturbation 10 %

Fig.2 : variation of error on stiffness.

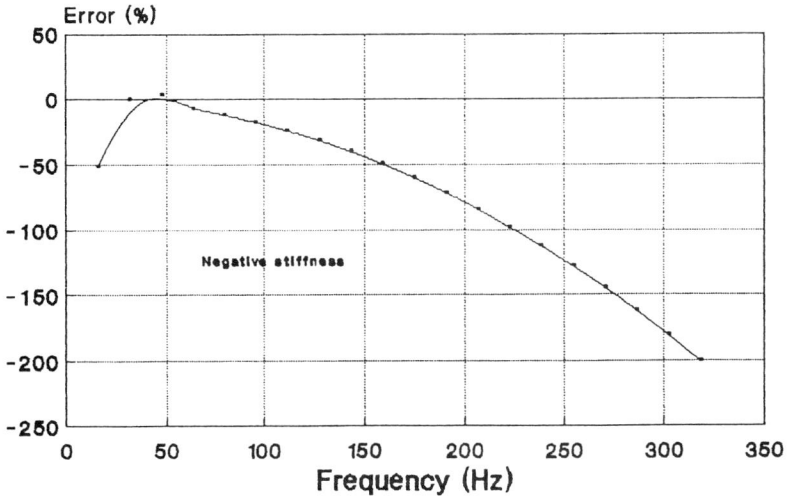

Perturbation 10 %

Waves and Vibrations in Elastic Fiber-composite Medium and Plates

P. NAVI

Ecole Nationale des Ponts et Chaussées,
Centre d'Enseignement et de Recherche en Analyse des Matériaux,
Central IV, 1 Avenue Montaigne, 93167 Noisy-le-Grand, France

ABSTRACT

Dispersion relations and modal displacement patterns of harmonic plane waves propagating in a periodic medium-laminated composite plate, fiber composite and porous medium are studied by a numerical method based on finite element solutions. The dispersion curves for wave propagating in an unidirectional fiber composite and in an unidirectional hollow medium, in a plane perpendicular to the fiber or hollow axes are shown. The wave patterns corresponding to the acoustical and optical branches of the frequency spectra for different wave-numbers are plotted and the differences between these displacements are discussed. Then a frequency spectra for a reinforced composite plate is generated. It is observed that a composite material, for wavelengths greater than the periodical dimensions of the medium and for low frequencies, behaves like a homogeneous medium.

KEYWORDS

Composite material ; hollow medium ; composite plates ; wave propagation .

INTRODUCTION

The study of harmonic wave propagations in an inhomogeneous periodical medium supplies important informations both on the practical and theoritical plane. One of these practical importances is that, a periodical composite infinite medium shows the aptitude for selecting certain frequencies of harmonic waves to passing or stopping. This attitude of the periodical composite medium is related to the geometrical inhomogeneities and mechanical properties of its constituents. Another phenomenon which is known for a long time is that the phase velocity of a travelling harmonic wave in a heterogeneous medium is related to the overall mechanical properties and the mass density of the medium. On the theoritical plane it is known (Turbe, 1982) that, the displacement field in a large periodical medium can be expressed in the form of Bloch developpement on the basis of the modal displacement patterns and their corresponding frequencies.

In past the propagation of harmonic waves in an infinite laminated plate and composite medium has been studied extensively. Review of the literatures on exact and approximate analysis of this problem (Sun, 1968 ; Sve, 1971 ; Dong et al., 1972 ; Nelson and Navi,1975 ; Delph et al., 1978) and more recently (Datta et al., 1988), shows that in general the

135

matrix-fiber structure of layered medium is replaced first by a laminated medium in which each layer is considered as a homogeneous elastic continuum, then the problem of wave propagations has been investigated in this laminated medium. The dispersion relations obtained from this modeling ignore the physical effect of the periodical matrix-fiber structure of the medium. This can be important in dynamic problems for the wavelengths which approache the periodical length of the fiber-matrix medium and for the high frequencies.In addition, only few efforts (Minagawa et al, 1984 ; Navi, 1988) have been made to illustrate the different modal displacement patterns of travelling waves belonging to different acoustical and optical branches, in theBrillouinsense (1952), of frequency spectra. According to Brillouin, the acoustical branches are the ones passing through the origin, the optical ones being the ones that don't.

This paper study the dispersion relations of a plane wave in a periodic fiber-composite medium. This medium can be an infinite composite plate or a periodical hollow material. In this study the main attention is given to the illustration of the modal displacement patterns for both acoustical and optical branches of dispersion curves. The differences between the modal patterns is shown and the influence of discretization on the travelling optical modal displacements in a periodical composite medium is discussed.

In this paper we have applied a numerical method based on finite element solutions presented in (Navi, 1973 ; Nelson and Navi, 1975). In this approach a periodic composite is idealized as of a periodic lattice type structure and its response to harmonic waves is investigated as a problem in lattice dynamics, a subject detailed in (Brillouin, 1952).

ANALYSIS

Review of the method

A periodic continuous medium is replaced by an assembly of identical regions, each with displacement fields represented by a finite number of generalized displacements. The stiffness and mass matrices of the entire region are determined by Hamilton's principle and the equations of motion are written in the form (1).

$$[M] \{\ddot{q}\} + [K] \{q\} = \{o\} \tag{1}$$

Where [M] and [K] are the mass and stiffness matrices of the medium and {q} is a vector of generalized coordinates. This developement thus replace the periodic continuous medium with a periodic lattice type structure in which the characteristics of each nodes are calculated. Let's assume that, this lattice type structure being formed by a basic unit cell defined by k nodes with position vectors $\vec{r}_1$, $\vec{r}_2$, ...$\vec{r}_j$, ...$\vec{r}_k$ with respect to the origin of the cell. The dimensions of the unit cell are defined by three positive vectors $\vec{d}_1$, $\vec{d}_2$ and $\vec{d}_3$ in x_1, x_2 and x_3 directions respectively. Let's also assume that, the center of a unit cell (N) to be at point 0 , defined from system coordinate by a position vector $n_1\vec{d}_1 + n_2\vec{d}_2 + n_3\vec{d}_3$, where n_1, n_2 and n_3 are integer numbers.Then the position vector of a node (J) in the cell (N) can be expressed by (2).

$$\vec{r}_j^{'(N)} = \vec{r}_j + n_1\vec{d}_1 + n_2\vec{d}_2 + n_3\vec{d}_3 \tag{2}$$

Since we are interested in plane harmonic waves propagation in a discrete lattice type structure, we write the displacement of node (J) in cell (N) in the following form.

$$\psi_j^{(N)} = Re \left[A_j^{'(N)} e^{i(2\pi\vec{\alpha} \cdot \vec{r}_j^{'(N)} - \omega t)} \right] \tag{3}$$

in which Re [ ] design real part of [ ], $\vec{\alpha}$ is a vector which defines the direction of wave propagation in the medium and has a magnitude equal to the reciprocal of the wavelength λ. $|\alpha^2| = \alpha_1^2 + \alpha_2^2 + \alpha_3^2 = \left(\frac{1}{\lambda}\right)^2$, ω is the frequency of the harmonic wave and $A_j^{'(N)}$ is a complex function of position vector of node (j) in the cell (N). On the basis of the Floquet theory we can show that $A_j^{'(N)}$ is a periodic function and has the same periodicity of the medium.

$$A_j^{'(N)} = A_j^{'(N+M)} = A_j^{'} \qquad (4)$$

Substituting (2) in (3) and taking acount of (4), relation (3) becomes ;

$$\Psi_j^{(N)} = \text{Re} \left[A_j \ e^{i\left(2\pi\vec{\alpha}.\left(n_1\vec{d}_1 + n_2\vec{d}_2 + n_3\vec{d}_3\right)-\omega t\right)} \right] \qquad (5)$$

where $A_j = A_j^{'(N)} \ e^{i\left(2\pi\vec{\alpha}. \ \vec{r}_j^{'(N)}\right)}$ or,

$$\Psi_j^{(N)} = \text{Re} \left[A_j \ e^{i(n_1 K_1 + n_2 k_2 + n_3 k_3)-\omega t)} \right] \qquad (6)$$

in which $K_1 = 2\pi\vec{\alpha}_1.\vec{d}_1$, $K_2 = 2\pi\vec{\alpha}_2.\vec{d}_2$ and $K_3 = 2\pi\vec{\alpha}_3.\vec{d}_3$.

Relation (6) gives relationship among all the identical nodes in the different unit cells. Therefore by evaluating just the displacements of k nodes belonging to a one unit cell at a time t, we can determine all the displacements of the entire region. This phenomena reduces the solution of the equations of the entire system to a small region consisting of a basic unit cell only defined by k nodes. Substituting (6) in (1) gives a set of equation in the form :

$$[\bar{K}] \ \{\Psi\} - \omega^2 \ [\bar{M}] \ \{\Psi\} = \{0\} \qquad (7)$$

where $\{\Psi\}$ is a vector of all the amplitudes of A_j (j = 1,..k) which belongs to one unit cell. The (-) appearing with mass and stiffness matrices indicates that the boundry conditions related to the adjacent nodes to the considered unit cell are included.

The effectiveness and the accuracy of this method have been shown in (Navi, 1973 ; Nelson et al., 1975 ; Navi, 1988). In the following parts we will apply this method to illustrate the dispersion curves and modal displacement patterns for a wave propagating in a unidirectional fiber composite and hollow medium and in a laminated composite plate.

EXAMPLES

Infinite Fiber Composite Medium

An infinite unidirectional fiber composite medium with double periodicity of d in x_1, and x_2 directions is considered. The elastic constants of the fibers are E_f = 26.6 GPa and ν_f = 0.33, and of the matrix are, E_m = 9.16 GPa and ν_m = 0.35. The mass density of the fibers and matrix are ρ_f = 2000 Kg/m^3 and ρ_m = 500 Kg/m^3. Fig. 1 shows the geometry of a periodic

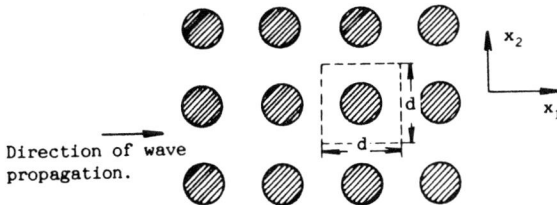

Fig. 1 Infinite fiber composite medium

medium and a basic unit cell. In this example the volume concentration of the fibers is $C_f = 0.35$. The frequencies and modal displacements of the medium for a cell represented by 32 nodes are calculated and the dispersion curves are plotted for the waves propagation in the x_1 direction. These curves are shown in Fig. 2. Then four lowest modal displacement patterns

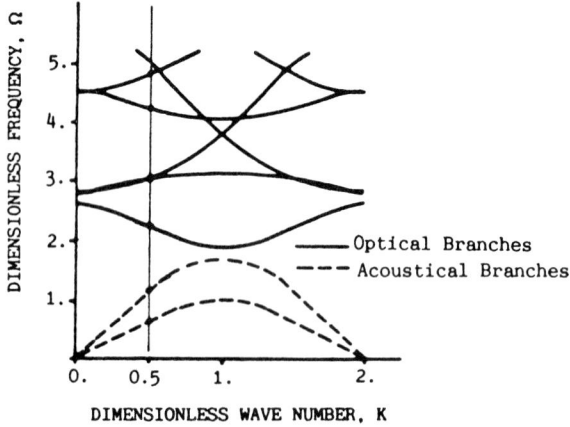

Fig. 2 Dispersion curves in a fiber
composite, wave propagating in
x_1 - direction

for wave number $K = 2d/\lambda = 0.5$ for three adjacent unit cells in the direction of the wave propagation are illustrated is Fig.3. The first two modal displacements are belonging to the acoustical branches.

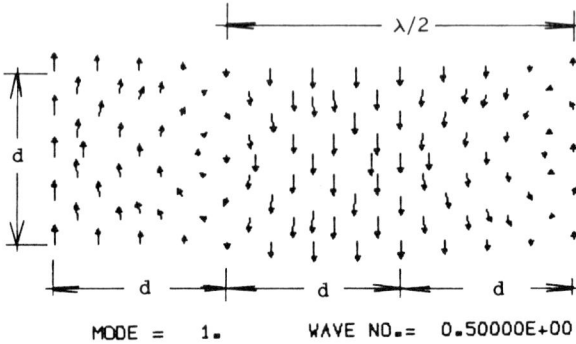

MODE = 1.   WAVE NO.= 0.50000E+00

.../...

139

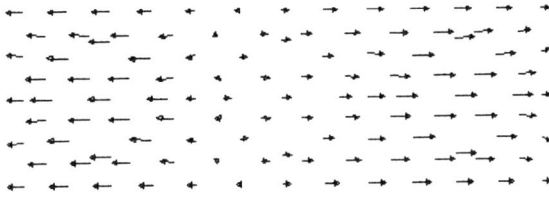

MODE = 3.   WAVE NO.= 0.50000E+00

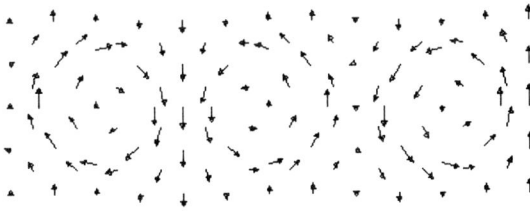

MODE = 5.   WAVE NO.= 0.50000E+00

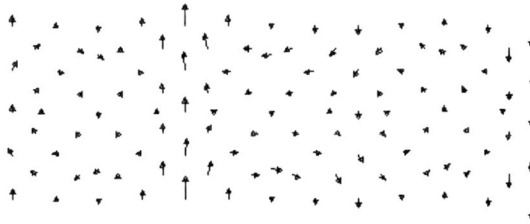

MODE = 7.   WAVE NO.= 0.50000E+00

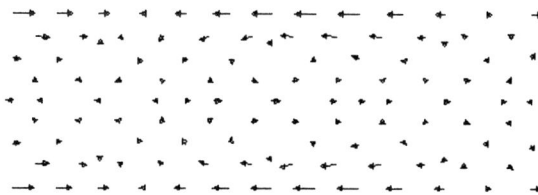

MODE = 9.   WAVE NO.= 0.50000E+00

Fig. 3 Modal displacement patterns of waves in a fiber composite medium at K = 2d/λ = 0.5

Infinite Hollow Fiber Medium

The propagation of a harmonic wave in an infinite hollow medium with 35 per cent porosity in the x_1 direction is studied. The geometry of the medium and a unit cell is shown in Fig. 4.

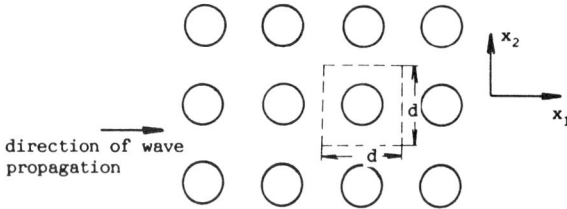

Fig. 4 Infinite hollow fiber medium

The dispersion curves are calculated with unit cell represented with 27 nodes and plotted in Fig. 5. Four lowest modal displacement patterns for a wavenumber K = 0.5 for three adjacent cells are shown in Fig.6.

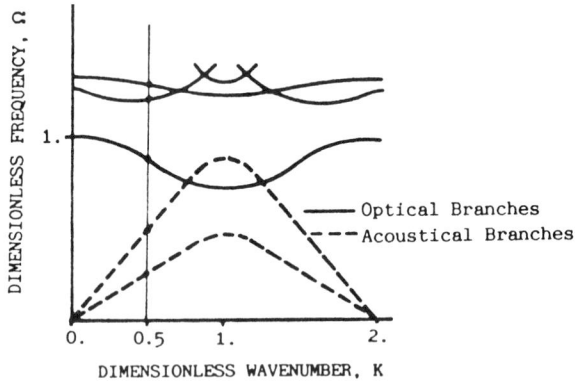

Fig. 5 Dispersion curves in a hollow
medium, wave propagating in
x_1 - direction

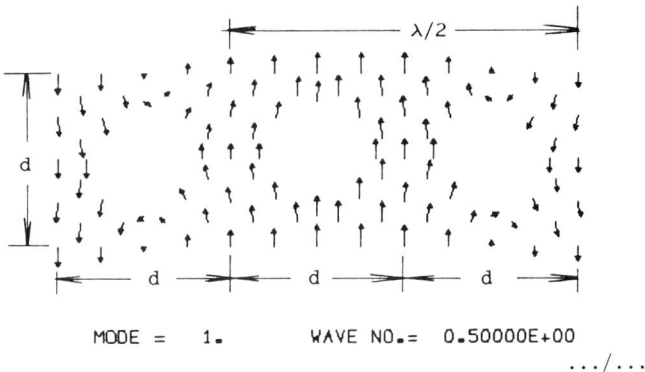

MODE = 1.   WAVE NO.= 0.50000E+00

.../...

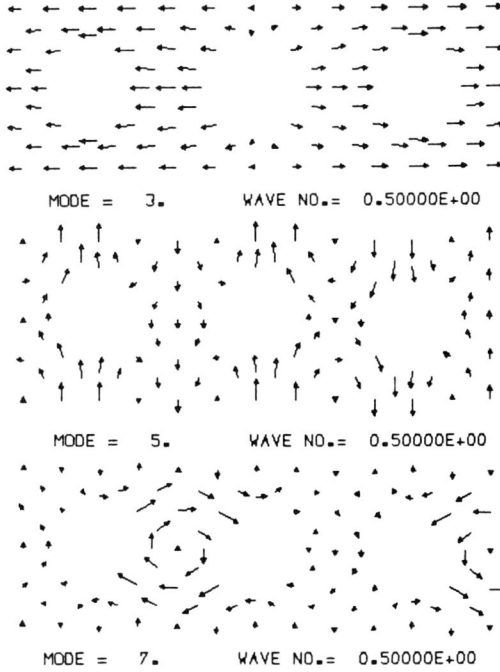

MODE = 3.   WAVE NO.= 0.50000E+00

MODE = 5.   WAVE NO.= 0.50000E+00

MODE = 7.   WAVE NO.= 0.50000E+00

Fig. 6 Modal displacement patterns of waves in a hollow fiber medium at $K = 2d/\lambda = 0.5$

Infinite Fiber Composite Plate

We have considered an infinite fiber composite plate with $H = d$, shown in Fig. 7. The six lowest frequency for various values of the wavenumber $0.01 \leqslant K \leqslant 1$ are computed and shown in

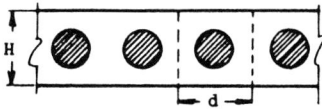

Fig. 7 Infinite composite plate, H = d

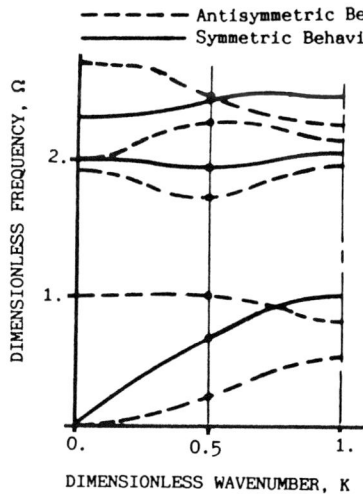

Fig. 8 Frequency spectrum for
a fiber-reinforced
plate, H = d

Fig.8. The first four symetric and antisymetric wave patterns for K = 0.5 are illustrated in Fig. 9. The unit cell is represented by 36 nodes.

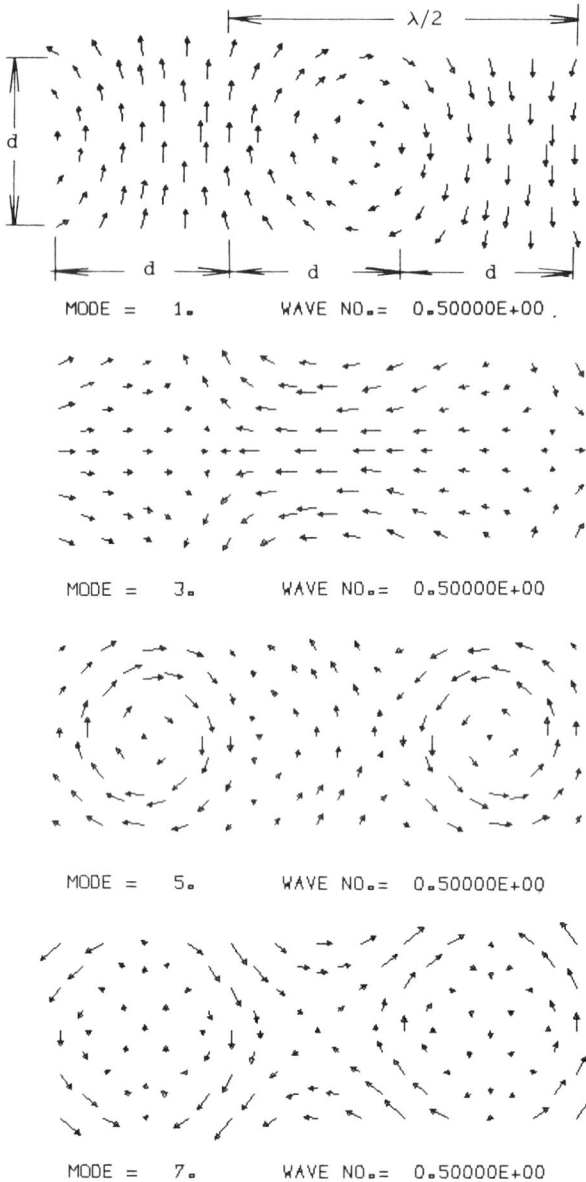

Fig. 9 Modal displacement patterns of waves in a composite layered medium at K = 2d/λ = 0.5 and H = d

DISCUSSION

We have studied the problem of the propagation of plane harmonic waves in infinite periodic elastic medium. The dispersion curves and modal displacement patterns of waves in a fiber composite, porous medium and a fiber composite plate, in a plane perpendicular to the fiber axes, were illustrated. These modal displacements show that :

(1) - In composite materials, the modal displacement patterns belonging to the (Brillouin, 1952) acoustical branches of dispersion curves essentially are different from those of optical branches.In the acoustical behaviour, at a given time t, the movements, (polarizations) of entire physical points inside a $\lambda/2$ distance of medium (in the direction of propagation), are unidirectional. There are two such a displacement patterns in a two dimensional problem. These acoustical behaviours are almost the same for both homogeneous and inhomogeneous periodic medium for the long waves (compared to periodical distances of the medium) and for the low frequencies.

(2) - Under higher frequencies, the periodic inhomogeneous medium reaches the optical branches . In this case the movements (polarizations), of the physical points in the $\lambda/2$ distances of the medium, are not all in the same direction. The relative directions of the polarization between neighbouring points and thus the modal patterns strongly related to the detailed configuration, and to the relative distribution of mass and mechanical characteristics in the basic unit cell.

(3) - The modal displacement patterns of optical behaviour obtained by a numerical procedure for a periodic inhomogeneous medium is also strongly related to the discretization pattern of the continuum medium. More a basic unit cell is represented with "identical" subregions and more the number of such subregions are increased and more the results are accurate.

(4) - On the contrary to the optical behaviour of the heterogeneous medium, the calculated acoustical modal patterns are less influenced by the uniformity and number of subregions representing the basic unit cell.

REFERENCES

Brillouin, L. (1952). Wave Propagation in Periodic Structures, Dover, New-York.

Datta, S.K., A.H. Shah, R.L. Bratton and T. Chakraborty (1988). Wave Propagation in Laminated Plates. J. Acoust. Soc. Am., 86, 2020-2026.

Delph, T.J., G. Herrmann and R.K. Kaul (1978). Harmonic Wave Propagation in a Periodically Layered, Infinite Elastic Body : Antiplane Strain. J.Appl. Mechanics, 45, 343-349.

Dong, S.B. and R.B. Nelson (1972). On Natural Vibrations and Waves in Laminated Orthotropic Plates. J. Appl. Mechanics, 39, 739-745.

Minagawa, S., S. Nemat Nasser and M. Yamad (1981). Finite Element Analysis of Harmonic Waves in Layered and Fiber Reinforced Composites. Int. J. Num. Meth. Engineering, 17, 1335-1353.

Navi, P. (1988). Harmonic Waves in Layered and Fiber Composite Materials. Proceeding of the International Conference on Impact Loading and Dynamic Behaviour of Materials. Oberurse (RFA), 2, 985-992.

Navi, P. (1973). Harmonic Wave Propagation in Composite Materials. Ph.d Thesis U.C.L.A., Los-angeles.

Nelson, R.B. and P. Navi (1975). Harmonic Wave Propagation in Composite Materials. J. Acoust. Soc. Am., 57, 773-781.

SVE, C. (1971). Time Harmonic Travelling Obliquely in a Periodically laminated Medium. J.Appl. Mechanics, 38, 477-482.

Turber, N. (1982). Application of Bloch Expension to the Periodic Elastic and Viscoelastic Media. Math. Meth. in the Appl. Sci., 4, 433-499.

On the Harmonic Waves Propagation in Multilayered Viscoelastic Media

T. NACIRI, P. NAVI and O. GRANACHER

Ecole Nationale des Ponts et Chaussées,
Centre d'Enseignement et de Recherche en Analyse des Matériaux,
Central IV, 1 Avenue Montaigne, 93167 Noisy-le-Grand, France

ABSTRACT

An analytical method is presented to study the propagation of plane harmonic waves in an infinite periodicaly laminated viscoelastic medium. The dispersion and the damping relations for a periodic layered medium, when the wave propagates in the direction perpendicular to the layers, are illustrated. From these curves, one can establish the static and dynamic characteristics of the homogenized medium. It is shown that when the half of the wave length approaches the periodic length of the medium, the medium becomes extremely dissipative. A comparison between an elastic and a viscoelastic multilayered medium is made. It is illustrated that, in the elastic medium at the zone of forbidden frequencies no waves penetrate into the medium, whereas for the multilayered viscoelastic medium the waves can penetrate but dissipate very fast.

KEYWORDS

Wave propagation. Composite materials. Viscoelasticity.

INTRODUCTION

The study of the plane harmonic waves propagation in an infinite inhomogeneous medium has received considerable attention the last two decades. In fact the phase velocity of a plane harmonic wave in a medium is correlated to the mechanical properties of the medium. The dispersion relation, which relates for a given propagation direction the wave numbers to the frequencies, and the damping relation, which is a relation between the damping coefficients and the frequencies, provide important information on the overall elastic and viscoelastic behaviour of the medium. These relations can also give informations on the aptitude of the periodic medium to transmition or absorbtion of certain frequencies.

The harmonic wave propagation in the composite materials with elastic constituents have been investigated extensivly in the past. This phenomenon has been studied analytically by Sun et al (1968), Sve (1971) for elastic laminated medium. Latter, the dispersion relations in a fiber composite have been investigated numerically by Navi (1973,1987), Mingawa et al (1984). The verification of dispersion relations for layered composites have been made experimentaly by Robinson and Leppelmeier (1974). In many situations the composite materials have been considered like a viscoelastic medium. Stern et al (1971) considered a composite material like a laminated medium consisting of alternating layers of elastic and viscoelastic materials. They have analyzed the wave propagation in the direction of the layers by an effective elastic theory. Mukherjee and Lee (1975) used a finite difference method to study the dispersion relations in the direction perpendicular to the layers. Ma et al (1980) used a finite element method to investigate the wave propagation in a fiber-composite material when the matrix is viscoelastic in shear. Sutherland (1975) illustrated some experimental studies on the dispersion of waves in a fiber-composite materials. However these works do not give a quantitative dispersion relations for a viscoelastic composite medium in which one can calculate the damping coefficients in the pass and stop regions of the wave frequencies

In this paper we have shown an analytical method which permits us to calculate the velocity and the damping coefficient for a plane harmonic wave propagating perpendicularly to an infinite periodic viscoelastic medium. The dispersion and damping relations for a two layered periodic structure, in which the layers are isotropic and viscoelastic, are determined.

In the first part of this paper, for a two layered periodic viscoelastic medium, the hypothesis are made and the equations for a wave propagating perpendicularly to the layers are established and the dispersion relations are derived. In the second part, a comparative study between an elastic and a viscoelastic multilayered medium is made.

ANALYSIS

An unbounded laminated periodic medium consisting of two alternative layers A and B is considered. The laminate has a periodic structure in the x_3- axis with periodic length $H = a + b$ shown in fig.1.

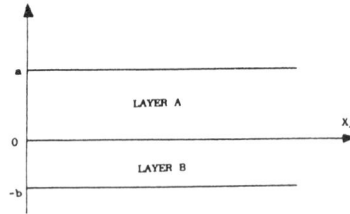

fig.1 : Geometry of the periodic laminate

We also consider the components of the stress tensor $\sigma_{ij}(M,t)$, $u_i(M,t)$ the components of the displacement vector and $\rho(M)$ the mass density of a point $M(x_1,x_2,x_3)$ at time t .

The equation of motion is :

$$\frac{\partial \sigma_{ij}}{\partial x_j}(M,t) - \rho(M).\frac{\partial^2 u_i}{\partial t^2}(M,t) = 0 \qquad (1)$$

Let $\lambda^1(t)$ and $\mu^1(t)$ designate the Lamé's relaxation functions for the layer A and $\lambda^2(t)$ and $\mu^2(t)$ the lame's relaxation functions for the layer B. Then in each layer the constitutive relation can be written in the following form :

$$\sigma_{ij}(M,t) = \lambda^m(o)\,\varepsilon_{kk}(M,t).\delta_{ij} + \int_{-\infty}^{t} \dot{\lambda}^m(t-\tau)\,\varepsilon_{kk}(M,\tau).\delta_{ij}d\tau \qquad (2)$$

$$+ 2\,\mu^m(o)\,\varepsilon_{ij}(M,t) + 2\int_{-\infty}^{t} \dot{\mu}^m(t-\tau)\,\varepsilon_{ij}(M,\tau)d\tau \qquad \text{where } m = 1,2$$

For each layer $\lambda^m(t)$ and $\mu^m(t)$ are determined by the parameters of the model which represent the viscoelastic behaviour of the medium. Here we assume that such a model corresponds to a

Zener model for both material A and B shown in fig.2. Then the relaxation functions can be expressed in the form :

$$\lambda^m(t) = \frac{\nu^m}{(1 + \nu^m)(1 - 2\nu^m)} \; (Ed^m + (Ei^m - Ed^m) \; e^{-t/\tau^m}) \; Y(t) \tag{3}$$

$$\mu^m(t) = \frac{1}{2(1+\nu^m)} \; (Ed^m + (Ei^m - Ed^m) \; e^{-t/\tau^m}) \; Y(t) \tag{4}$$

in which $Ei^m = E^m$, $Ed^m = (E^m + E_1^m)/(E^m.E_1^m)$, $\tau^m = \eta^m/(E^m + E_1^m)$,ν^m is the Poisson's ratio and $Y(t)$ the Heavyside function.

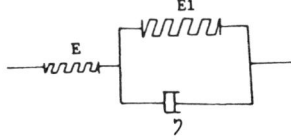

fig.2 : Zener's model

Since we are interested in the propagation of a harmonic wave, we can then write :

$$u_i \; (M,t) = Re \left[u_i^* \; (M) \; e^{i\omega t} \right] \tag{5}$$

where $Re \; [x]$ is the real part of x and $u_i^*(M)$ are the complex components of the amplitude displacement vector.

The strain tensor associated to the displacements given in (5) are :

$$\varepsilon_{ij} \; (M,t) = Re \left[\varepsilon_{ij}^* \; (M) \; e^{i\omega t} \right] \tag{6}$$

where

$$\varepsilon_{ij}^* \; (M) = \frac{1}{2} \left[\frac{\partial u_i^*}{\partial x_j} \; (M) + \frac{\partial u_j^*}{\partial x_i} \; (M) \right] \tag{7}$$

Substituting of the equation (5) into (2) gives us the constitutive relation (8).

$$\sigma_{ij}(M,t) = Re \left[(\lambda^*(M,i\omega) \; \varepsilon_{kk}^*(M) \; \delta_{ij} + 2\mu^*(M,i\omega) \; \varepsilon_{ij}^*(M)) \; e^{i\omega t} \right] \tag{8}$$

where

$$\lambda^*(M,i\omega) = \lambda^m(o) + \int_0^\infty \dot{\lambda}^m(u) \; e^{-i\omega u} \; du \tag{9}$$

and

$$\mu^*(M,i\omega) = \mu^m(o) + \int_0^\infty \dot{\mu}^m(u) \; e^{-i\omega u} \; du \tag{10}$$

where $m = 1$ if the point M is in the layer A and $m = 2$ if M is in the layer B

We can also introduce the complex numbers K_p and K_s such that :

$$K_p^2 = \frac{\rho(M).\omega^2}{\lambda^*(M,i\omega) + 2\mu^*(M,i\omega)} \tag{11}$$

$$K_s^2 = \frac{\rho(M).\omega^2}{2\mu^*(M,i\omega)} \tag{12}$$

Moreover, we know from the Poisson theorem (see Mandel 1966) that every small perturbation in an isotropic viscoelastic medium can be decomposed into an irrotational motion (progressive wave u^P) and an isochoric motion (shear wave u^s).Therefore the equation of motion (1) with the aide of (8), (11) and (12) and by applying the Poisson theorem can decompose into the two following wave equations :

$$\frac{\partial^2 u_k^P}{\partial x_j^2} (M) + K_P^2 \, u_i^P \, (M) = 0 \; \textit{for progressive waves} \tag{13}$$

$$\frac{\partial^2 u_k^S}{\partial x_j^2} (M) + K_S^2 \, u_i^S \, (M) = 0 \; \textit{for shear waves} \tag{14}$$

We know that the two polarized progressive ans shear waves are perpendicular to each other and in addition a progressive in a isotropic motion polarize in the same direction of the wave propagation. Then we can write :

$$\vec{u}^P(M) = u^P(x_3) \; \vec{x}_3 \tag{15}$$

and

$$\vec{u}^S(M) = u^S \, (x_3) \; \vec{x}_2 \tag{16}$$

Where $\vec{x}_2$ and $\vec{x}_3$ are unit vectors. Then, the relations (15) and (16) become :

$$\frac{\partial^2 u^P}{\partial x_3^2} (x_3) + K_P^2 \, u^P \, (x_3) = 0 \; \textit{for progressive waves} \tag{17}$$

$$\frac{\partial^2 u^S}{\partial x_3^2} (x_3) + K_S^2 \, u^S(x_3) = 0 \; \textit{for shear waves} \tag{18}$$

We can see that although p-wave and s-wave move in two different directions the relations (17) and (18) have the same form. We can then take u for u^P or u^S and K for K_P or K_S and solve the following equation :

$$\frac{\partial^2 u(x_3)}{\partial x_3^2} + K^2(x_3).u(x_3) = 0 \tag{19}$$

In (20) the equation (19) is written for two alternative layers A and B.

$$\begin{cases} \dfrac{\partial^2 u_A}{\partial x_3^2} (x_3) + K_1^2 u_A(x_3) = 0 & \text{for the layer A} \\[4mm] \dfrac{\partial^2 u_B}{\partial x_3^2} (x_3) + K_2^2 u_B(x_3) = 0 & \text{for the layer B} \end{cases} \tag{20}$$

Interfaces conditions

At the interface between the two layers, the displacements field and the stress vector are continuous. Moreover, by FLOQUET theory (Stocker 1950), there exists a complex number K^* such that a solution of the motion equation (19) is :

$$u(x_3) = P(x_3).e^{iK^* x_3} \text{ with } P(x_3 + a + b) = P(x_3)$$

Hence we have the four conditions :

$$u_A(0) = u_B(0) \tag{21}$$

$$\sigma_{33}^A(0) = \sigma_{33}^B(0) \tag{22}$$

$$u_A(a).e^{iK^*(a+b)} = u_B(-b) \tag{23}$$

$$\sigma_{33}^A(a).e^{iK^*(a+b)} = \sigma_{33}^B(-b) \tag{24}$$

If we introduce the complex numbers :

$$Z = e^{iK^*(a+b)} \; , \; S = e^{iK_A(a+b)} \; , T = e^{iK_B(a+b)} \; , \delta = \frac{K_A(\lambda_A^* + 2\ \mu_A^*)}{K_B(\lambda_B^* + 2\mu_B^*)} \text{ for P-waves and } \delta = \frac{K_A\mu_A^*}{K_B\mu_B^*}$$

for S-waves, the equation of motion (20) and the interfaces conditions (21),(22),(23) and (24) give us the relation :

$$2\delta\ (Z^2 + 1) + \frac{Z}{2}\left[(1-\delta)^2(\frac{S}{T} - \frac{T}{S}) - (1+\delta)^2(ST + \frac{1}{ST})\right] = 0. \tag{25}$$

where the only unknown parameter is Z (or K^*). The resolution of (25) gives us the value of K^* for each value of the pulsation ω. The real part of K^* is $2\pi/\lambda$ (where λ is the wave length) and the imaginary part of K^* is A (where A is the damping coefficient). We can then exhibit the dispersion and the damping relations.

EXAMPLES AND DISCUSSION

Comparison between elastic and viscoelastic multilayered media

Two different two-layered periodic composites are considered. The first medium is composed of two different elastic materials where the second one is constituted of two linear viscoelastic ones. The dispersion and the damping relations are calculated and plotted in fig.3-6 . Fig.3 shows the damping relation for a multilayered elastic medium. This curve illustrates the allowed propagation frequencies for which the damping coefficient is zero and the forbidden frequencies where the damping coefficient is not zero in the curve. This phenomena has already been shown before by Brillouin (1953) for a discrete periodic crystals, verified experimentally on layered composite material by Robinson and Leppelmeier (1974) and by Navi (1987) for the fiber composite materials.

On the other hand, Fig.4 gives the relationship between frequencies and wave number. For the frequencies in which the damping coefficient is not zero, the group velocities $V_g = \frac{d\omega}{dk}$ (for $k = \frac{2\pi}{\lambda}$) become infinite. To explore the phenomenon of forbidden frequencies zone, we have calculated the energy flow for a period of periodic composite material. It is shown that when the half of the wave length approches the periodic length H the energy flow becomes zero. That means a signal with frequencies in the forbiden zone, when propagating in an elastic multilayered medium decreases very fast in the depth of the media and no energy penetrates in the composite.

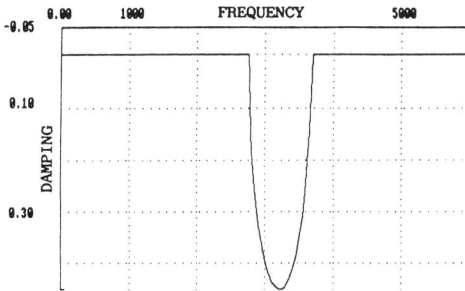

fig.3 : Damping curve for an elastic laminate

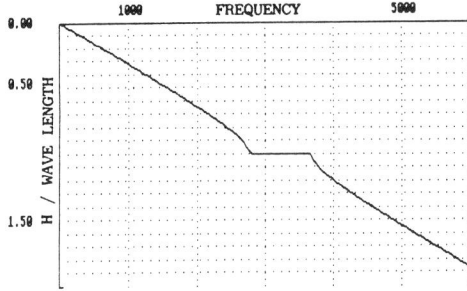

fig.4 : Dispersion curve for an elastic laminate

For the viscoelastic multilayered medium the damping and dispersion curves are shown in Fig.5-6 respectively. These curves for viscoelastic media illustrate the forbiden frequencies phenomena. But in this case the group velocity for forbidden frequencies in the media is no longer infinite. We have analyticaly shown that for the layered viscoelastic medium the energy flow for a period of periodic structure is not zero. That means that a wave in the forbidden frequencies can propagate in the viscoelastic periodic multilayered and carry energy with dissipation along the media. Of course the Fig.5-6 show that the amplitude of the waves which propagate with the allowed frequencies also decreases along the media but the damping coefficient are much less than the waves with forbidden frequencies.

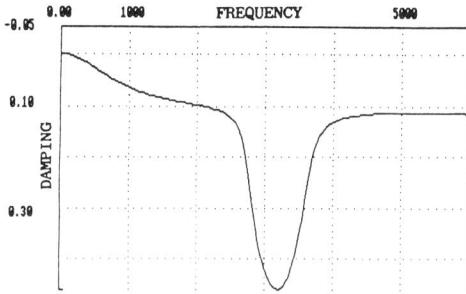

fig.5 : Damping curve for a viscoelastic laminate

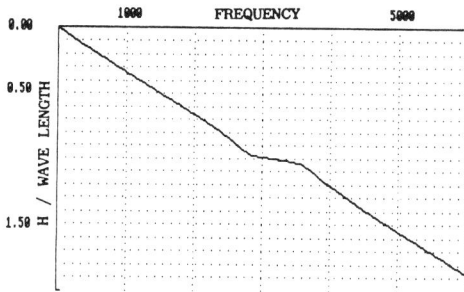

fig.6 : dispersion curve for a viscoelastic laminate

REFERENCES

Sun, C.T., Achenbach, J.D. and Herrmann, G. (1968). Time harmonic waves in a stratified medium propagating in the direcction of the layers. J. Appl. Mechanics, 408-411

Sve, C. (1971). Time harmonic waves travelling obliquely in a periodically laminated medium. J. Appl Mechanics, 38,477-482

Navi, P. (1973). Harmonic wave propagation in composite materials. Ph.d Thesis U.C.L.A., Los Angeles

Navi, P. (1987). Harmonic waves in layered and fiber composite materials. Proceeding of the international conference on impact loading and dynamic behaviour of materials, 2, 985-992

Minagawa, S., Nemat Nasser, S. and Yamad, M. (1981). Finite element analysis of harmonic waves in layered and fiber reinforced composites. Int. J. Num. Meth. Engineering, 17, 1335-1353

Robinson, C.W. and Leppelmeier, G.W. (1974). Experimental verification of dispersion relations for layered composites. J. Appl. Mechanics, 89-91

Stern, M.,Bedford, A. and Yew, C.H. (1971). Wave propagation in viscoelastic laminates. J. Appl. Mechanics,448-454

Mukherjee, S. and Lee, E.H. (1978). Dispersion relations and mode shapes for waves in laminated viscoelastic composites by varaitional methods. Int. J. Solids & Structures, 14, 1-13

Ma, T.C., Scott, R.A. and Yang, W.H. (1980). Harmonic wave propagation in an infinite viscoelastic medium with a periodic array of cylindrical elastic fibers. J. Sound & Vibration, 69 (2), 257-264

Sutherland, H.P (1975). Dispersion of acoustic waves by fiber rienforced viscoelastic materials. J. Acoustical Soc. Am., 57, 870-875

Mandel, J. (1966). Mécanique des milieux continus. Gautier Villars Editeur, Paris

Stocker, J.J. (1950). Nonlinear vibrations in mechanical and electrical systems. Interscience publishing Inc.

Brillouin, L. (1953). Wave propagation in periodic srtuctures. Dover, New York

Vibration Analysis Techniques Applied to a Large Turbocharger Rotor

G. M. CHAPMAN and J. TURNBULL

Department of Mechanical Engineering,
Loughborough University of Technology,
Loughborough, Leics. LE11 3TU, UK

ABSTRACT

Improved designs for large turbocharger rotors require both a thorough understanding of the modal properties of the system and a knowledge of how the rotor will respond to typical force excitation. This paper describes the approach taken to develop a mathematical model using finite element analysis to represent a complete turbocharger rotor based upon verification by experimental data obtained using conventional modal analysis and electronic speckle pattern interferometry (ESPI) techniques. The relative merits of each technique are discussed.

KEYWORDS

Turbocharger, transient vibration, modal analysis, ESPI, finite element.

INTRODUCTION

Turbochargers use the exhaust gas from an engine to supply air to the engine intake. The exhaust gas drives a gas turbine mounted on a short rotor with a compressor attached to the other end. Rotors of large turbochargers are subjected to a variety of vibration excitations as a consequence of out-of-balance, changes in rotor running speed and the vibration content of the exhaust gas stream from the engine. Each blade as it passes a nozzle, is subjected to a variation in force excitation resulting from the local gas stream pressure and the phasing of the engine exhausts into the turbocharger entry ports (Connor et al, 1985). The rotor will therefore be stimulated into vibration at many frequencies instantaneously and each of its characteristic modes will either increase in amplitude or decay depending upon whether the excitation overcomes the inherent damping in the system.

A typical example of the force exerted by the engine exhaust pressure on a single turbocharger blade is shown in Fig.1a together with an estimate of the frequency content in the gas stream shown in Fig.1b.

153

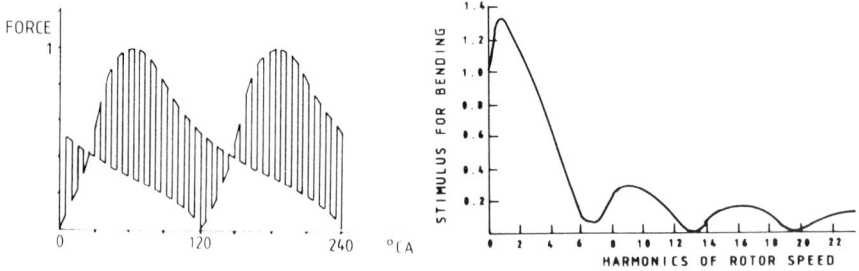

Fig. 1. Force experienced by a single turbocharger blade
a) time history
b) frequency content

As each blade passes behind a nozzle blade and back into the gas stream it is instantaneously subjected to pulsations of force modified by the instantaneous pressure associated with the pressure level in the entry port. Hence pulses at the blade pass frequency are continuously being generated but modified in amplitude of excitation. The rotor running speed can be continuously varying to match engine exhaust conditions to air intake demand upon the compressor. Hence the integrity of components to withstand failure caused by fatique as a consequence of vibration is a major design requirement which requires knowledge of the transient response characteristics of the rotor.

One approach to assist with the prediction of response previously reported (Turnbull and Chapman, 1988) would be to develop a mathematical model of the rotor which truly represents all the dynamic characteristics of the system. Such a model could then be adjusted geometrically to optimise the parameters to either limit the onset of sustained vibration modes or reduce the amplitude of vibrations. Clearly such a task is difficult if the model is to represent all the intricacies of the design and is required to yield response behaviour with a high degree of accuracy for a realistic usage of computer time. The rotor used as a model for this investigation is illustrated in Fig.2.

The rotor shaft is comparatively rigid but it carries a fairly heavy gas turbine disc at one end and a light compressor at the other. The geometric shape particularly of the gas turbine and compressor blading is extremely complex in relation to the rotor shaft and discs. A true geometric mathematical model of such a rotor would be expensive to develop and require considerable computation to evaluate its modal parameters. Such an approach would be excessively complex for the information required by designers about rotor response. However it would be of significant benefit for a full understanding of the vibration performance of the rotor's blading.

Fig. 2. Turbocharger rotor

There is therefore a need to develop simplified mathematical representations of the real rotor which contain sufficient mathematical integrity to model the critical modal characteristics over the frequency range of excitation. To check the accuracy of the model it is useful to be able to identify the rotor modal characteristics experimentally and to modify the mathematical model accordingly. This approach assures realistic values for damping and realistic boundary conditions.

EXPERIMENTAL METHODS

Several experimental methods exist to determine vibration characteristics of systems. The two methods adopted for this analysis were modal analysis to identify natural frequencies and damping values, and electronic speckle pattern interferometry to view mode shapes of specific natural frequencies. The information obtained was then used to improve simplified models of the rotor assembly for later theoretical simulation tests for response evaluation.

Modal analysis technique

Vibration characteristics can be determined by investigating the response of the system to a known force excitation. Typically the rotor can be excited by an impulse hammer blow or by an electromagnetic shaker with either sinusoidal or pseudo-random white noise excitation. Whatever the excitation form the force must be monitored and compared to a measurement of response, which is commonly obtained by attaching an accelerometer to the system. By monitoring response at strategic locations on the surface of the rotor it is possible to build up transfer functions for the vibration response at these locations. The data obtained can be manipulated to identify natural frequencies, damping values at specific natural frequencies and the characteristic mode shapes. The technique is detailed in Ewins (1984).

Electronic speckle pattern interferometry technique

Vibration mode shapes can be observed by making use of the technique of interference between a reference beam and an illuminating beam viewing the vibrating surface. The interference patterns can be scanned in a video system and displayed on a monitor. This technique, described by Jones and Wykes (1983) offers the opportunity of observing complete surface areas of such objects as rotor discs etc. The object under investigation must be excited by a constant frequency sinusoidal excitor, which is tuned to each natural frequency in turn. Detection of vibratory motion is currently restricted to a uni-directional axis and in this investigation was restricted to out-of-plane motion of the vibrating surface. It is now possible to extend the scope of ESPI to monitor separately, in-plane motion along any two perpendicular planes (Shellabear and Tyrer, 1987)

Merits and disadvantages of experimental methods

The relationship between finite element and modal analysis has been presented previously (Rieger, 1986). Methods for experimental determination of modal characteristics of systems are continually improving. Each method has its own merits with accompanying disadvantages.

Electronic speckle pattern interferometry

The major advantage of this technique is that vibration amplitudes for specific natural frequencies can be viewed in real-time with a full view image of all points on the surface of the object under investigation. The method is comparatively fast as observation of vibration amplitude can also be viewed whilst the frequency of excitation is varied. A considerable amount of data is generated for each mode as it is possible to interrogate the computer stored image and to extract the displacements at all points by counting the number of interference fringes generated at each point. At present the technique is restricted to measurement of single plane motion but research is currently investigating the possibility of combining several images obtained in each of the three axes. The opportunities of using pulsed lasers as the illumination source offer the possibility of vibration monitoring of moving (rotating) objects. The method offers a useful first stage investigation of vibrating objects and is excellent for eventual verification of vibration modes identified by other methods.

Among its draw-backs ESPI can only view modes at prescribed excitation frequencies. It is therefore possible to miss natural frequencies by poor scanning of the frequency range and by problems caused by coupled frequencies distorting the response characteristics. In particular care has to be exercised with selection of excitor and excitor location as these decisions can influence the detected mode shape. Complex structures can present difficulties as the surface to be investigated must be visible in the plane of viewing; hidden surfaces cannot be observed. Curvature introduces difficulties with interpretation because the technique is restricted to single displacement plane measurement. The object being viewed must be rigidly attached to a boundary as the technique measures the relative motion between the stationary object referenced to the instrument and the maximum deflected amplitude during vibration. The method is noisy and very often, for good images, it is necessary to paint the surface to obtain good quality speckles.

Modal analysis

Experimental modal analysis offers a highly accurate and fast method for identifying natural frequencies. The data obtained is in a digital form and thus the method lends itself to computational techniques for later manipulation to identify modal characteristics of the system. Estimates of damping at each natural frequency can be obtained comparatively easily by using logarithmic decrement or quadrature techniques on curve-fitted data. Modern developments in computer techniques have helped to develop highly sophisticated systems capable of vibration mode simulation with the potential for structural modification for vibration characteristic alteration.

The method is however time consuming when applied to the analysis of a complete structure. It suffers from the disadvantage that its basic monitoring devices i.e. accelerometers and force excitors are primarily uni-directional and the excitation system can influence the response characteristics. Impulse excitation using a force instrumented hammer is highly convenient but lacks sufficient accuracy when frequencies in excess of 7 kHz are of interest. Electromagnetic vibration excitors can be used with pseudo-random white noise excitation but allowance for the mass of the armature must be considered. Mode shape identification requires many monitored points and often it is necessary to be aware of the likely mode shape in order to be able to specify the monitoring points for measurement.

THEORETICAL METHOD

Mathematical modelling of the rotor was undertaken using a conventional finite element package PAFEC with an interactive graphics suite (PIGS) for viewing deformed mode shapes.

Finite element package

To enable the best correlation between experimental and theoretical analyses it was decided to first model the rotor in its free-free mode, as this configuration is least effected by boundary restraints.

INITIAL RESULTS FOR TURBINE DISC

The initial results obtained by finite element, ESPI and modal analysis of for the gas turbine disc natural frequencies are illustrated in Table 1. The rotor was held at its bearing locations by simulated flexible elements using rubber. This constraint was not expected to effect the turbine disc diametral modes although it was accepted that the circle modes would be highly damped as the circle modes are intrinsically linked.

It can be clearly seen that the first attempt at a finite element model of the rotor was far from a reasonable representation of the actual system. Increasing the number of master degrees of freedom and decreasing the mesh size failed to improve the frequency predictions. in particular the failure of the model to predict the strongly detected OC-2D mode highlighted the poor correlation.

Table 1. Initial frequency and damping determination
for gas turbine disc.

Mode	Finite Element nat. frequency	ESPI nat.frequency	Modal Analysis nat.frequency (damping)	
OC-OD	-	-	2275	(0.0007)
OC-2D	5540	3133/3157	3125/3175	(0.0002)
1C-OD	7450	-	4175	(0.00035)
OC-3D	9290	4646/4727	4650/4725	(0.0003)
OC-4D	-	6208	6200	(0.00025)

This correlation between the experimental and theoretical techniques was not considered adequate to progress further and so simplified rotors were investigated to establish the relevance and accuracy of each technique.

TEST RESULTS ON SIMPLIFIED ROTORS

In order to build a simplified model of the turbocharger rotor it was necessary to consider its construction. The gas turbine disc was shrunk into a collar integral with the shaft whilst the compressor disc was secured with a through bolt at the other end.

A solid shaft with three discs attached (Fig. 3) using shrink-fits to simulate the assembly technique employed on the rotor for the gas turbine, compressor and gas turbine location collar was constructed and tested using modal analysis. The location collar rested up against the gas turbine disc to simulate the actual rotor assembly.

Fig. 3 Simplified rotor

The results obtained were compared to natural frequencies predicted for the free-free modes by finite element representation of an identical mathematical model. Large errors were detected between the natural frequencies predicted by finite element analysis and measured by experimental modal analysis. Increasing the mesh size and manipulating the master degrees of freedom in an attempt to obtain a better mathematical simulation failed to reduce these errors. ESPI was used to check the modal analysis results and served only to confirm the experimental data.

Closer investigation of the modelling of the boundary conditions suggested several possibilities for the anomalies between theoretical and experimental results. Firstly the gas turbine collar had been modelled as a solid boundary between disc and collar, whereas the construction of the simplified rotor could not guarantee this condition. Inspection of the rotor indicated that this interface was not solid and that a small gap existed between the surfaces. Thus vibration of the disc was less restrained than expected with a solid support collar. Secondly no guarantee could be placed upon the ability of the shrink-fit to maintain contact between disc and rotor along the whole length of their axial overlap. Two small stub shafts were fabricated; one with a shrink-fit disc attached the other machined out of solid. Modal analysis identified a small difference in the natural frequencies, thus indicating that the circumferential contact area was likely to be smaller than anticipated.

By modelling these two boundary conditions i.e. allowing both a gap between the collar and gas turbine disc and by reducing the circumferential contact area between discs and shafts, the frequency errors were reduced to a negligible level. Confirmation of these two effects was assured by welding the collar and simulated turbine disc together and by welding the simulated compressor disc to the shaft. A further modal analysis survey was conducted to verify results which are presented in Table 2.

Table 2.   Natural frequency comparison for simplified rotor

Mode	Finite Element Solid rotor	Experimental with shrink-fit	with welded joints	Finite Element with allowance for gaps
OC-2DT	4570	3700	4150	3670
OC-3DT	–	7990	8110	–

DEVELOPMENT OF FULL ROTOR MODEL

Having identified the most significant modelling parameters giving rise to the greatest error between experimental and theoretical natural frequency prediction, a finite element model was developed by manipulating the clearances between connecting components on the rotor.

Besides the clearance between thrust face and the reduced shrink-fit contact area, an allowance was made for a gap between the inner surface of the compressor disc bore and the rotor shaft. Attachment of the compressor disc was modelled by considering a through bolt acting on small thrust collars at either end of the compressor disc. This was included to give a better representation of the the real rotor situation but was found to have little influence upon the compressor disc natural frequencies. The compressor disc is comparatively rigid near to its centre and quite flexible towards its rim, so consequently restraints at the bore have little effect upon components towards the outer diameter of the compressor disc.

It was found necessary to make allowance for the effect of the mass and inertia of the gas turbine blading by simulating all the blades as a solid ring element attached to the circumference of the disc. An artificial inertia effect was provided by assigning a high value of density to this element.

The finite element finally developed for comparison with the experimental free-free modal analysis is shown in Fig 4. The model includes spring elements on the compressor disc to represent the effect of the compressor blades upon the disc. The compressor blades are extremely flexible in relation to the disc and need only represented as stiffness elements without mass for prediction of rotor performance.

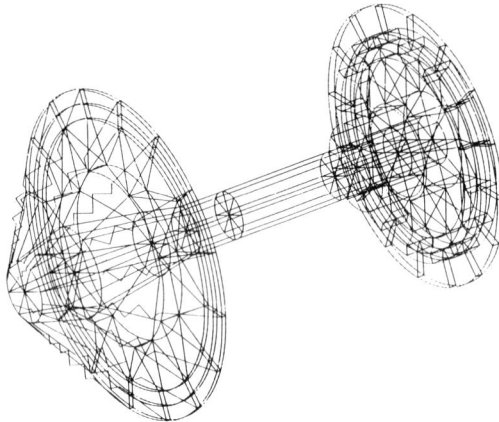

Fig. 4.   Finite Element model of turbocharger rotor

Further adjustments were required to accommodate discrepancies between rotor shaft natural frequencies by redistributing mass along the shaft axis. A comparison of predicted and experimentally determined natural frequencies for the rotor shaft assembly is shown in Table 3.

All disc diametral modes for the finite element model were predicted to within 6% of the experimentally measured natural frequencies, whereas the circle modes differed substantially. Work is continuing to develop the finite element model to simulate both these frequencies and the rotor shaft natural frequencies.

The relevant damping values obtained experimentally were assigned to each of the natural frequencies of the system.

Table 3. Comparison of predicted and experimentally determined natural frequencies for rotor.

Mode	Finite element predicted natural frequency	Modal analysis measured natural frequency
1DI-1DT-1B	950	438
1DI-1DT-2B	2539	3400
OCI-OCT-AV	2874	2275
2DT	3216	3150
1DI-1DT-3B	3703	4825
3DT	4611	4686
2DI	4860	4725
3DI	4970	5200
OCI-OCT-AV	4981	4715
4DI	5141	5400
5DI	5360	5700
4DT	6040	6200
5DT	7760	7403

CONCLUSIONS

The use of modal analysis and electronic speckle pattern interferometry has considerably assisted the development of a finite element representation of a turbocharger rotor for further analysis of its transient performance. It has been shown that attention must be paid to local boundary conditions as these can considerably influence the accuracy of the model. In particular component interfaces must be modelled in such a way as they truly represent the constraint in the real system.

The value of using several methods of analysis has been strongly identified as all currently available analytic techniques have their own specific advantages with associated disadvantages.

ACKNOWLEDGEMENTS

The authors wish to thank the Science and Engineering Research Council for funding this investigation, Napier Turbochargers for providing rotors and supporting data and the Department of Mechanical Engineering at Loughborough University of Technology for the research facilities.

REFERENCES

Connor W.A., Swain E., Bellamy A.G. and Chapman G.M., 1986, Excitation of turbine blade vibrations in large turbochargers, Conf. Proc.Turbochargers and Turbocharging, London, UK, I.Mech.E.

Ewins, D.J, (1984),Modal Testing: Theory and Practice, Research Studies Press, England

Jones R. and Wykes C., 1983, Holographic and Speckle Interferometry, Camb. Univ. Press, UK.

Rieger N.F., 1986, The relationship between finite element analysis and modal analysis, Sound and Vibration.

Shellabear, M.C. and Tyrer, J.R., 1988, Three dimensional vibration analysis using electronic speckle pattern interferometry (ESPI), Proc SPIE, vol 952.

Turnbull J. and Chapman G.M., 1988, Strategies for the vibration analysis of a large turbocharger, Conf. Proc. Vibrations in Rotating Machinery, Edinburgh, UK, I.Mech.E.

Prediction and Measurement of Rotor Blade/Stator Vane Dynamic Characteristics of a Modern Aero-engine Axial Compressor

R. J. WILLIAMS, K. L. JOHAL, H. A. BARTON and
S. T. ELSTON

Rolls Royce plc, Aero-engine Division,
P.O. Box 31, Derby DE2 8BJ, UK

ABSTRACT

Commercial pressures require ever more fuel efficient and light weight propulsion systems. Inevitably this results in an increased stage loading that can exacerbate vibration problems of stall flutter, acoustic resonance, and conventional mechanical resonances from the immediate upstream and down stream blade/vane rows. This paper traces the design of a rotor and stator from design to engine validation. The components are analysed at design stage using finite element models and modified to give acceptable dynamic characteristics. The predicted and measured results are compared and the method of assessment described.

KEYWORDS

Rotor, Stator, Vibration, Fatigue, Strain Gauge, Holography

THEORETICAL ANALYSIS

Figures 1a and 1b pictorially represent the mathematical idealisation of a typical rotor blade and a typical cantilevered stator vane. To produce accurate results it has been found necessary to simulate the entire component, aerofoil and root, earthed over the bearing surfaces. The discretisations were constructed using a semi-automatic procedure and the resulting finite-element models analysed using an in-house computer code SADIE. The models were composed entirely of twenty node isoparametric brick elements which can incorporate:-

1. geometric non-linearity
2. the stiffness generated by initial stress
3. mass unstiffening
4. anisotropic material properties

The dynamic characteristics were calculated using subspace iteration on the mass and stiffness matrices linearised at the steady stress condition. From the results the design was assessed against criteria relating to :-

1. steady stress
2. natural frequencies and corresponding mode shapes
3. flutter
4. modal stress distributions and corresponding HCF strength.

If the criteria were not satisfied an iterative loop was entered with the Compressor Aerodynamic Design until the optimum efficiency component, which satisfied the integrity constraints, was obtained.

163

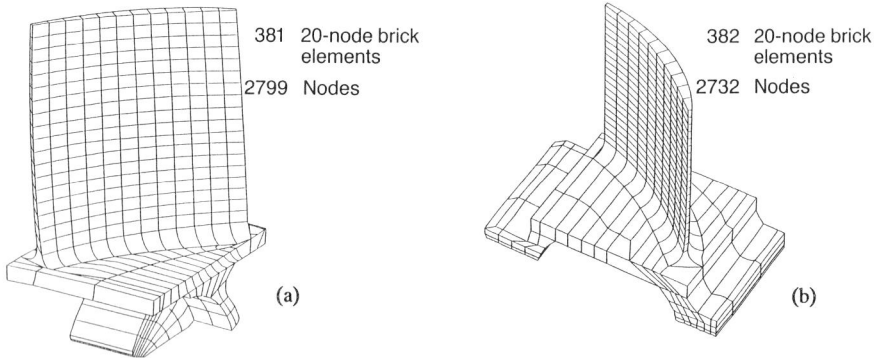

Fig. 1   HP Compressor mid-stage (a) rotor blade,
(b) stator vane finite element models.

EXPERIMENTAL COMPONENT TESTING

A thorough assessment of each aerofoil row with regard to its non-rotating:-
1. natural frequencies and corresponding scatter
2. mode shapes of all modes which resonate in the operating range
3. high cycle fatigue (HCF) strength of all relevant modes

was made by testing hardware. In the experimental rigs the components were restrained to reproduce the engine clamping conditions as closely as possible. It was discovered, however, that there was significant damping when the stator vanes were fatigue tested, and failure amplitudes could not be achieved. This problem was overcome by clamping the vanes across the platform faces. The resulting change in stress distribution was investigated using the theoretical models which indicated only minor differences in the peak stress/tip amplitude ratios. The subsequent experimental fatigue strengths were therefore representative of the engine.

Natural Frequencies

The frequency survey rig is a semi-automatic measurement system consisting of a computer controlled Fast Fourier Transform spectrum analyser operating on the output of a microphone which detects the component's response to a light impact. A large number of nominally identical blades and vanes were tested so that the true scatter was established.

Statistical analysis revealed a reasonable correlation between the natural frequencies of 1st Flap and/or 1st Torsion with high order modes. Consequently all blades/vanes were frequency checked in the two lowest frequencies and the results etched on the root. In the event of an HCF failure the frequency characteristics of damaged blades could therefore be inferred.

Mode Shapes

A mean frequency component from each aerofoil row was selected for holographic determination of its mode shapes. This was accomplished by forcing the blade or vane to vibrate, via an exciter contained within the holding fixture. This resulted in a very compact excitation system with negligible mechanical coupling which did not distort the aerofoil mode shape. Some difficulties were encountered in mounting the holding fixture "rigidly" to the optical table while providing sufficient isolation to prevent significant vibration of the optical hardware. This was largely overcome by using very thin layers of isolation material between the holding fixture and the table.

Holograms were produced by a time averaged technique using an Argon Ion 2 watt (CW) laser. The recording medium was a holographic camera incorporating a reusable thermoplastic plate. Conventional photographic techniques were employed for reconstruction. The optical layout was arranged so that the illuminating and viewing directions were approximately normal to the aerofoil chord. This ensured that the holographic fringes represented displacement normal to the chord thus allowing direct qualitative comparison with the standard finite element pictorial output.

An analysis technique was developed to measure "total amplitude" using a pair of holograms with off-axis sensitivity vectors. However this proved too time consuming for routine application. From the limited number of examples processed it was apparent that the vast majority of modes have very little motion tangential to the chord.

Figure 2 compares a sample of measured frequencies and modeshapes with theoretically predicted values.

Mode	Predicted (FE analysis)	Measured (Holography)
5	19710Hz	20330Hz
6	22710Hz	21716Hz
7	26817Hz	25548Hz

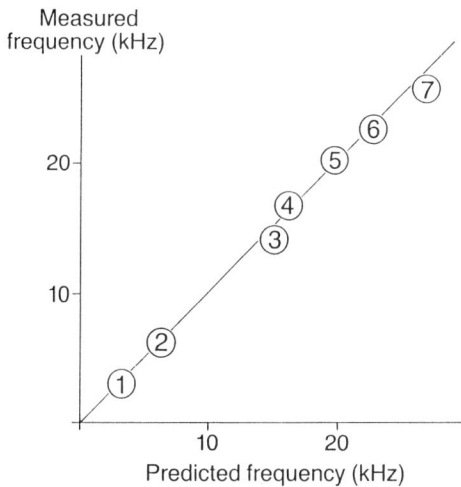

Fig. 2 Comparison of stator mode shapes and frequencies predicted by FE analysis with holographic measurements.

The fringes represent approximately linear contours of increasing modal displacement. The brightest fringes, zero fringes, represent nodal regions. A correction for non-linearity in the vicinity of zero fringes has not been applied because the main purpose was to produce data for a qualitative comparison with theory.

High Cycle Fatigue Testing

It is necessary to test hardware to:-
1. ensure the Design Intent has been satisfied
2. ensure that the manufacturing techniques have been optimised to produce maximum fatigue strength
3. allow rigorous interpretation of engine straingauge data
4. allow comparison with theoretical predictions.

For ease of testing this work is usually carried out at room temperature. Some hot testing is necessary however to validate the techniques used to convert to engine conditions.

In the lower modes tip "af" (amplitude x frequency) gives a convenient way of comparing similar blades/vanes as described by Armstrong (1988). Over the years a considerable databank has been established which is used to assess components and thus highlight material and/or geometry problems prior to engine run.

For engine integrity assessment the fatigue characteristics are more conveniently expressed in terms of the 10^7 cycle endurance tip amplitude. Initially a large number of components were tested at constant amplitude levels. From this data it was recognised that in the 10^6 - 10^8 cycle range the log (tip amplitude) against log (life) plot was sensibly linear. The constant of proportionality C was broadly similar for components of the same material in all modes of vibration. However the value of C was significantly different between forged and machined aerofoils. An appreciable disparity also exists between polished specimen data and blade hardware.

Using the appropriate value of C permitted a compressed programme of fatigue testing to be devised. This involved fatigue testing a small quantity of components at approximately the 10^7 cycle strength (determined by trial and error). Each individual result was extrapolated to the 10^7 strength as shown in fig. 3b. A statistical analysis of these results was then used to determine the 10^7 cycle endurance. The extrapolation errors were minimal because the fatigue curve is typically very shallow in the 10^6 - 10^8 cycle region and the trial and error choice of test level ensured that the extrapolation was small.

In practice a full interpolation, Fig. 3a, was only performed on a limited number of occasions for each material in order to establish the material constant C. Subsequent testing was restricted to very small samples as indicated in Fig. 3b.

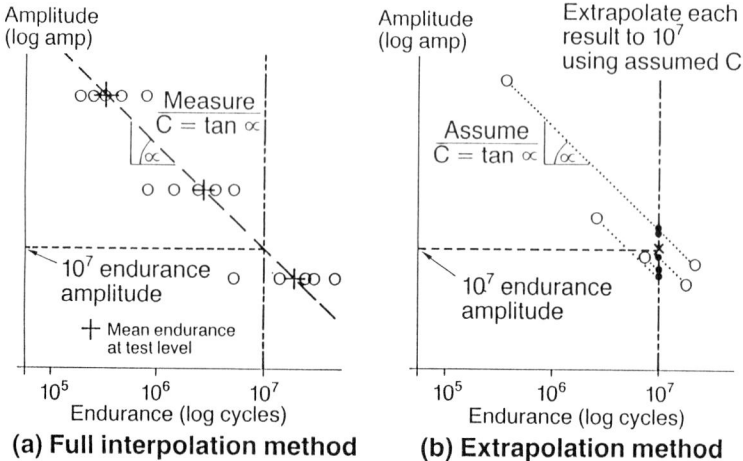

Fig. 3   Measurement of 10^7 cycle endurance amplitude.

Strain Gauge Calibration

It was found that the strain gauge reading per tip amplitude varied considerably on nominally identical components. However very little scatter was measured in the fatigue strength expressed as a 10^7 cycle tip amplitude. Consequently each instrumented component which ran in an engine was calibrated in a rig prior to the engine test. The rig measured the strain gauge/tip amplitude ratio in all relevant modes. In this way the integrity of the blades and vanes could be rigorously assessed as described in the next section.

INTEGRITY ASSESSMENT ENGINE STRAIN GAUGE TESTING

Rigorous interpretation of engine strain gauge results requires considerable preparatory work. In particular:-

1. individual calibrations for each instrumented component in all relevant modes of vibration
2. corresponding 10^7 cycle fatigue strengths at room temperature and at the highest engine metal temperature for the row
3. fatigue strength against temperature relationship
4. metal temperature, direct stress and modal stress distributions as a function of engine rotational speed

are required. Relating the calibration to the fatigue strength provides an unsteady stress level for a cold blade. By scaling for temperature and using a Goodman diagram a modal strain gauge endurance for 10^7 cycles and at maximum stage temperature and steady stress is obtained. An aerofoil responding below this level would be expected to last indefinitely where as one exceeding the limit would fail.

In practice to save time and hardware only a small number of modes are fatigue tested prior to engine test. The theoretical model is used to supply the missing strengths. This is considered a safe option because in numerous instances the theoretical predictions have never exceeded the measured strengths.

Figures 4a and 4b present the strain gauge records, measured during a slow uniform acceleration, of a typical rotor blade and stator vane respectively transformed into the frequency domain. The vibration modes are discernible with the response excited by the immediate up stream and down stream aerofoil rows clearly visible. From these plots the peak response in each mode, usually the resonance excited by the upstream row, and the corresponding engine speed can be extracted. Using strain gauge limits permits these values to be expressed as a percentage of the appropriate 10^7 cycle endurance. It is general practice, where possible, to use two strain gauges on each aerofoil. This provides a useful check because the endurance levels calculated from each can be compared. A number of blades/vanes are instrumented from each row and the maximum percentages of endurance for each mode determined. Table 1 presents this data for the rotor blade and stator vane rows discussed earlier.

In addition to recording the stress through uniform accelerations and decelerations, strain gauge readings are taken with different bleed offtakes and varying vane schedules. This ensures that the complete engine operating envelope is covered in the integrity assessment. It is possible for a number of modes to respond at the same engine speed. Consequently it is necessary to allow for this in the integrity assessment. This has been accomplished using the theoretical modal stress distributions and the corresponding measured endurance levels.

Due regard is also taken of frequency scatter which could increase the total vibratory stress on an adverse tolerance component.

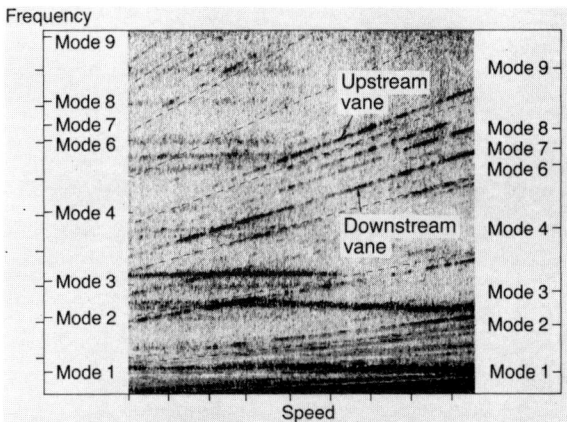

Fig. 4a Engine Campbell Diagram for rotor blade with
theoretical static and dynamic modal frequencies.

High Cycle Fatigue Engine Endurance Running

The final proof of integrity is engine running over the complete operating envelope. In development testing it is accomplished by 50 rpm incremental testing from ground idle to redline speed. The time at constant condition is selected to ensure all relevant modes exceed 10^7 vibration cycles.

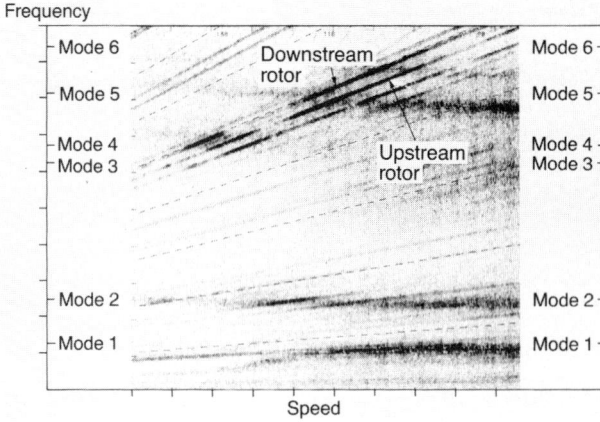

Fig. 4b Engine Campbell Diagram for stator vane with theoretical static modal frequencies.

Mode	Rotor blade	Stator vane
1	11	7
2	31	5
3	–	–
4	(37)	4
5	–	7
6	29	25
7	(24)	–

( ) Based on theoretical fatigue strength

Table 1 Maximum percentage of endurance from engine running.

CONCLUSIONS

Having shown by means of holography, that the finite element analysis technique adequately predicted the modal characteristics of the complex aerofoils, the technique was used with confidence to influence aerofoil design at an early stage. In particular, frequency modifying changes could be introduced to ensure that components were resonance free at crucial engine speeds. Subsequent laboratory testing of engine components demonstrated whether the design intent had been satisfied, and whether the required metallurgical strength had been achieved. Calibration of strain gauged components prior to the test, assured that the complex response in the engine could be interpreted and assessed against realistic integrity criteria. Finally, once the design had been optimised for all components, an incremental endurance test was used to demonstrate that all components had an adequate fatigue strength throughout the engine life.

ACKNOWLEDGEMENTS

The authors are grateful to Rolls-Royce plc for their permission to publish this paper. Any views expressed herein are the authors' own and not necessarily those of the company.

REFERENCES

Armstrong EK (1988). Fatigue and assessment method of blade vibration. In **AGARD-AG-298** Manual on Aeroelasticity in Axial Flow Turbomachinery. Vol. 2 Structural dynamics and Aeroelasticity. Chapter 12.

Theoretical and Experimental Determination of Natural Frequencies of Laced Blading

G. M. CHAPMAN and X. WANG

*Department of Mechanical Engineering,
Loughborough University of Technology,
Loughborough, Leics. LE11 3TU, UK*

ABSTRACT

It is common practice to introduce a lacing wire into the gas turbine blading of large turbochargers in order that vibration of the blades can be kept to a minimum.

This paper describes theoretical and experimental investigations aimed at developing an understanding of the inter-relationship between the blading and the lacing wire. It discusses the experimental procedures developed and reinforces the results obtained by comparison with a finite element analysis of the blade assembly.

The investigation progresses from a full study of single unlaced blades to single laced blades and finally reports on the vibration behaviour of multiple laced blade assemblies.

KEYWORDS

Turbine blades, vibrations, mode shapes, lacing wire, ESPI, finite element.

INTRODUCTION

Turbocharger blading is subjected to vibration excitation as a consequence of blade/nozzle passing frequency and as a result of disturbances in the gas stream entering the gas turbine disc from the engine exhaust. The turbocharger rotor is a free running rotor which varies in speed as the demand on the turbocharger changes. This results in vibration excitation of the blades with the inherent risk of fatique failure.

Lacing wires are one method of coupling blades and reducing vibration amplitude. They are threaded through blades and their function has in the past been likened to a damping device which operates as the blade attempts to slide on the wire during vibration. However at the rotational speeds commonly experienced by turbocharger rotors, the centrifugal force generated would almost certainly make this behaviour unlikely as the sliding forces required would be extremely high. Evidence of fretting between the wire and the blade hole is not commonly experienced. Distortion in the wire between consecutive blades as a consequence of the centrifugal force is often identified and would almost certainly indicate an almost positive connection between blade and wire at the lacing wire/blade hole.

Based upon these considerations, a series of experiments to investigate the characteristics of the relationship between lacing wire and turbocharger blading were conducted. Three techniques were employed to verify the vibration behaviour; modal analysis to identify natural frequencies, electronic speckle pattern interferometry (ESPI) to observe mode shapes and finite element analysis to build up mathematical models of the systems.

Fig. 1.   Single turbocharger blade

The turbocharger blade selected for this investigation is shown in Fig. 1. It represents a single blade of approximately 91 mm height with a lacing wire location at approximately 78 mm from the top of the blade root. There is almost a 50° chord twist along the blade length.

INITIAL ANALYSIS

The investigation initially concentrated on identifying the natural frequencies of a single unlaced blade without the effects of centrifugal loading. To arrange for blade root loading to approximate the situation experienced by a rotating blade, the blade was held in a solid block machined with a fir tree root slot with two bolts reacting on the base of the root of the blade.

Initial experimental determination of natural frequencies

An initial modal analysis investigation was conducted using a sub-miniature accelerometer attached to the surface of the blade and applying impulse excitation to a mesh of points on the surface of the blade by a force instrumented hammer. This served to excite frequencies up to about 7 kHz and gave an insight into mode shapes. The results obtained were always considered with suspicion as the accelerometer added mass to the system and because the hammer and monitoring accelerometer recorded data along their polarised axes.

Initial mathematical model of turbocharger blade

The initial mathematical model for the blade was developed using the PAFEC finite element package. To faithfully represent the geometry of the blade, whilst meeting the geometric constraints of the elements offered by the package it was found necessary to model the blade by 120 twenty noded elements. This rather large number of elements for a single blade was dictated by the requirement to meet length/width/depth requirements for all elements including those representing the highly tapered trailing edge and those required for the thick leading edge. The results for natural frequencies did not compare well with the experimentally determined data, and so attempts to elongate the blade model to cater for flexibility in the root section were investigated. Whilst this approach lead to a better model it was considered necessary to reduce the number of elements to reduce the computer time required for solution.

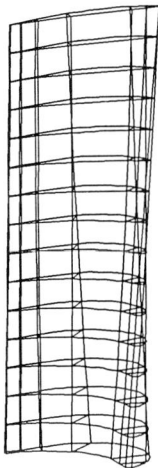

Fig. 2.   Finite element of blade using 90 twenty noded elements.

A considerable improvement in the use of computer time was achieved by reducing the model to 90 elements (Fig.2), but still the predicted results did not fully correlate with the results obtained by ESPI and modal analysis.

Second experimental investigation of blade

To check the discrepancy between the finite element analysis and initial modal analysis, a survey was conducted of the natural frequencies by exciting the blade with an electromagnetic exciter in front of an electronic speckle pattern interferometer (ESPI). This technique described by Chapman and Wang (1988) uses an interference technique based upon the difference in path length of a laser beam travelling a fixed path and a second beam which effectively illuminates the peak displaced image of the vibrating object.

Vibration excitation techniques

Preference for the use of an electromagnetic exciter was based upon its flexibility, as the ability to move a point excitation probe across the surface of the blade outweighed the disadvantages of the added mass caused by the vibrator armature. In general at lower frequencies it was found necessary to excite at low amplitude near to the nodal lines of modes to avoid mode shape distortion, whereas at high frequency it was necessary to excite at the antinodes of modes in order to inject sufficient energy.

Initial ESPI results

The results obtained were interesting in that some natural frequencies were in agreement with the modal analysis results, whilst others were detected by only one of the experimental methods. The finite element analysis however identified most of the frequencies detected experimentally and indicated others.

A common difficulty experienced with the experimental determination of natural frequencies was found with modes which had close natural frequencies. For such cases it was found relatively easy to excite combined modes which appeared to have amplitudes of vibration equivalent to true natural frequencies. Figure 3 shows the ESPI images obtained by exciting the turbocharger blade at three frequencies around the first torsional (1T) and second flap (2F) mode.

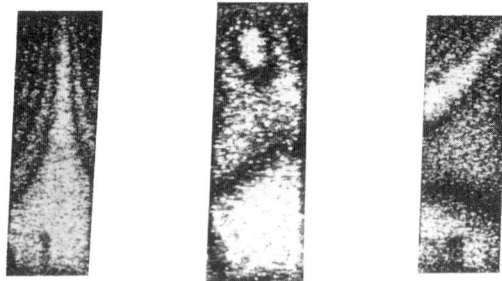

a) 1T mode   b) coupled mode   c) 2F mode

Fig. 3.   Examples of combined mode excitation.

It was found that only by excitation at the node of one frequency could the other frequency be stimulated independently. Care therefore was required with interpretation of the ESPI images to ensure correlation with the finite element predictions.

SIMPLIFIED SINGLE BLADE ANALYSIS

Interpretation of the mode shapes for the turbocharger blade was also difficult because of the high degree of curvature. To solve this problem, several simplified blades were designed to investigate the effect of parameter changes on natural frequencies. The blades tested are shown in Fig. 4. Each blade has approximately the overall height, width and thickness of the turbocharger blade. The parameters investigated were based upon adjusting a straight, rectangular cross-sectioned cantilever blade by introducing taper along blade length, taper across blade width, curvature across blade surface and twist along blade length.

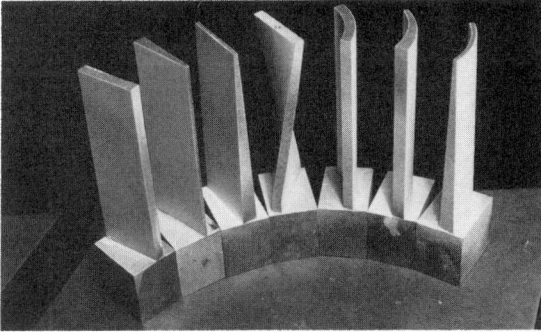

Fig. 4.   Simplified blades for mode classification analysis

To verify the experimental results, the finite element model shown in Fig. 5 was developed using twenty noded elements : 8 elements to represent the blade and 6 elements for the root assembly.

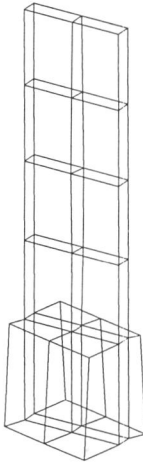

Fig. 5.   Finite element model of simplified blade.

This analysis was the basis for building up a blade vibration mode classification system reported in Chapman and Wang (1988). The results obtained showed good correlation between theoretical and ESPI derived results and identified modes which had been previously missed experimentally because of the viewing axis of the out-of-plane detecting ESPI facility.

Based upon these results a mathematical model simulating the curved turbocharger blade was developed using 21 sixteen noded elements plus 1 twenty four noded element to represent the root configuration. By tuning this model it was possible to represent the vibration characteristics of all modes with natural frequency below 10 kHz.

176

The initial results obtained for the single blade analysis are shown in
Table 1.

Table 1. Initial single blade results

Vibration mode	ESPI	Modal Analysis	Finite Element (90 elem)	Finite Element (21 elem)
1F	1255	1050	1290	1202
1E	2392		2960	2431
1T	4510	4250	4700	4534
2F	4594		5610	4829
3F	7246	6850	9030	7991
2T	9824	9600	10800	10295

INVESTIGATION OF EFFECT OF LACING WIRE ON A SINGLE BLADE

In order to simulate the effect of the lacing wire upon the vibration
characteristics of a group of blades, investigations were conducted on a
loaded single blade. A rig was developed in which a vertical force could be
applied to a short length of lacing wire passing through the lacing wire
hole of single blades. The loading structure is shown in Fig. 6

Fig. 6   Loading rig for simulating
centrifugally loaded lacing wire

Load was applied to the blade assembly by dead-weight using a pulley and
once induced into the system was maintained by locking the assembly with
bolts. Typical values of load were determined by considering the effect of
centrifugal force experienced in a rotating turbocharger on the element of
the lacing wire acting on the blade. Care was taken to reduce the effect of

the loading structure upon the vibration characteristics of the blade.

Results obtained by finite element analysis and by experimental determination of natural frequencies by ESPI are shown in Table 2.

Table 2. Natural frequencies of laced and unlaced single turbocharger blade obtained by ESPI.

Mode	1F	1T	1E	2F	2T	3F
unlaced	1290	4700	2960	5810	10280	9031
laced	2575	4845	4181	7777	7881/10032	

The blade investigated had an inclination between lacing wire axis and the blade chord section of approximately 50°, this tended to cause coupling between modes. The results indicate that the lacing wire had little effect upon the characteristics of the 1T mode and only marginal influence upon the 2F, 2T and 3F mode. The theoretical analysis modelled the lacing wire by simulating the direct stiffness along the axis of the wire, by using beam bending theory to predict stiffness in the edgewise mode and using bending theory to predict torsion stiffness about the axis of the blade. The geometries selected were based upon the actual wire dimensions and the clamping conditions on the rig. Good correlation was achieved between the theoretical and mathematical representation, although it was difficult to assign a mode shape to the laced blade response for the equivalent 2T and 3F mode.

MULTI-BLADES

The investigation was developed to identify the effect of coupling several blades together, firstly by ESPI investigation of four linked blades and then by mathematical modelling of multi-bladed assemblies. Simple rectangular blades were used with the lacing wire fitted differently for three tests; a) passing freely through the lacing wire holes, b) attached by super-glue and c) brazed into the holes. To ensure a typical representation of the root fixture each blade was machined with an integral rectangular sectioned root block and the whole assembly was held in place by using a screw clamp and wedges between blades. This assembly is shown in Fig.7 where it can be seen that the inclination of the wire axis has been increased to an angle of 35° to assist with viewing by ESPI.

In the above Figure, there are four coupled blades and a single reference blade used for mode identification.

Experimental tests

For the cases of the loose and super-glued lacing wires the results indicated that the blades had very weak coupling between each other. Each assembly exhibited four natural frequencies (a family of frequencies) associated with each fundamental blade mode. Whereas the brazed blading exhibited the same number of natural frequencies on the frequency range under investigation but in this case it was extremely difficult to identify a logical mode classification because of the coupling between modes and the overlap of families of frequency modes. The coupling resulted directly from the cross stiffness effects induced by the inclination of the lacing wire.

Fig.7   Multi-bladed tests system

To clarify the vibration behaviour, the finite element analysis using the simplified blade model was extended to simulate the laced four blade system and reasonable correlation was obtained between FE and ESPI analysis. The analysis was extended to predict the natural frequencies for up to 10 laced blade assemblies. The results obtained are based on specific eigenvalue solutions and do not therefore suffer from the coupling experienced by the effects of damping in the practical tests. Hence easily identifiable modes can be predicted. Figure 8 shows the mode shape detected for a natural frequency in the 1T family where there is 1 node along the lacing wire axis. This mode was classified as the 1T1 mode.

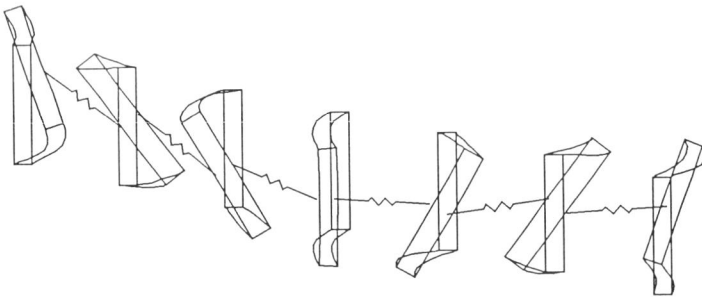

Fig.8   1T1 mode for laced seven bladed assembly

The theoretical analysis was extended to investigate the effect of the number of blades upon the natural frequencies of multi-bladed assemblies.

Figure 9 illustrates how the first flap (1F) family of frequencies develop with increase in the number of blades.

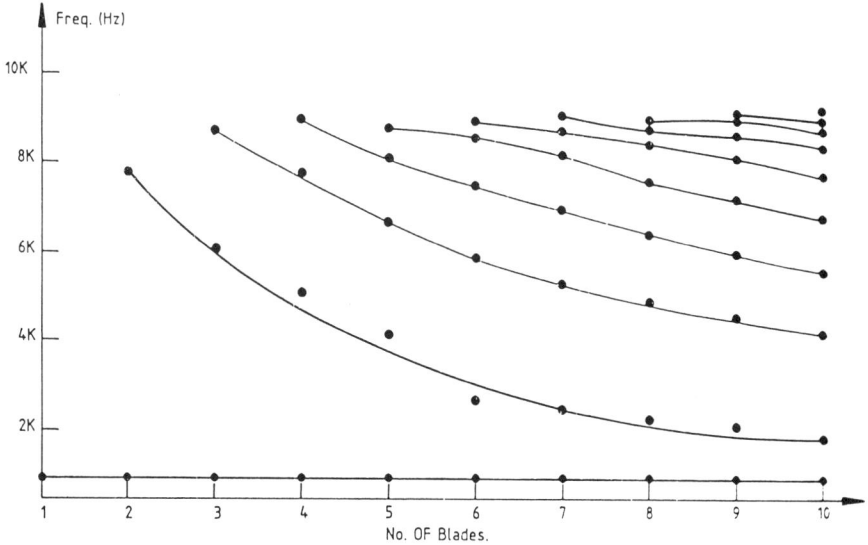

Fig.9   Development of 1F family frequencies with increase in number of blades.

The 1F0 mode remains constant for all assemblies, whereas all other modes i.e. 1F1, 1F2 etc start at higher levels and decrease as the blade number increases.

CONCLUSIONS

For a full understanding of the vibration performance of laced turbocharger blading, it is necessary to employ experimental and theoretical techniques to verify results. The effect of the lacing wire is to alter the vibration characteristic of multi-bladed assemblies by coupling via the inclined lacing wire. Complication arise with mode identification because the wire assist coupling between families of frequency modes.

ACKNOWLEDGEMENTS

The authors wish to thank Napier Turbochargers Ltd for financial support and Loughborough University of Technology for the research facilities.

REFERENCES

Chapman G.M. and Wang X., 1988, Interpretation of experimental and theoretical data for prediction of mode shapes of vibrating turbocharger blades, Jl of Vibration, Acoustics, Stress, and Reliability, ASME, vol 110, No.1, pp 53-59

Theoretical Evaluation of Experimental Techniques for the Determination of Structural Dynamic Properties

C. GENATIOS*, O. LOPEZ* and M. LORRAIN**

*IMME/UCV, Aptdo 50361, Caracas 1050A, Venezuela
*INSA, Dept. Genie Civil, Av. Rangueil, Toulouse 31077 Cedex, France

ABSTRACT

Experimental dynamics techniques are of great importance to improve or corroborate structural models. More realistic models lead to higher confiability designs. This paper presents a summary of a detailed evaluation of the experimental techniques used to identificate structural systems properties. Free vibration, harmonic vibration and random vibration procedures are discussed. Identification methods are presented for each technique based on single and multi degree of freedom formulations. The parameters that condition modal participation are identified. Fourier Transform methodology is presented to improve results for free vibration techniques.

KEYWORDS

Experimental structural dynamics; dynamic properties; free vibration; harmonic loading; shaking table loading; ambient vibration.

INTRODUCTION

For traditional design purposes, dynamic properties (modal frequency w_i and modal shape ϕ_i) are defined by mathematical models, or proposed upon experience (modal damping coefficients ξ_i). Experimental techniques are necessary in order to improve or comprobate systems idealizations, in the search of high confiability designs. Specially for highly irregular structures, composite material or largely repeated constructions, experimental validation becomes indispensable,because umprecise modelling might lead to less secure or more expensive constructions. This paper summarizes a detailled theoretical evaluation of structural dynamics experimental techniques (largely discussed on Genatios 1985, 1988; Lopez 1975). Its main objective is to present for each technique efficient identification procedures and the parameters conditioning modal participation, and to justify the use of Fourier Transform methodology for free vibration.

FREE VIBRATION PROCEDURE

This procedure is achieved by imposing an initial condition (displacement or velocity) on the structure, generating its free vibration. The response for each vibration mode is :

$$\eta_i(t) = \rho_i \, e^{-\xi_i \omega_i t} \cos(\omega_{di} t - \psi_i) \tag{1}$$

Generally, it allows the identification of the fundamental periode (T_1) and modal shape (ϕ_1) by direct observation, and damping coefficient (ξ_1) by the logarithmic decay formula. The answer of multi-degree of freedom (DOF) systems is obtained by modal superposition. Time registers analysis is usually limited to the evaluation of first modes properties. Higher modes contributions usually represent perturbations for first modes time domain studies. The variables that condition the results are (Genatios 1985): (1) Structural systems characteristics. For equally damped modes, higher frequency modes will decay faster than lower frequency ones, thus frequencies separations will condition relative modal contributions. Beam-column (shear behavior) buildings have closer frequencies than tall shear-wall (flexural behavior) buildings, so their higher modes interference will be more important. (2) Initial conditions imposed to the structure. The relative initial condition of each mode depends on the shape of the systems initial condition. If an initial deformation similar to the ith mode is imposed, the system will only vibrate with that forme. (3)Transducers location. In order to magnify ith mode's contribution, it is necessary to obtain the register on its largest relative coordinate. Consequently, to eliminate a certain modes contribution, measurements must be taken on its minimal coordinate.(4) Measured variable. Displacement or acceleration registers are generally obtained on structural dynamic testing. Oscillatory phenomenum shows that the amplitude of modal acceleration is equal to the displacement amplitude multiplied by w_i^2. Thus acceleration registers will magnify higher frequency modes contributions, facilitating higher modes identification.

Fourier Transform Applications

Time domain studies will generally be limited to first modes evaluation. Fourier Transform allows higher and lower modes evaluation(Genatios1988). Systems frecuency can be easily identified, from direct inspection. In the case of modulus of the Fourier transform of the response, the damping coeficient can be obtained applying the bandwidth method :

$$\xi = \frac{1}{2} \frac{p_2 - p_1}{\omega} \tag{2}$$

p_1 and p_2 must be evaluated at $\sqrt{2}/2$ times the maximal modulus value. For the real component, equation (2) can be applied, but p_1 and p_2 must be evaluated a 1/2 times the maximal value. For

the imaginary component, equation (2) can also be applied by
assigning p_1 to the maximal value and p_2 to the minimal one
(Genatios, 1988).For systems with several degrees of freedom,
the relative contributin of each mode is determined by the
variables already described on the previons section.

Example (free vibration)Four resulting Fourier Transforms
registers are presented on Fig.1 They represent different
loading or measuring conditions applied to a four level ideal
beam-colum building. The comparison between Fig. 1.a and 1.b
shows the transducer locations influence:the 4th floor is the
1st modes largest coordenate, so it magnifies its relative
influence and the 1st. floor magnifies higher modes
influence.The comparison between 1.a and 1.c shows that the
acceleration registers magnifie higher modes influence, due to
w_1^2. Finally, 1.a and 1.d illustrate initial conditions
influence by showing how a load applied on the 4th floor
produces a deformation that is similar to the 1st modes one,
thus reducing higher modes influence, that can be increased if
the load is applied on the 1st level.

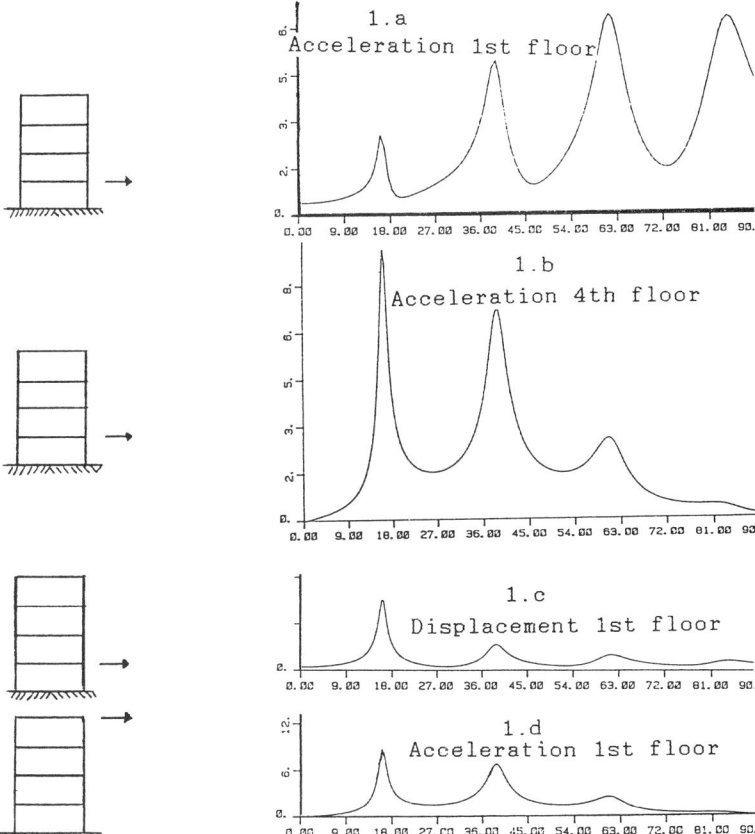

Fig.1   Four level beam-column building
        Free Vibration Procedure by
        Initial Lateral Load

FORCED HARMONIC VIBRATION

Structures are usually subjected to harmonic excitations. If the acting frecuency can be controlled, frecuency response curves can be obtained, allowing resonance determination. Modal properties can be identified from the resonance zone of these curves. Harmonic loads are obtained by rotating-mass vibration generators or hydraulic actuators. The first can be located on different parts of the structure and its force is proportional to mep^2 (m=rotating-masses value, e= rotating-masses excentricity and p= rotating frequency). Hydraulic actuators can be directly applied to the structure, but in general are connected to a shaking table acting on its base.

Rotating-mass vibration generator essay

The structural displacement of a single DOF system is:

$$u(t) = D \, sen(pt-\psi) \; ; \; D = \frac{m \, e}{M} \frac{(p/\omega)^2}{\sqrt{(1-(p/\omega)^2)^2+(2\xi p/\omega)^2}} \qquad (3)$$

This kind of expressions allow frequencies evaluation (by direct observation) and ξ_i determination (by the bandwidth method, eq.2) (Nielsen, D. 1964). This method gives exact results for constant amplitude harmonic loading. In the case of rotating-masses essay, displacement amplitudes are proportional to p^2 so they should be divided by p^2, but eq.2 can directly be applied with very small errors. For accelerations, it is recommended to divide the obtained amplitudes by p^4 or at least by p^2. Other methods are proposed for damping evaluation (Nielsen, D.1964), but the bandwith one leads to more accurate results (Genatios, C.1985). For N DOF systems, the response can be obtained by adding modal contributions. The displacement response for jth DOF, due to a rotating-mass device acting on the kth, is

$$u_j(t)= \sum_{i=1}^{N} \phi_{ji}\phi_{ki}\frac{m \, e}{M_i^*} \frac{(p/\omega_i)^2}{\sqrt{(1-(p/\omega_i)^2)^2+(2\xi_i p/\omega_i)^2}} sen(pt-\psi_i) \qquad (4)$$

where u_{ji} is the ith modal contribution to u_j displacement, M_i^* is the ith mode normalized mass and ψ_i the phase angle. Eq.4 shows that there will be resonance whenever $p=w_i$, allowing w_i and ξ_i identification (direct observation and bandwidth method). Modal shapes can be obtained from observation of resonance amplitudes and phase angles. Variables conditioning modal participation are:(1)Structural systems characteristics. As structural frequencies lie closer, modal interference is stronger, difficulting modal properties identification. Easier identification will then be possible for flexural type buildings than for shear type.(2) Rotating-Mass devices location. To obtain the largest contribution of a choosen mode, it is necessary to generate its response on its largest relative coordenate, i.e. higher levels for 1st modes evaluation.(3) Transducers location. Similarly, the largest modal contribution must be recorded in order to magnify a certain mode presence (4) Measured variable. For each mode contribution, acceleration amplitudes will be equal to displacement amplitudes multiplied by p^2; this produces a stronger interference on acceleration registers of lower modes on higher modes amplitudes, difficulting its identification. Generally, it will be recommendable to use displacement amplitude registers or acceleration amplitude registers divided by p^2, for low or high modes evaluation.

Example (harmonic vibration). The same ideal four floor building is submitted to the action of a rotating-masses device placed at the 4th level.Fig 2.1 shows 4th levels displacement and acceleration amplitudes, with its corresponding modal components; first modes contributions difficult higher modes identification from acceleration amplitudes. Fig.2.2 shows first levels amplitudes, allowing 2nd and 3rd modes evaluation.

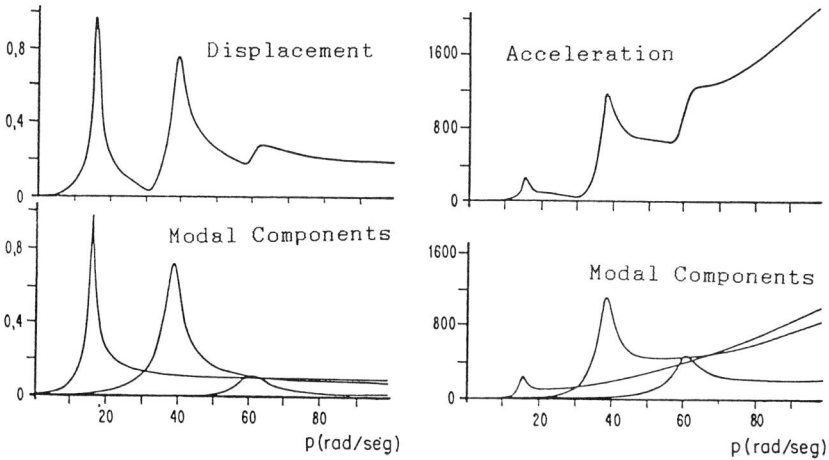

Fig. 2.1 4th Level Response

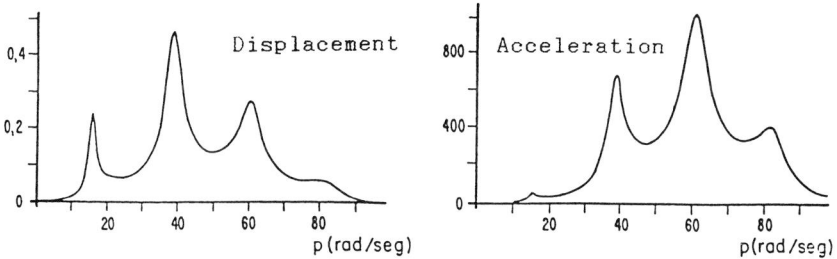

Fig. 2.2 1st Level Response

Fig. 2 Four level beam-column building
rotating-masses device applied on 4th level

Harmonic Loading by Shaking table

Shaking table harmonic tests are usually carried out with constant amplitude displacement or acceleration base movements. The relative responses amplitude for the jth DOF is :

$$u_j(t) = \sum_{i=1}^{N} \phi_{ji} \cdot \sum_{k=1}^{N} \phi_{ki} \frac{\ddot{u}_g(p)}{M_i^*} \frac{1}{\sqrt{(1-(p/\omega_i)^2)^2+(2\xi_i p/\omega_i)^2}} sen(pt-\psi_i) \quad (5)$$

with $\bar{u}_g(p)$=constant (constant amplitude acceleration base movement) or $\bar{u}_g(p)=\bar{u}_g(p)p^2$ where $\bar{u}_g(p)$=constant (constant amplitude displacement base movement). p^2 multiplying factors allow transition from one type of register to the other. Direct observation and bandwidth methods are used for properties identification. For $\bar{u}g(p)$=constant, acceleration amplitudes are more convenient for lower and higher modes studies. For $\bar{u}_g(p)$=constant, displacement amplitudes are more convenient; they can also be obtained dividing acceleration amplitudes by p^2.Practical variables are: structures characteristics, monitoring devices location and measured variable. They all lead to the same considerations as in the rotating-masses essay.

RANDOM BASE MOVEMENT

This technique consists on applying an irregular movement to the base of the structure. The corresponding records can be generated by hydraulic actuator movements (shaking table tests) or by actual seismic movements. The Transfer Function of the structure, can be obtained from the applied base movement and the registered response, allowing thus dynamic properties evaluation. From structural dynamics theory, we know that the transfer function of absolute acceleration (Ha(p)) is (Single DOF systems):

$$Ha\ (p) = \bar{a}\ (p) \Big/ \bar{a}_g(p) \ = \Big\{ \ \frac{1 + (2\xi p/\omega)^2}{(1-(p/\omega)^2)^2+(2\xi p/\omega)^2} \ \Big\}^{1/2} \ . \ e^{i\psi} \qquad (6)$$

with $\bar{a}(p)$= absolute acceleration Fourier Transform,$\bar{a}_g(p)$=ground acceleration Fourier Transform. Transfer function of the relative response is:

$$H_u(p) = (1 - H_a(p))/p^2 \qquad (7)$$

Transfer functions are complex values; their modulus represent the amplitude of the response due to a constant amplitude load. This allows w_1 and ξ_1 identification, by direct observation and by the bandwidth method (Lopez,O.1975). For small damping values, the absolute acceleration modulus also leads to appropriate results. For the kth DOF of a multi-DOF system, the transfer function will be:

$$H_{ak}\ (p) = \sum_{j=1}^{N} \ \gamma_j \phi_{kj} \Big\{ \ \frac{1+(2\xi_j p/\omega_j)^2}{(1-(p/\omega_j)^2)^2+(2\xi_j p/\omega_j)^2} \ \Big\}^{1/2} \ . \ e^{i\psi_j} \qquad (8)$$

with γ_j= jth mode participating factor= $\phi_j^t \underline{M} \underline{b}/\phi_j^t \underline{M} \phi_j$ .Eq.8 allows properties evaluation in resonance, especially for relative responses. Practical variables are: structures characteristics, monitoring devices location and measured variable, leading them all to the same considerations as in the preceeding cases.

Example (random base movement).Fig.3 shows the modulus and the phase angle of the absolute acceleration Transfer Functions of the first and last floor of an 8-level building. Transfer Functions where obtained from (a) direct calculation (eq.8) and (b) use of acceleration records obtained from time integration (eq.6). Registers differences are due to time integration procedures and Fourier Transform algorithms. The 1st floor transfer function allows high modes evaluation, while the 8th

level transfer function outlines first mode contribution
(Lopez,O.1975).

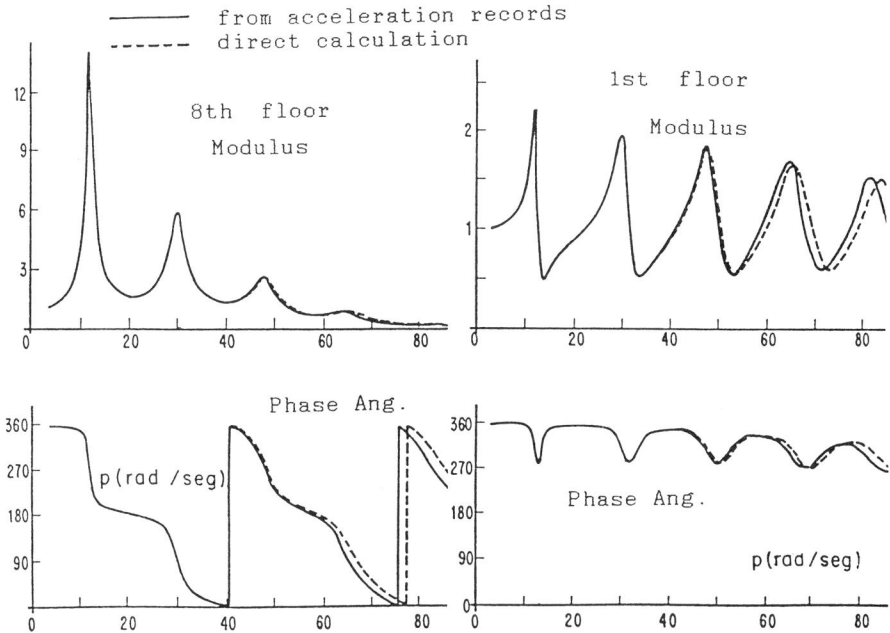

Fig 3   Eight-level beam-column building
         Absolute acceleration transfer functions

AMBIENT VIBRATION

This technique allows dynamic properties identification from
the analysis of registered vibrations due to environmental
agents such as wind, traffic, machinery, etc. Even if the
vibration levels are about 50 times lower than the ones
normally obtained on free or harmonic vibration essays, dynamic
properties can be precisely determined. Designating by F(p) the
vector containing the Fourier Transform of the applied load
vector, the Fourier Transform of the response of the kth mode
is given by (Cascante,1985)

$$\overline{u}_k (p) = \sum_{i=1}^{N} \phi_{ki} H_i (p) \underline{\phi}_i (p) \underline{\phi}_i^t \ \overline{F} (p) \tag{9}$$

where $H_i(p)$=transfer function of the ith mode. If the load has
the "white noise" shape (same intensity for all sampled
frequencies), the Fourier transform of the structural response
will have a shape very similar to the transfer functions, thus
allowing properties identification by direct observation (w_i)
and by applying the bandwith method (ξ_1). Due to the fact that
the actual load almost never has the "white noise" shape, an
averaging technique together with long enough recordings are

usually employed in order to compensate this effect. This is the reason why 10 to 20 minutes monitoring time is usually recommended for structural dynamics applications. In order to obtain modal shapes, a reference accelerometer (or any employed vibration transducer) must be maintained on the same place, and a second measuring device shall be moved along different structural locations. For each mode, modal coordenates are calculated by dividing the Fourier transform modulus corresponding to the sampled coordenate, by the one corresponding to the reference coordenate.

Example (Ambient Vibration).This example is included in order to show the importance of averaging frequency response modulus. A two DOF system is submitted to an irregular vibration, similar to those generated by the environment. Fig.4.1 shows the Fourier transform of the load and response signals, corresponding to three averages; also the transfer function is shown; due to load irregularities, the Fourier transform of the obtained response does not aproach the systems transfer function, thus not allowing properties identification. Fig.4.2 shows the excitation and the response averaged 500 times. The excitation Fourier transform modulus is now very close to the "white noise" shape, and the structural response represents the transfer function modulus, allowing correct properties identification.

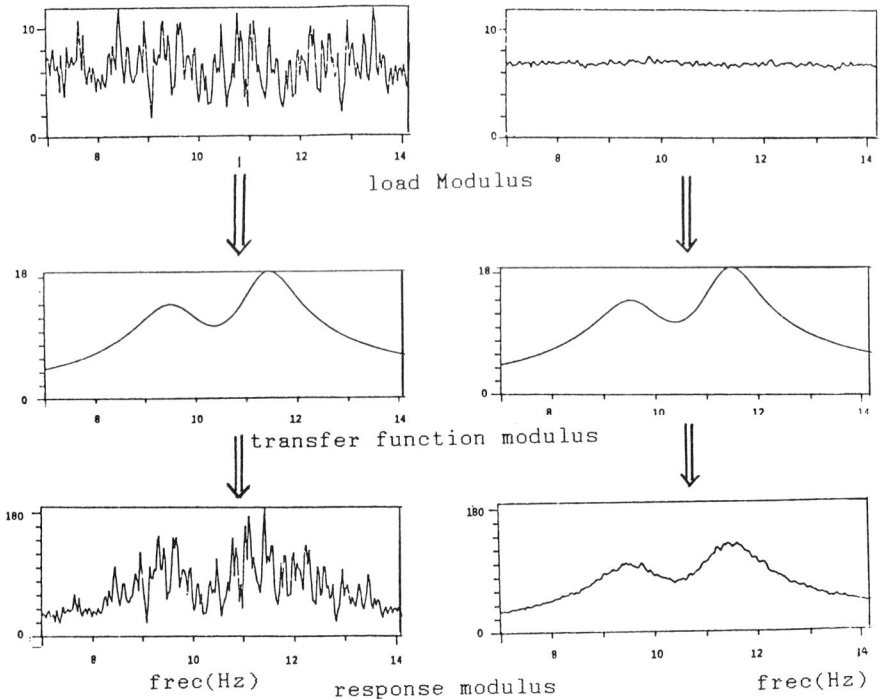

load Modulus

transfer function modulus

response modulus

frec(Hz)

frec(Hz)

Fig 4.1   two DOF system
3 averages

Fig 4.2   two DOF system
500 averages

CONCLUSIONS

Four kinds of structural dynamic experimental techniques were reviewed: (1) Free vibration studies based on time domain registers lead generally to 1st mode identification. In order to decrease higher modes perturbations, displacement records must be used, the load must be applied on the highest levels and the measurements must also be recorded on the higher levels. For higher modes identification, Fourier Transform analysis must be employed; acceleration records shall be used and should be obtained from lower floors, and the load shall also be applied on the lower levels. Original bandwidth and two modified bandwidth methods can be used to evaluate damping capacities. Direct observation allows natural frequencies and modal shapes identification. (2) Forced harmonic vibration tests allows lower and higher modes properties identification. The bandwidth method gives the most accurate results. First modes magnification requires displacement amplitude registers obtained from the higher levels, and if it is the case, the load applying device should be placed on the higher levels. For higher modes, it is recommended to use displacement or p^2 normalized acceleration records obtained from lower levels (with the loading device acting on the lower floors). (3)Random base movements usually give absolute acceleration results that should in general be transformed to relative displacement registers. Monitoring levels recommendation coincide with previous cases. (4) Ambiental vibration tests do not demand special load generating mechanisms. Enough time sampling and averaging techniques improve significantly obtained results Thus low and high modes can be properly evaluated.

REFERENCES

Cascante,G. (1985). "Determinacion de Propiedades Dinamicas de dos Estructuras".Instituto de Materiales y Modelos Estructurales. (IMME) Universidad Central de Venezuela (UCV) Caracas, Venezuela.

Genatios,C. (1985)."Evaluacion de Procedimientos Experimentales para la Determinacion de Propiedades Dinamicas de Estructuras". Instituto de Materiales y Modelos Estructurales Universidad Central de Venezuela, Caracas

Genatios,C. (1988). "Identificacion Estructural por Procedimientos de Vibracion Libre". Instituto de Materiales y Modelos Estructurales, Universidad Central de Venezuela, Caracas

Lopez,O. (1975) "Propiedades Dinamicas de Edificaciones obtenidas de lo Acelerogramas".Boletin Tecnico del IMME, N°52 Instituto de Materiales y Modelos Estructurales, Universidad Central de Venezuela, Caracas

Nielsen,N. (1964) "Dynamic Response of Multistory Buildings" Ph.D. Thesis, California Institute of Technology, Pasadena, California

Comparison among Modal Analyses of Axial Compressor Blade using Experimental Data of Different Measuring Systems

G. D'EMILIA*, C. SANTOLINI** and E. P. TOMASINI**

*Dipartimento di Energetica, University of L'Aquila, 67100 L'Aquila, Italy
**Dipartimento di Meccanica, University of Ancona, 60131 Ancona, Italy

ABSTRACT

The aim of this work is to compare the results of modal analyses performed on a mechanical structure in the same testing conditions, with different measuring systems.

The experimental data have been successively measured on the compressor blade by an eddy current probe proximity transducer, a laser Doppler vibrometer and a very light accelerometer; the same software package has been used for the modal analysis. The Authors were interested both in checking the differences in modal parameters estimation and in finding right working limits in the examined frequency range (0-5000 Hz). For each transducer all the presumable causes of bias error have been taken in to account and, if possible, eliminated, in order to obtain the best results for each measuring system. The excitation of the structure was through use of an impact hammer. The measuring mesh was the same in all cases.

With respect to the blade modal parameters estimation, the Authors examined the effect of analyzing experimental data of displacement, velocity and acceleration, also with reference to the results dispersion of a modal analysis. The correspondence among the measuring systems is in general quite good except for the accelerometer which put in evidence some problem in measuring highest frequency vibrations near the blade tip.

KEYWORDS

Modal Analysis, Accelerometer, Laser Doppler Vibrometer, Proximity Probe, Errors.

NOMENCLATURE

$H_{pq}(\omega)$	inertance FRF function		j	imaginary unit
ω	angular frequency of vibration		ω_r	natural ang. frequency, mode r
$r^A{}_{pq}$	modal constant (mode r FRF $_{pq}$)		ξ	critical damping ratio
$M^R{}_{pq}$	residual mass		$K^R{}_{pq}$	residual stiffness
η_r	damping loss factor for mode r		p, q	point coordinates

191

INTRODUCTION

The extension of the modal analysis to more and more specialized applications, makes the requests to the used dynamic transducers more demanding. In fact the use of the "classical" accelerometers often turns out to be unsatisfactory for some applications, mainly in the field of the smallest mechanical structures: there, the modification of the structure dynamic characteristics may turn out to be intolerable.

In the tests with modal analysis techniques, different Authors have therefore used either displacement or velocity or acceleration tranducers, having non contact and high frequency response characteristics, very good spatial resolution, and other interesting performances.
As a matter of fact, different Authors have used optical fibre displacement sensors (Patton and Trethewey, 1985, 1988), (Radwan and Chohshi, 1985), (Castagna, 1986), strain piezoelectric sensors (Brown, 1986), optical fibre and not laser Doppler vibrometers (Laming *et al., 1985)* and torsion meters. The study of these systems generally is a check of their transfer function but, in most cases, the applications we find in literature are more oriented to prove the possibility of using the said systems than to verify the errors deriving from their application in the modal analysis.

The characterization of the dynamic behaviour of such transducers on standard laboratory applications is evidently important. However, we must notice that the error propagation in the different steps of the modal analysis of non simple structures, such as those on which we generally must carry on the analysis, needs to be checked and experimentally quantified. That, too, in order to verify the real importance for each of the measuring systems of the different, possible, causes of bias and random errors. The possibility of using several measurement systems (and also different phisical quantities) consequentely sets the need of checking the consistency of the results obtained.

As a first step to the identification of the most important causes of error in the subsequent stages of the modal analysis, we have carried out a comparison among the obtained results by using experimental data of different kinematic quantities, indipendently measured; the mechanical structure on which the experimentations have been performed is a blade of a gas turbine compressor. We want to point out that the application has been selected in such a way as not to categorically "a priori" leave out the use of each of the selected systems; in particular we have tried to limit the load error due to the accelerometer, using a 2.4 g mass one. The used methodology, the software and the selection of the reference point for the processing of the experimental raw data, have been the same for the three different analyses. The purpose of the tests has also been to point out, for each measurement system, the particularly important bias error causes and to identify the possible right working limits of each system, obviously referring to the tested mechanical structure (see also D'Emilia *et al*, 1989). All the procedures for a correct experimental practice, largely recommended in literature (Ewins, 1984, 1986; Van Dessel and Snoeys, 1983) have been used with each measurement system (dynamic and static calibrations, right positioning checks, and so on), even if they were particularly laborious. In the first stage, in fact, the comparison among the different systems involved more the repeatibility of the results than the use ease and flexibility.

Further on, we will then describe the used measurement chains, we will point out some of the behaviour limits and measure the differences among the various systems with respect to the different modal parameters of interest. We will critically discuss the used comparison methods in order to point out as well some techniques that allow the elimination of further, little, systematic errors. We also mention some additional tests to be carried out in order to furtherly improve the comparison among the different measurement systems of modal quantities.

EXPERIMENTAL TEST RIG

Structure to be examined: grid of the measurement points.

Fig. 1a shows the blade on which all the measurements have been carried out. In the graph of fig. 1b is described the mesh, formed by 84 measurement points; the grid has been defined in such a way

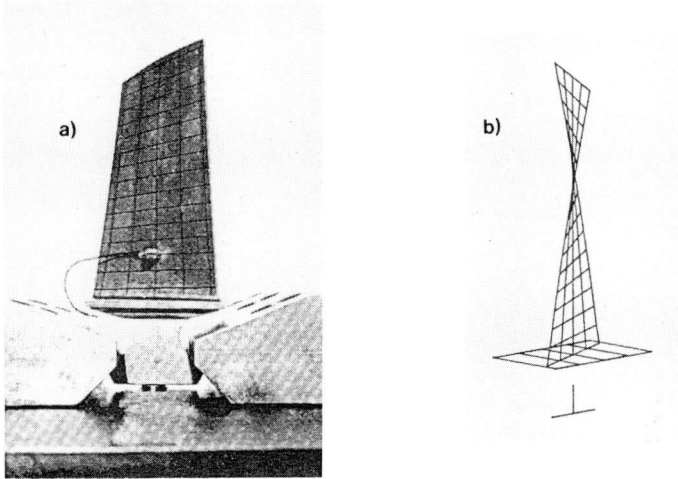

Fig. 1   a) A view of the blade and of the grounding frame;
         b) the experimental mesh with the driving point.

as to be able to properly describe the modal shapes of the studied frequency range. However, we have verified that a further increase of the spatial resolution did not produce remarkable improvements in the quality of the results.

Problems Referring to the Boundary Conditions

The blade has been fixed to the frame, trying to reproduce the boundary conditions of the structure in the operative assembling situation; they have been kept identical for the three series of tests. It has to be pointed out that , at this stage of study, the Authors were much more interested in the repeatability of the boundary conditions than in their perfect correspondence to the working situation.

Measurement Systems

The structure excitation, identical in all tests, has been of impulse type, by means of an impact hammer, instrumented with a force transducer. The diagram of fig. 2 summarizes the used measurement chains. The procedures of static and dynamic calibration for the different measurement systems are described in details in (D'Emilia et al., 1989). In the same reference (D'Emilia et al.,1989) are fully reported the problems relating to the positioning of the sensor and the effect of the difference in the measurement area for each of the sensors.

Acquisition and Processing of the Experimental Data

The tests have been carried out by keeping fixed the driven point and moving the measurement one along the mesh. This procedure and the selection of the driven point near the blade bottom have been suggested by the following criteria: maximum relative stiffness of the excitation area, its distance from the nodal lines, constant excitation direction. As to this last matter, it is important to point out the remarkable warpage of the tested blade.

We have checked that the excitation system in the frequency range (0-5000 Hz) has a satisfactory

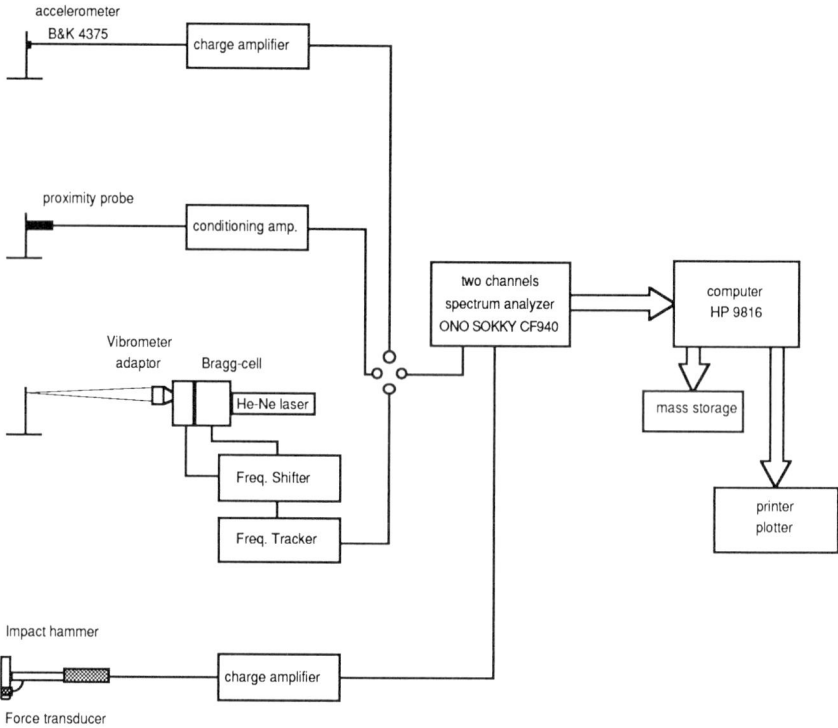

Fig. 2   Schematic of all measurement, excitation, data acquisition and processing systems.

frequency content (attenuation at 5000 Hz within 15 dB). Because of the very low damping of the structure, we have used an exponential window, the same for the three systems: the selection of the imposed damping, as we will subsequently show, turned out to be very important, mainly for the first modes. The FRF of each point has been obtained by averaging on eight acquisitions, checking that the coherence function did not point out positioning or orientation remarkable errors of the impact hammer. In order to get the wanted modal parameters (natural frequencies, dampings and modal shapes), we have used the software package ENTEK EMODAL 3.12, which utilizes for the curve fitting the algorithm of the "Rational Fraction Least Square Method": where imposed by the the FRF form, the analysis has been carried out in MDOF. We have finally checked the phisical meaning, with reference to the real test conditions, of the residual terms used in the reconstruction of the frequency response function.

RESULTS

Table 1 shows the comparison among the different measurement systems related to the natural frequencies for the modes identified in the inspected frequency range. This table also shows the differences existing among the damping evaluations.

Tab.1 - Comparison of Natural Frequencies and Damping Ratios

Mode	Accelerometer		Proximitor		L.D.Vibrometer	
	Freq.(Hz)	ξ (%)	Freq.(Hz)	ξ (%)	Freq.(Hz)	ξ (%)
1	289	0.10	288	0.16	290	0.14
2	640	0.27	637	0.24	642	0.27
3	1310	0.08	1315	0.10	1317	0.12
4	1432	0.27	1440	0.33	1443	0.29
5	2157	0.23	2173	0.19	2184	0.19
6	2893	0.08	2913	0.05	2916	0.04
7	3824	0.07	3829	0.06	3829	0.06
8	3997	0.16	4001	0.11	4006	0.08
9	4453	0.15	4470	0.15	4475	0.09

The main elements that can be pointed out are:
- the correspondence among the modal frequencies is very satisfactory; the difference between the two systems without contact (eddy current proximity probe and LD vibrometer) is generally within 0.5%; only for the first two flexural modes (mode 1 and 2) the difference is 1%; for them the analysis carried out by using the vibrometer data, sistematically tends to provide higher frequency values; particularly satisfactory is the correspondence for the modes at higher frequency;
- equally the dampings are in good agreement, also considering the type of low damping structure and the use of an impact hammer; the difference is in general within 20%; the average uncertainties that can be conservatively attributed to these measurements are $\pm$ 1% for natural frequencies and $\pm$ 15% for damping ratios;

- also the correspondence of the modal shapes is good; their correlation for different modes has been quantified by means of parameters M.S.F. and M.A.C. (Ewins, 1984); figg. 3 point out the correlation degree of the obtained modal shapes for each measuring system, for some of the modes.

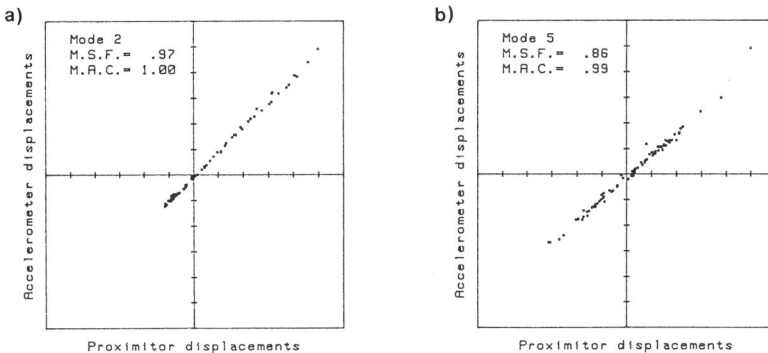

a)

b)

Fig. 3   Diagrams of M.S.F. and M.A.C. parameters (Ewins, 1984),
comparing data from different measuring systems:
a) proximity probe versus accelerometer (mode 2)
b) proximity probe versus accelerometer (mode 5)

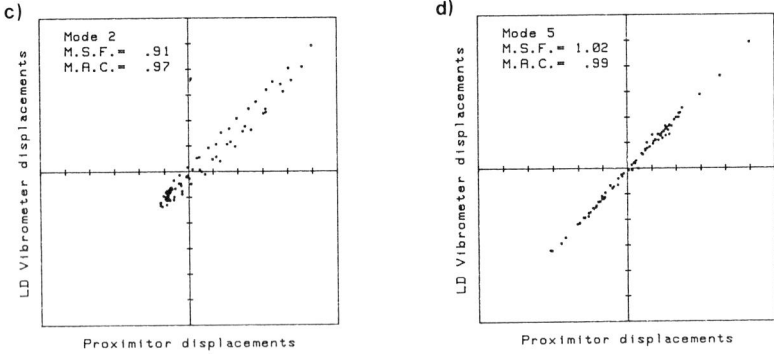

Fig. 3   Diagrams of M.S.F. and M.A.C. parameters (Ewins, 1984),
          comparing data from different measuring systems:
          c) proximity probe versus LD vibrometer (mode 2)
          d) proximity probe versus LD vibrometer (mode 5)

For all the individuated modes, the correlation between the vibrometer and eddy current probe measurements is very good (M.A.C. near 1); also the reconstruction of the nodal lines shows results in very good agreement (fig. 4). As to the accelerometer, this is valid only for the first modes.

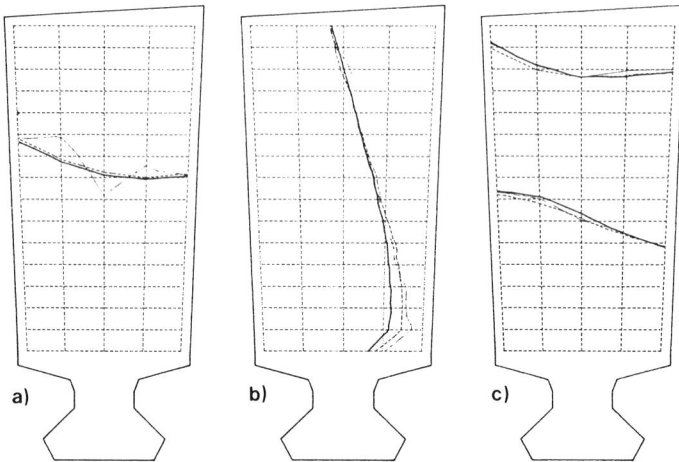

Fig. 4   Comparison among nodal lines, reconstructed starting from
          experimental data.
          a) mode 2   _ _ _ :   accelerometer
          b) mode 3   _____ :   LD vibrometer
          c) mode 5   _____ : proximity probe

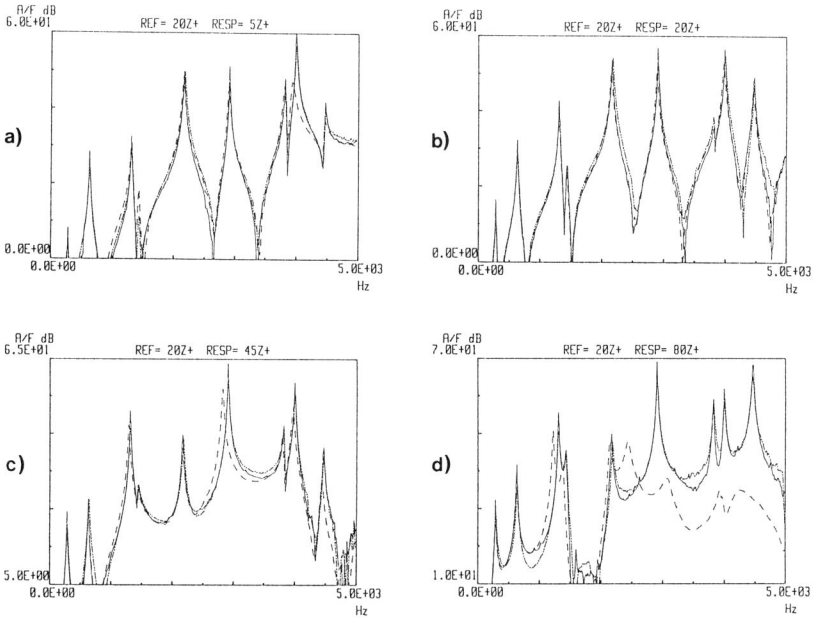

Fig. 5   Plots of the experimental FRF magnitude curves (1/kg) at the varying of the measuring point:

a) blade bottom;   _ _ _ : accelerometer
b) driving point;   - .. -- : LD vibrometer
c) half blade;   _____ : proximity probe
d) blade tip;

- the comparison between the results of proximity probe and vibrometer does not point out any problem for the modal analysis in the frequency range 0-5000 Hz; this means that the proximity sensor, even if it measures displacements, allows the valid identification of the modal shapes relating to the higher frequencies. On the other hand, the accelerometer causes an appreciable load effect on the structure; fig. 5 show the obtained FRF for different points on the grid: (a) blade bottom, (b) driving point, (c) half blade, (d) blade tip. It has to be noticed that the measurements with accelerometer show a progressive drop of the resonance frequencies and a flattening of the curves, also pointing out a damping effect of the sensor. This behaviour causes a worsening of the evaluation of the structure global response in the higher part of the frequency range. On the contrary, it is particularly satisfactory the comparison between the experimental FRF curve, obtained by LD vibrometer, and the one reconstructed using the measurements on the different points of the mesh, according to the following algorithm : (Ewins, 1984), (fig. 6).

$$H_{pq}(\omega) = \sum_{r=1}^{n}\left(\frac{-\omega^2 \, _rA_{pq}}{\omega_r^2-\omega^2+j\eta_r\omega_r^2}\right) + \frac{1}{M_{pq}^R} - \frac{\omega^2}{K_{pq}^R}$$

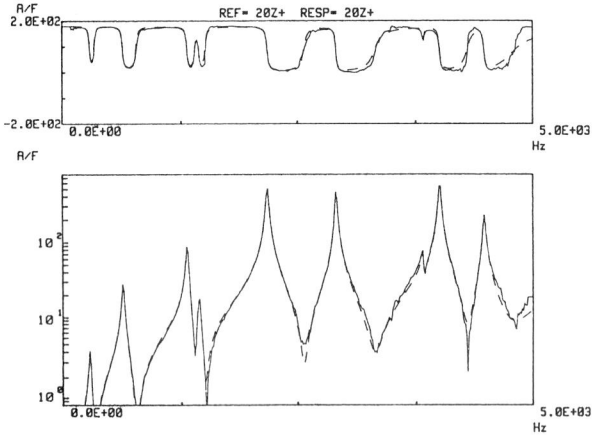

Fig. 6   Comparison between the experimental FRF curve (by LD vibrometer) and synthesized one.

a) phase (degrees)   b) magnitude (1/kg)

As previously pointed out, the highest discordances, though limited, exist for the first two modes of the structure (first and second flexural): the found difference is not due to the selection of the point in which the the experimental FRF is choosen for the curve fitting (fig. 7). On the lower frequency modes, mainly on a low damping structure like the described blade, the right selection of the exponential window has a great influence. Further tests will be carried out in order to check the influence of the exponential window on the results referring to each system for ther first two modes; the acquisition time lenght will be increased, limiting the investigation frequency range.

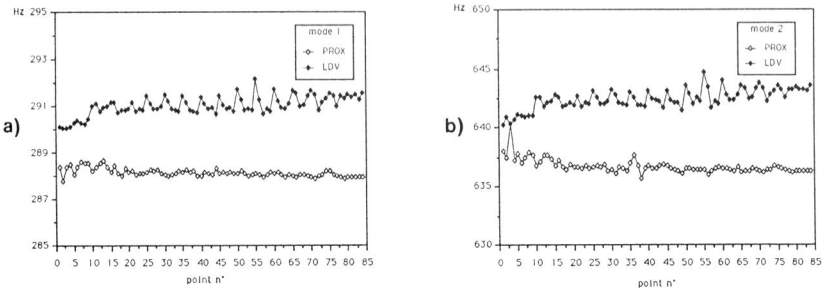

Fig. 7   Computed mode frequecy behaviour, varying measuring point on the mesh, using experimental data of proximity probe and LD vibrometer for mode 1 (a) and 2 (b).

CONCLUSION

Non contact displacement and velocity measurements, respectively carried out with eddy current proximity probe and laser Doppler vibrometer, have shown the possibility to be efficaciously used in the modal analysis of a blade of a gas turbine compressor, in the frequency range 0-5000 Hz. The tests with accelerometer have pointed out that on this structure we may have problems in identifying the modal parameters for the higher frequency modes. The comparison among the different systems has been performed after having calibrated them and provided for the elimination of all the predictable systematic causes of error.

The agreement among the modal parameters evaluation carried out by using each of the above mentioned transducers with those obtained by means of piezoelectric accelerometers, is generally fully satisfactory, except for the higher frequencies modes , where the accelerometer, although of very limited mass (2.4 g), has shown heavy load effects on the structure. This is pointed out with the lowering of the resonance frequencies of the structure and the damping effects. The differences between the non contact measurement systems are, on the contrary, very small for the higher frequencies. The experimental dispersion of natural frequencies and dampings at the varying of the mesh point, resulted negligible (in absence of systematic errors !). If for the resonance frequencies we assume the uncertainties usually accepted in the practice of modal analysis for average complex structures(2-3%), the measurements of the different systems look in perfect agreement; the same is true also for the dampings, taking in to account the impact excitation and the low damping (<0.3 %) of the examined structure modes. However the prosecution of this study will deal with both the investigation about the causes of other bias error now considered less important, and the acquisition of useful information, in order to set up a complete analysis procedure of the error propagation in the modal analysis.

ACKNOWLEDGMENTS

The Authors wish to thank Nuovo Pignone S.p.A. of Florence, Italy for releasing publication of experimental information about the compressor blade.

REFERENCES

Brown, R.H. (1986). Piezo-film Vibration Sensors: New Techniques for Modal Analysis. 6th Int. Modal Analysis Conference, Orlando, 1986.
Castagna, J. (1986). A Method of Determining the Modes of Vibration of Disc Drive Heads and Suspensions During Operation. 6th Int. Modal Analysis Conference, Orlando, 1986.
D'Emilia, G., Santolini, C., Tomasini, E.P. (1989). A Contribution to the Evaluation of the Uncertainty of Modal Parameters Estimation, Starting from Experimental Data. Internal Report of Dipartimento di Meccanica, University of Ancona, 1989.
Ewins, D.J., Griffin, J. (1981). A State of The Art Assessment of Mobility Measurement Techniques - Results for the Mid Range Structures (30-3000 Hz). Journal of Sound and Vibration, vol. 78, pp. 197-222.
Ewins, D.J. (1984). Modal Testing: Theory and Practice. Research Studies Press, Letchworth Hertfordshire.
Ewins, D.J. (1986). Theory and Practice of Modal Testing: Additional Notes. Research Studies Press, Letchworth Hertfordshire .
Laming, R.I., Gold, M.P., Payne, D.N., Halliwel, N.A. (1985). Fiber-optic Vibration Probe. Proc. of the Int. Conf. on Fiber Optic Sensors, SPIE Vol. 586, Cannes 1985.
Patton, M. E., Trethewey, M. W. (1985). A Technique for Non Intrusive Modal Analysis of Ultralightweight Structures. 5th Int. Modal Analysis Conference, London, 1985.
Patton, M. E., Trethewey, M. W. (1988). External intensity Modulated Fiber Optic Sensors for Structural Vibration Measurements. Journal of Modal Analysis, 1988.
Radwan, H.R., Chohshi, J. V. (1985). Use of Non-Contact Measurements for modal Analysis of Disk Drive Components. 3th Int. Modal Analysis Conference, Orlando, Florida, (January 1985).
Van Dessel, H., Snoeys, R. (1983). Accuracy of Modal Parameters: Survey. KUL, Dept. of Mechanics, Leuven, Int. Note.

Application of Acceleration and Strain Measurements in Forced Vibration Testing

W. F. TSANG, E. RIDER and P. GEORGE

Royal Naval Engineering College, Plymouth, UK

ABSTRACT

This paper presents the results of the application of an experimental modelling technique on a beam structure using acceleration and strain data. It has been shown that valid low order mathematical models of the beam can be derived. A further application of the derived models for force prediction has also been given as an example.

KEYWORDS

Modelling; force prediction; spatial models; forced vibration.

INTRODUCTION

Forced vibration is a very useful experimental technique in the study of dynamics of structures. The method is based on the characterisation of a structural system by its force/response characteristics. With the advent of modern experimentation and computer systems, such characteristics can often be measured experimentally.

A structure responds to the external disturbances (forces) with internal stresses, strains and external kinematic deformations. Therefore the response of a structure can be measured in terms of its kinematic and/or rheological deformations. Acceleration measurement using accelerometers is well established in modal testing whereas strain measurement using strain gauges is less common. This is substantiated by the scarcity of literature on this subject [References 1,2,3]. The use of strain gauges can offer advantages such as light construction and cheapness in terms of hardware costs. These features can be very beneficial from the experimental point of view. An investigation into their usage has been undertaken based on these premises.

201

This research is largely a follow-on of the work by Hillary and Ewins [3] who have used the Strain Frequency Response technique to derive modal models . At the Royal Naval Engineerimg College (RNEC) , research on modal analysis techniques has been carried on for quite some time using both frequency as well as time domain methods on a variety of structures [4,5,6,7,8]. However in all these works , a response measurement system based on piezo-electric accelerometers was used exclusively . Current research at RNEC is aim at investigating the technique of experimental dynamic modelling and force prediction. The recent method proposed by Tsang and Rider [10] to derive spatial models of structures using a non-modal approach has proved to be a substantial departure from the traditional modal approach adopted in previous investigations. The interest in the use of strain gauges was generated quite recently therefore the programme is still very much in the learning stage. It is our ojective to explore what strain measurement can offer as an experimental modelling tool and this is exactly what this paper is concerned with. The results of the experimental modelling of a simple beam structure (shown in Fig.1) using both acceleration and strain measurement methods, are reported in this paper.

THEORETICAL APPROACHES

Description of the general theory will not be given and only a particular loading case will be considered ; i.e. the flexural vibration of a beam ,under a harmonically varying unit concentrated excitation force ($e^{-j\omega t}$) applied at position x=a . This is because the more general loadings in both time and space can be treated similarly using the principle of integral superposition. The governing equations of the problem concerned is given by :

$$ EI \frac{\partial^4 V}{\partial x^4} + \rho A \frac{\partial^2 V}{\partial t^2} = \delta(x-a)\, e^{-j\omega t} \qquad (1)$$

E : YOUNG'S MODULUS $\qquad \rho$: MASS DENSITY $\qquad \delta$: DIRAC DELTA FUNCTION

I : MOMENT OF INERTIA $\qquad A$: AREA OF SECTION $\qquad \omega$: ANGULAR FREQUENCY

The complete solutions of Eqn. (1) have been given in reference [9] and some of them are reproduced below :

for x < a, the solution is :

$$ V1(x) = A1\, \sin\beta x + B1\, \cos\beta x + C1\, \sinh\beta x + D1\, \cosh\beta x \qquad (2a)$$

for x > a, the solution is :

$$ V2(x) = A2\, \sin\beta x + B2\, \cos\beta x + C2\, \sinh\beta x + D2\, \cosh\beta x \qquad (2b)$$

constants A1, B1,...C2, D2 can be solved depending on the specifed boundary conditions. What is important in the solution is that the displacement function can be represented by the product of a time independent function $G(x,a,\omega)$ and a time dependent function $e^{-j\omega t}$. Until the analytic displacement function V(x,t) is solved, the bending moment at location x on the beam can be obtained by partial differentiation of V(x,t) *twice with respect to x.*

i.e.
$$B(x,t) = EI \frac{\partial^2 V}{\partial x^2} \qquad (3)$$

Knowing the bending moment distribution function B(x,t), the strain distribution function S(x,t) can be found from Equation (2)

$$S(x,t) = y * \frac{\partial^2 V}{\partial x^2} \qquad \text{where } y \text{ is the distance of the point from the neutral plane of the beam.} \qquad (4)$$

The SFRF S(x,w) is then given by :

$$S(x,\omega) = \frac{F(S(x,t))}{F(e^{-i\omega t})} \qquad \text{where } F \text{ is the Fourier Transform operator.} \qquad (5)$$

This is the usual approach in the analysis of flexural vibration of a beam in which the displacement function is determined first, and the strain function derived later. However the same problem can also be approached using the theory of elasticity which deals with stress and strain directly. In this approach, the governing equations are formulated for each individual infinitesimal particle rather than for each section of the beam. The spatial translation of each particle is completely defined by the three displacement components U, V and W in the x, y and z axes respectively. The relationship between the various strains and the displacements U, V and W can be derived purely from the consideration of geometry.

$$\varepsilon_x = \frac{\partial u}{\partial x} \qquad \gamma_{xy} = \frac{\partial v}{\partial x} + \frac{\partial u}{\partial y}$$

$$\varepsilon_y = \frac{\partial v}{\partial y} \qquad \gamma_{yz} = \frac{\partial w}{\partial y} + \frac{\partial v}{\partial z} \qquad (6a - 6b)$$

$$\varepsilon_z = \frac{\partial w}{\partial z} \qquad \gamma_{zx} = \frac{\partial u}{\partial z} + \frac{\partial w}{\partial x}$$

There are six independent strain components to specify the strain state :

$$\varepsilon_x, \ \varepsilon_y, \ \varepsilon_z, \ \gamma_{xy}, \ \gamma_{yz}, \ \gamma_{zx}$$

which are compelely defined by three displacent components U, V and W of each particle. This implies that other fixed relationships exist between the strain components. These equations are called the compatibility equations. The motion of the particle is governed by the differential equations called the equilibrium equations. The complete solution of the stresses and strains within the material requires the satisfaction of all these equations simultaneously for all the particles within the material. In addition, the boundary conditions detailing the stresses at the boundary surfaces have to be satisfied as well. Typically, these equations were solved by proposing, with some intelligent guesses, some stress functions Q(x,y,z) with a hope that the stresses and strains derived from these stress functions will satisfy all the equilibrium, compatibilty and boundary conditions. Hence by virtue of the Uniqueness Theorem these stresses and strains form the unique solution. Even for a particular case which involves stress and strain analysis of a simple cantilever beam in flexural vibration (which is essentially a two dimensional plain stress problem) the problem is still not easy to solve. Therefore it will be beyond the scope of this paper to describe their analytical solution as this can be a subject in its own right and merits a separate paper for its complete desciption. These theories are quoted here in order to illustrate the complexity and difficulty involved in the direct derivation of the analytical strain-force relationship.

EXPERIMENTAL APPROACH

It has been shown that analytical solution of vibration of structures can be very involved mathematically and sometimes even impossible for more complicated situations. This paper is concerned with the application of experimental rather than theoretical approaches. It will be shown later how valid low order mathematical models can be derived from experimental forced vibration data. Although high order models such as those created by Finite Element Analysis are sophisticated, low order models are often found to be superior in terms of accuracy in most cases. The experimental approach being described in this paper is a frequency domain method using data either in terms of accleration or strain frequency response functions . The basic instrumentation used is very much the same as those in modal testing i.e. a shaker, signal analysers, force and response transducers, a computer and so on. Interested readers are referred to the equipment check list in Ref. 5 for further details. The forced vibration tests reported here were performed using both step-sine (SS) and the pseudo-random (PRBS) excitation methods.

EXPERIMENTAL MODELLING OF STRUCTURES USING ACCELERATION MEASUREMENTS

Instead of dealing with the beam structure as a continous system which requires a solution of the partial differential equations ; a discrete system is dealt with. The governing equations of a discrete system can then be represented by a system of linear equations as depicted below :

$$(-\omega^2 [M] + \omega j [C] + [K]) \{ V(\omega) \} = \{ f(\omega) \} \qquad (7)$$

The task of the experiment is to measure the input force and the acceleration response. In the frequency domain, there is a simple relationship between the various kinematic quantities given by :

$$\dot{V}(\omega) = \frac{\ddot{V}(\omega)}{j\omega} \quad and \quad V(\omega) = \frac{\ddot{V}(\omega)}{(j\omega)^2} \qquad (8)$$

therefore the need to measure all three kinematic quantities is alleviated. It has been shown in reference [10] that the system matrices [M], [C] and [K] can be derived from the measured experimental data and that the assumption of zero damping can be justified for this beam system since it was very lightly damped. Therefore the [M] and [K] matrices alone constitute an undamped spatial model.

EXPERIMENTAL MODELLING OF STRUCTURES USING STRAIN MEASUREMENTS

Foil gauges are relatively cheap devices and can be used in large quantities without the problem of mass-loading the structure, because of their light-weight construction . The associated bridge amplifying and conditioning units can often be obtained at a fraction of the price of charge amplifers for piezo-electric accelerometers. However it is also true that strain gauges are generally not as sensitive as accelerometers. The other reason for using strain gauges is that there are many situations in which kinematic responses are small while strain responses are significant. Therefore strain gauges are an invaluable supplement to acclerometers as structural response measurement devices.

By analogy with Equation (7), it is postulated that Equations (9) and (10) apply based on the fact that the two systems of linear equations possess the same eigen-values.

$$\{ V(\omega) \} = [T] * \{ S(\omega) \} \qquad (9)$$

$$(- \omega^2 [M] [T] + [K] [T]) * \{ s(\omega) \} = \{ f(\omega) \}$$

or $[D] = [M] * [T]$ *and* $[E] = [K] * [T]$

$$then \quad (-\omega^2 [D] + [E]) \{ s(\omega) \} = \{ f(\omega) \} \qquad (10)$$

Hence, The task of the experiment is then to determine the unknown system matrices [D] and [E] . Unlike the mass [M] and stiffness [K] matrices which carry the familiar meanings of inertia and stiffness, the physical interpretation of the [D] and [E] matrices is not readily apparent. In view of the fact that Equation (10) is an appropriate form of a second order dynamic system, then the derivation of the [D] and [E] matrices can be seen merely as a curve-fitting process in which the experimental strain data are made to fit this equation.

FORCE PREDICTION AS A PARTICULAR APPLICATION OF EXPERIMENTAL STRUCTURAL MODELLING

The derived models can be utilized in a number of ways. For instance one can predict the response of a structural system which is subjected to a specified loading. Alternatively one can predict the imposed loads on a structure if the responses are known. The latter is often called 'force prediction'. There are many situations in which the measurement of forces on a structure is difficult if not impossible. Distributive loadings and inaccessibilty of the placement of force measurement transducers are typical situations . As an illustration of this application, the derived models are used to predict the forces imposed on the beam. In the experiment, the single input force and the response auto-spectra at all five measurement points were measured over a frequency range. The predicted force vector {f} was then calculated based on the measured response data using equations (7) and (10).

RESULTS

Typical experimental and data analysis results of the following models are given :

- Model A1 : acceleration data from SS tests (5 DOF model) (see Fig. 2   and Table 1)

- Model S1 : strain data from PRBS tests (4 DOF model) (see Fig. 3)

- Model S2 : strain data from SS tests (5 DOF model) (see Fig. 4   and Table 1   )

- Model S3 : Strain data from SS tests (4 DOF model) (see Fig. 5)

DISCUSSION

The [M] and [K] matrices of model A1 have been given in Table (1) and they convey information on the inertial and stiffness properties of the beam. However those [D] and [E] matrices asssociated with Model   S2   as shown in Table (1) are difficult to interpret physically.

Generally speaking, the derived models A1, S1, S2 and S3 are good, as proven by the coincidence of the predicted and the measured FRF plots in which the shapes of the curves bear close resemblance and the positions of the resonant frequencies are correctly predicted. However , considerable errors did occur in regions near resonance and regions beyond the coverage of the chosen frequency points upon which the models were derived.   The errors due to the former were unavoidable and were inherited in the initial assumption of undamped models.

The force prediction results   show that the position and magnitude of the applied force have   been predicted with some accuracy. This was especially so over the frequency range in which good models had been obtained.

The strain models do not account well for mode 1 which was at about 7 Hz . This was probably because the radii of curvature   throughout the beam were small when the beam was vibrating at its first mode, . Hence the responses detected by the strain gauges were also small.

While it has been shown that both acceleration and strain frequency response data are applicable to derive mathematical dynamic models of structures (a beam in this case), the two measurement methods are very different in nature.   The acceleration measurements provide temporal information with regard to the rate of change of kinematic quantities with respect to time.   Strain measurements , however, provide spatial information with regard to the rate of change of displacment with respect to space. The full potential of the use of strain gauges in the application mentioned, can only be realized with further research.

REFERENCES

1.   A   R   Walker.   "Experimental determination of rotating helicopter blade deformations using strain pattern analysis". Proceedings of the Conference On Stress And Strain Measurement In Aeronautics, Naval Architecture And Offshore Engineering, 6-8 Sep 1988.

2. S S Liu, G A O Davies. "The determination of rotor blade loading from measured strains". Paper presented at the 13th European Rotorcraft Forum Arles, France, Sep 8-1, 1987, paper No. 6.9.

3.   B Hillary, D J Ewins.   "The use of strain gauges in force determination and frequency response function measurements". Proceedings of the 2nd International Modal Analysis Conference (IMAC), Orlando, Florida, 1984.

4.   C S Hallett.   "Force predictions in a torsional system from limited /remote response data". Msc Dissertation. Royal Naval   Engineering   College.   Internal   Report   No. RNEC-TR-86006.

5.   S J Lloyd. "An assessment of modal analysis techniques for force deterimation in open-coupled systems".   Msc Dissertation. Royal Naval Engineering College, July 1987.

6.   B J Dobson, D G Dubowski. "Computation of excitation forces using structural response data".   Paper presented at

the 56th Shock and Vibration Symposium, Monterey, cailfornia, Oct 1985.

7.   R Randall.   "Modal analysis of a cylindrical structure Immersed in water". Proceedings of the 3rd IMAC, Florida, 1985.

8.   C P Ratcliffe.   "Dynamic structural modelling for time domain analysis". Ph.D thesis, Sep 1985. RNEC.

9.   K F Graff. "Wave motion in elastic solids". 1975, Oxford University Press.

10.   W F Tsang, E Rider. "The technique of extraction of structural parameters from experimental forced vibration data". Proceedings of the 7th IMAC , Las Vegas, 1989.

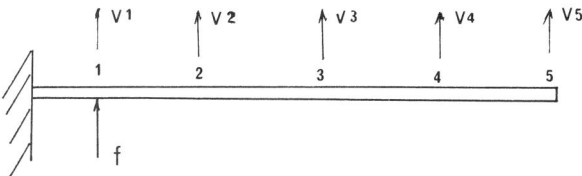

Figure 1. SCHEMATIC DRAWING OF THE BEAM STRUCTURE SHOWING MEASUREMENT AND FORCE LOCATIONS.

[M] model A1	$\begin{pmatrix} .16634 & .054195 & -.025742 & -.050307 & -.060382 \\ .054195 & .30393 & .10499 & .35592 & .24276 \\ -.025742 & .10499 & .25705 & .32474 & .28282 \\ -.050307 & .35592 & .32474 & 1.1263 & .8047 \\ -.060382 & .24276 & .28282 & .8047 & .7437 \end{pmatrix}$
[K] model A1	$\begin{pmatrix} 422070 & -294020 & 687.37 & -242420 & -181570 \\ -294020 & 441490 & -284350 & 192390 & -7060.1 \\ 687.37 & -284350 & 447770 & -119050 & 201680 \\ -242420 & 192390 & -119050 & 255330 & 35948 \\ -181570 & -7060.1 & 201680 & 35948 & 175640 \end{pmatrix}$
[D] model S2	$\begin{pmatrix} -1.6631E\text{-}6 & -7.1977E\text{-}6 & -1.9666E\text{-}5 & -5.0638E\text{-}6 & 2.7124E\text{-}5 \\ -7.1977E\text{-}6 & -7.6863E\text{-}5 & -4.4963E\text{-}5 & -.00010417 & .00035209 \\ -1.9666E\text{-}5 & -4.4963E\text{-}5 & 7.4086E\text{-}5 & -3.021E\text{-}5 & .00030839 \\ -5.0638E\text{-}6 & -.00010417 & -3.021E\text{-}5 & -.00014835 & .00074134 \\ 2.7124E\text{-}5 & .00035209 & .00030839 & .00074134 & -.0017914 \end{pmatrix}$
[E] model S2	$\begin{pmatrix} 81.86 & -2.6442 & 25.679 & -2.0403 & 49.047 \\ -2.6442 & 65.328 & -104.12 & 70.062 & -290.71 \\ 25.679 & -104.12 & 154.82 & -107.76 & 264.87 \\ -2.0403 & 70.062 & -107.76 & 46.548 & 180.28 \\ 49.047 & -290.71 & 264.87 & 180.28 & -1163.1 \end{pmatrix}$

Table 1 : TABLE OF THE DERIVED SYSTEM MATRICES OF MODELS A1 AND S2.

Figures 2 - 5. OVERLAID PLOTS OF THE MEASURED AND THE PREDICTED SPECTRA.
where R_{ij} : Receptance FRF plots of the response (displacement) at i and force at j.
S_{ij} : Strain FRF plots of response (strain) at i and force at j.
F_j : Auto-spectra of force at j.

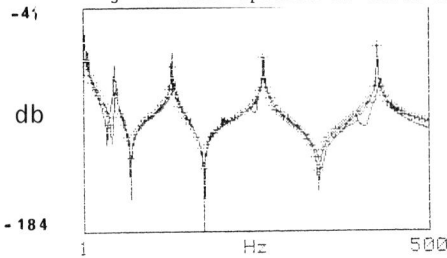

Fig. 2a : R_{11} (model A1).

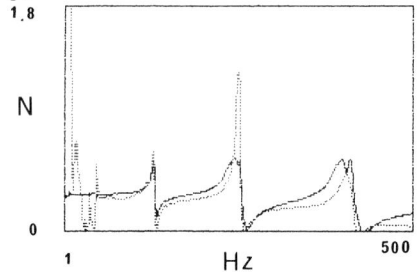

Fig. 2b : F_1 (model A1).

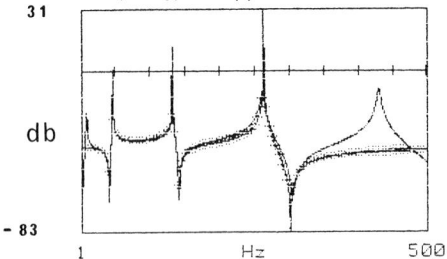

Fig. 3a : S_{11} (model S1).

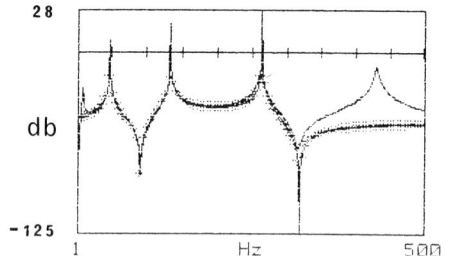

Fig. 3b : S_{21} (model S1).

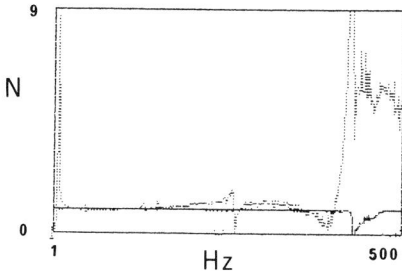

Fig. 3c : F_1 (model S1).

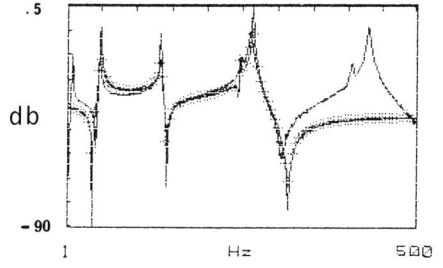

Fig. 4a : S_{11} (model S2).

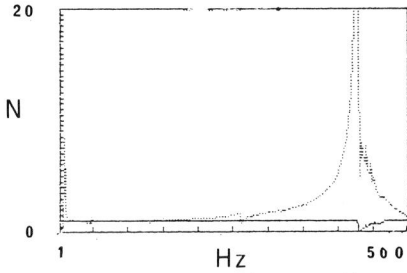

Fig. 4b : F_1 (model S2).

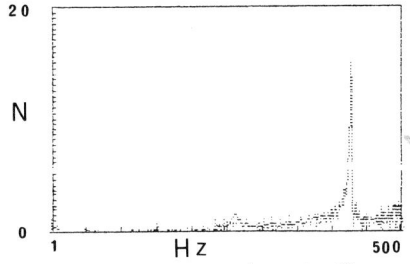

Fig. 4c : F_2 (model S2).

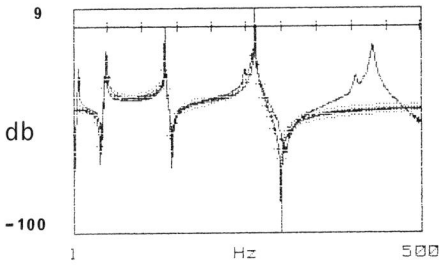

Fig. 5a : S_{11} (model S3).

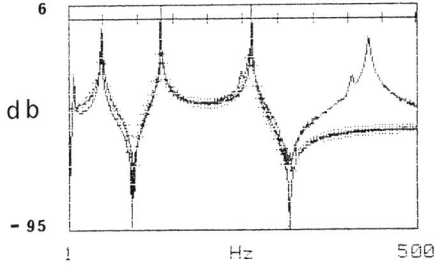

Fig. 5b : S_{21} (model S3).

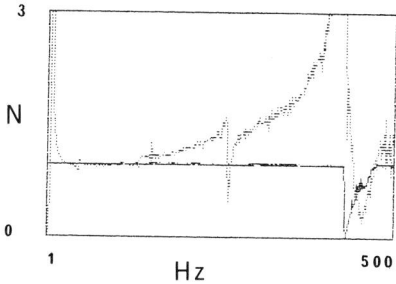

Fig. 5c : F_1 (model S3).

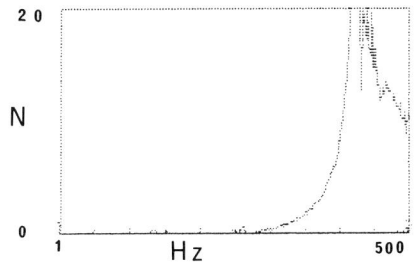

Fig. 5d : F_3 (model S3).

Vibratory Bowl Feeder Design using Numerical Modelling Techniques

D. MORREY* and J. E. MOTTERSHEAD**

*School of Engineering, Oxford Polytechnic,
Headington, Oxford OX3 0BP, UK
**Department of Mechanical Engineering, University of Liverpool,
P.O. Box 147, Liverpool L69 3BX, UK

ABSTRACT

This paper describes how a numerical model developed and running on a
mainframe system has been modified and adapted to run on a spreadsheet
package. The earlier model considered eight degrees-of-freedom of the
feeder structure, and enabled predictions of the dynamic characteristics of
bowl feeders with a range of geometric parameters to be made. The later
version of the model, which runs on an Apple Macintosh, treats the structure
as a four degree-of-freedom system, allowing for vertical, rotational and
tipping movements of the bowl, but neglecting the motion of the base. In
addition to calculating the dynamic characteristics, the package enables the
aceleration of the bowl to be determined in response to a particular forcing
function. Earlier work by other authors has shown that there is a direct
relationship between bowl acceleration and component feed-rate.

The model uses many of the standard features available on a spreadsheet-
graphics-database package in order to investigate changes in geometric
parameters and to find optimum values that result in the maximum feed-rate
obtainable. A programme of experimental work has been undertaken in order to
verify these values.

The development needs of Bowl Feeder Manufacturers are considered,and
further work being undertaken to enable the package to be used as a
'portable', user-friendly design tool allowing development engineers to
investigate the performance of different configurations of feeders is
described.

KEYWORDS

Vibratory Bowl Feeders; Mechanical Handling; Automation; Mechanical
Vibrations; Spreadsheet Packages.

211

INTRODUCTION

Vibratory bowl feeders are widely used in automated processes where high-volume handling and orientation of components is required. The use of dedicated tooling within a 'blank' bowl feeder enables identical parts to be oriented and delivered at a steady rate for presentation to a workstation or further tooling.

The basic structure of the feeder as illustrated in Fig.1 consists of a bowl separated by sets of inclined springs from the base. The driving force is provided by an electromagnetic coil mounted between the bowl and the base.

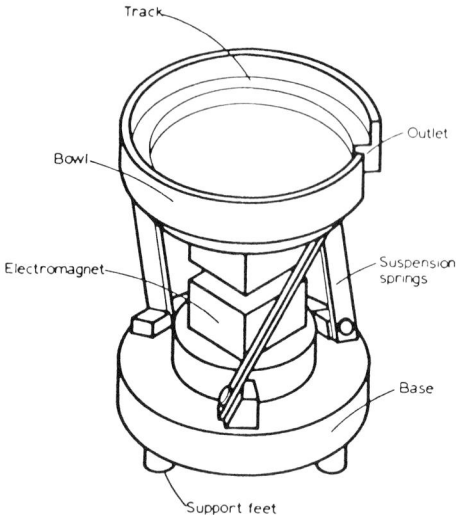

Track

Outlet

Bowl

Electromagnet

Suspension springs

Base

Support feet

Fig.1. Vibratory Bowl Feeder Structure

The oscillatory motion of the bowl causes the components to move along the spiral delivery track inside the bowl with a 'hopping' motion. Tooling is fixed at suitable positions along the track in order to reject those parts which have an incorrect orientation, or may be defective or out-of-tolerance.

The design of vibratory bowl feeders has been directed at producing smooth-running, quiet feeders which give a uniform flow of components whilst minimising the tendency to 'jam' during operation. Other important factors are the power consumption of the feeder, and the ease of tuning of each individual feeder. In addition, the exit velocity or feed rate of components is a critical parameter in the design of automated manufacturing systems; it is necessary that bowl feeders can match the operating speeds of the wide range of machinery which they may be used in conjunction with. It is therefore desirable that a working model of the bowl feeder structure is available to enable the behaviour and performance of the feeder structure to be predicted during the development process.

The work described here has been aimed at developing a numerical model of
the feeder structure which can be used as a design tool. This is based on
the Microsoft EXCEL Spreadsheet package running on an APPLE MACINTOSH
microcomputer. It uses standard database and graphics facilities in order
to investigate the variation of the geometric parameters of the feeder.
The final stage of this project is the addition of 'menus' and 'dialog-
boxes' to the package in order to make it more 'user-friendly'.

EARLIER WORK

An eight degree-of-freedom model of the bowl feeder structure has been
developed by the authors (Morrey and Mottershead, 1987). This takes into
account vertical motion in the z-direction, rotation about the z-axis, and
tipping about the x- and y-axes of both the bowl and the base, as shown in
Fig.2. It has been coded in FORTRAN and uses NAg Library Subroutines to
calculate the eigenvalues and eigenvectors of the model. The mode shapes
are plotted using GHOST graphics routines.

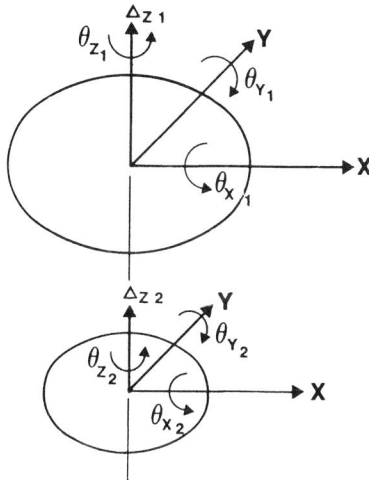

Fig.2. Eight Degree-of-Freedom Bowl Feeder Model

The natural frequencies and mode shapes were verified by a programme of
experimental modal analysis undertaken by suspending and exciting the feeder
as a free structure, and processing data from a dual-channel spectrum
analyser with a microcomputer-based system. There was good correlation
between the two sets of results at lower frequencies in the operating range
of the feeder.

DEVELOPMENT NEEDS OF THE MANUFACTURERS.

Having developed a model which described the dynamic behaviour of the
feeder, and which enabled a better understanding of its operation, it was
then necessary to ask some quite fundamental questions about the future
direction of this particular project. In order to do this, it was decided
that the needs of the development engineers working in the field should be

addressed. Their objectives for the product can be summarised as four main points:

(a) an increase in the feed-rate of components is desirable. There is a limit to this however, since very high feed-rates can cause jamming in the tooling, and components cannot be presented at workstations at a rate faster than that which they can be handled.

(b) a reduction of the power input and runing costs is desirable for both the end-users and the manufacturers.

(c) a quieter and smoother running feeder is desirable in view of noise legislation and the fact that this further reduces susceptibility to jamming.

(d) a feeder which is easier to tune i.e.the matching of spring sets to a particular bowl and base. This can often represent one of the largest costs of manufacture, particularly when a 'troublesome' bowl exhibits 'dead spots', where components remain stationary or feed backwards.

Although it has not been possible to address all of these objectives in the next stage of the project, their consideration has been essential in order to focus the direction of the programme of work. The objectives were therefore:

(a) optimisation of the geometric parameters of the bowl feeder structure in order to achieve an increase in feed-rate, smoother, quieter running and lower power consumption.

(b) the production of a 'portable', user-friendly design tool which would enable development engineers to investigate the performance characteristics of alternative structures and arrangements.

THE SPREADSHEET MODEL

Having developed an eight degree-of-freedom model which described the dynamic behaviour of both the bowl and the base, it was decided to adapt this for use in a spreadsheet package which could run on a microcomputer. This offered the following advantages:

(a) It would allow the package to be easily transferred for use by engineers involved in the design and development of bowl feeders.

(b) A spreadsheet package allows calculations of the'what-if?'type, where parameters are varied over a certain range, to be easily and readily performed.

From the work carried out on the eight degree-of -freedom model, it was decided that the model used in the spreadsheet package would be constrained to a four degree-of-freedom system, allowing for vertical motion, rotation about a vertical axis and in-plane tipping of just the bowl. Since it was the motion of the bowl which provided the driving force for the components, and that every effort was made with feeders operating in the field to prevent any motion of the base by rigidly clamping the structure to a supporting stand, it was felt that it was valid to neglect the movement of the base.

Because of the user-friendly facilities offered by the Apple Macintosh, and its ease of use, it was decide to use the Microsoft Excel spreadsheet package which runs on the Macintosh. This offers a wide range of graphics and database facilities, but also allows high-level programs to be written for use within the spreadsheets using 'Macros', and special-purpose menus and dialog-boxes to be added to enhance a package for a particular requirement.

At this stage of the project, the package has three main elements:

(a) A Basic 'Worksheet' which calculates the dynamic characteristics of the model given the input parameters of the feeder structure.

(b) High-level 'Macros' have been written in order to calculate the acceleration of the bowl using the technique of modal superposition. This has assumed a forcing function of a pure sine wave, which was that used in the programme of experimental work described in the next section.

Previous work undertaken by Redford and Boothroyd (1967-68) shows that bowl acceleration is directly related to feed-rate, and therefore maximising the acceleration maximises the feed-rate.

(c) The 'Table' function of the package allows the input parameters of the worksheet to be varied over a selected range in order to investigate how these affect acceleration values. The results of these calculations are given in the next section.

EXPERIMENTAL WORK

In order to verify the results from the spreadsheet model, a programme of experimental work has been undertaken. In this, the electromagnetic coil of the feeder was replaced by an electromagnetic vibrator. This was connected to the bowl of the feeder via a drive rod which incorporated a thrust bearing allowing both rotation and the vertical motion of the bowl. A piezoelectric force transducer was placed in the drive rod directly below the bowl in order to measure the input. The control to the vibrator allowed a range of frequencies to be input, and the resulting acceleration of the bowl was measured using an accelerometer. Initial measurements were taken at various locations across the bowl to ensure that the structure had been properly 'tuned', and that the acceleration was constant across the bowl.

The first programme of experimentation was designed to verify the modal superposition model. The input force was varied over a range of 350N to 650N, representing the maximum force input to a medium-sized feeder. This was repeated over a range of frequencies around the normal operating frequency of 50Hz. The results obtained from this and a comparison with the predicted values are shown in Fig.3.

The second programme of experimental work was designed to verify the use of the spreadsheet package in investigating the variation of the geometric parameters of a feeder structure. 'Blocks' were manufactured to allow the spring banks of the feeder to be set at different angles; this was varied in the range of 55 to 75 degrees at 5 degree intervals. The input force was maintained at 650N. Graphs comparing experimental and predicted results at 45 and 50Hz are given in Fig.4 and Fig.5 respectively.

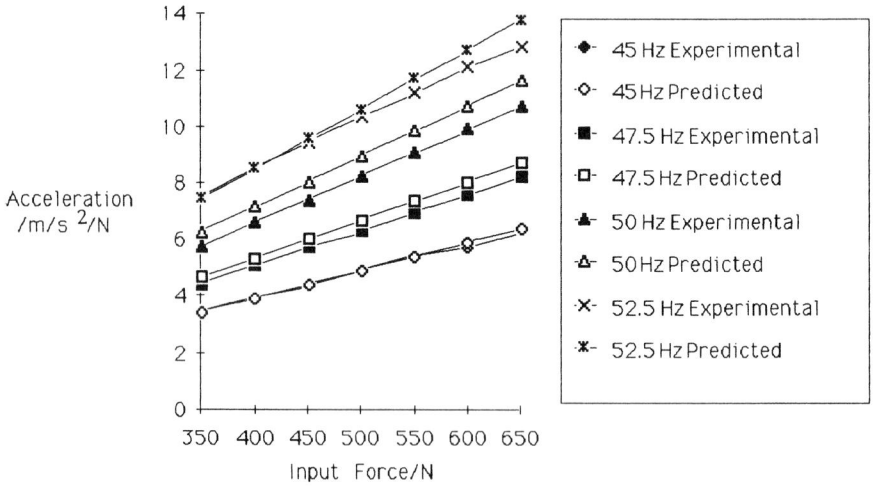

Fig.3 Graph of bowl acceleration against input force.
(Spring angle = 65°)

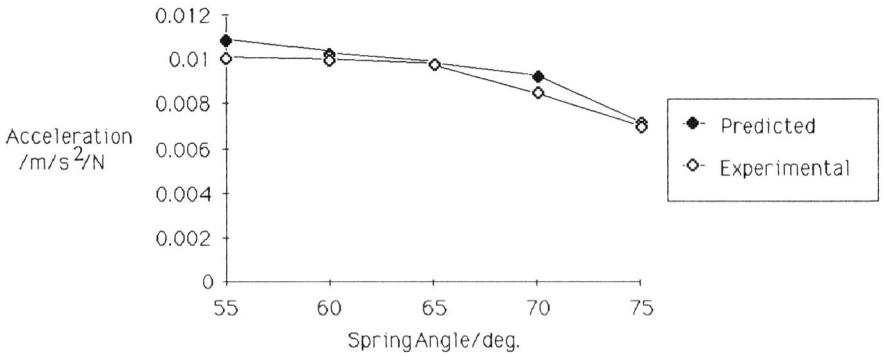

Fig.4 Graph of bowl acceleration against spring angle at 45 Hz.

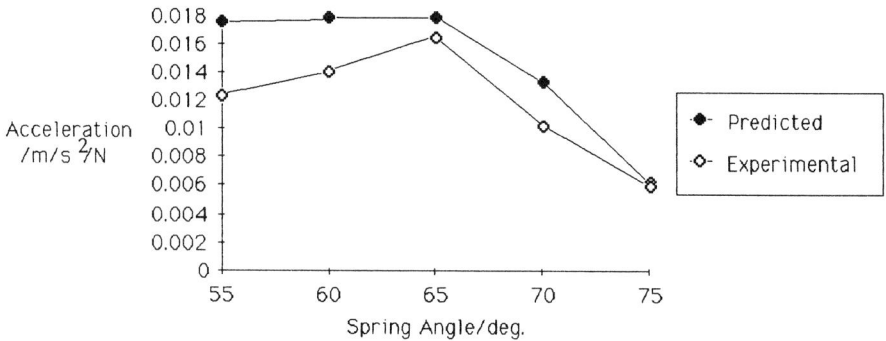

Fig.5 Graph of bowl acceleration against spring angle at 50 Hz.

217

DEVELOPMENT OF A DESIGN TOOL

Work has already been undertaken to add 'menus' and'dialog-boxes' to the
spreadsheet package so that it can easily be used by development engineers
working in the field to investigate the performance characteristics of
alternative arrangements of bowl feeder structures. The resulting package
will prompt the user through the required input values, and calculate
dynamic characteristics and bowl acceleration values. When used in
conjunction with the trends predicted by the spreadsheet package described
earlier, this will enable design and development work in this area to be
approached in a more systematic manner.

REFERENCES

Morrey,D. and J.E.Mottershead (1987). Modelling of Vibratory Bowl Feeders.
Proc. Inst. Mech. Engrs.: Part C, 200, 431-437.

Redford,A.H. and G. Boothroyd (1967-68). Vibratory Feeding. Proc. Inst.
Mech. Engrs., 182, 135-146

Non-linear Modeling of Structures undergoing Free Vibrations

A. F. MOUSTAFA* and M. A. EL-GEBEILY**

*Department of Mechanical Engineering,
Jordan University of Science and Technology, Irbid-Jordan
**Department of Mathematics, King Fahd University,
Dhahran, P.O. Box 1777, Jordan

ABSTRACT

In this work the modeling problem of a structure supported by a soil medium exhibiting nonlinear behaviour is considered. The upperstructure is represented by a linear model with translational degrees of freedom. Static condensation is used to eliminate the rotational degrees of freedom. The model of the interaction allows for soil nonlinearity which is represented by nonlinear springs and dampers connecting the upperstructure foundation to the ground. Free vibration characteristics of the structure interaction model is analysed. A quantitative investigation of the effect of the nonlinear behaviour of the soil on the natural frequencies of the structure model is presented

KEYWORDS

Modeling; Interaction; Nonlinearity; Perturbation; Free Vibration.

INTRODUCTION

The modeling problem of structures undergoing free vibrations and supported on soil medium exhibiting nonlinear behaviour is considered. The upperstructure is modeled by lumping its mass at a number of nodes. The degrees of freedom considered are the horizontal translational displacements at each of the nodes. The resulting mass matrix is diagonal with zero diagonal elements for the rotational degrees of freedom. The rotational degrees of freedom are eliminated from the stiffness matrix by static condensation. The soil supporting the structure is represented by nonlinear springs and dampers connecting the foundation to the ground.

A quantitative analysis of the effects of the nonlinearities of the interaction on the natural frequencies of the structure was recently carried out (Moustafa and El-Gebeily 1989). A brief review of the results of this analysis as applied to the problem at hand is given. It is shown that the natural frequencies are significantly changed. A method to calculate the modified frequencies is presented.

219

UPPERSTRUCTURE MODEL

The upperstructure is idealized as a planer beam with its mass being concentrated at a finite number of nodes at which the translational displacements
are defined. The mass to be lumped at each node is obtained by assuming
that the upperstructure is divided into segments where the nodes serve as
connecting points. The considered degrees of freedom are the translational
displacements, x_i, and the rotational displacements, θ_i, at each of the
considered n-structural nodes. The mass concentrated at the ith node is
denoted by m_i, and its rotational inertia is considered to be so small that
it can be neglected. The mass matrix of the upperstructure, M_S, is, thus,
diagonal and has zero diagonal elements for the rotational degrees of
freedom.

The stiffness matrix, K_S, on the other hand, is found by appropriately super-
posing the stiffness of each of the idealized segments. The resulting
stiffness matrix is symmetric and has coupling terms between the translational and rotational degrees of freedom. The damping matrix, C_S, could be
obtained by a superposition process similar to that of the stiffness matrix.
But since this method is impractical in practice (Clough and Penzien 1983),
the damping matrix will be assumed to be of the classical type. Classical
damping makes modal decoupling possible and, thus, structural analysis
becomes easier. The classical damping matrix is usually made up as a linear
combination of the mass and stiffness matrices as follows,

$$C_s = b_1 M_s + b_2 K_s$$

where b_1 and b_2 are constants.

Assuming that the damping forces associated with the rotational degrees of
freedom can be neglected, the equation of motion of the upperstructure, in
the absence of external forces, can be written as,

$$M_s \ddot{v} + C_s \dot{v} + K_s v = 0 \qquad (1)$$

where

$$M_s = \{m_1, m_2, \ldots, m_n, 0, 0, \ldots, 0\}$$

$$= \left[\begin{array}{c|c} M & 0 \\ \hline 0 & 0 \end{array} \right]$$

$$C_s = \left[\begin{array}{c|c} C & 0 \\ \hline 0 & 0 \end{array} \right]$$

$$K_s = \left[\begin{array}{c|c} K_{xx} & K_{x\theta} \\ \hline K_{x\theta} & K_{\theta\theta} \end{array} \right]$$

$$v^T = [x^T, \ \theta^T]$$

$$x^T = [x_1, x_2, \ldots, x_n]$$

$$\theta^T = [\theta_1, \theta_2, \ldots, \theta_n]$$

In the above, the notation $\{a_1, a_2, \ldots,\}$ is used to denote a diagonal matrix with diagonal elements $a_1, a_2, \ldots$, and T denotes the transpose operation.

Since the rotational degrees of freedom were not considered in the mass and damping matrices, it is necessary to exclude them from the stiffness matrix. This is to be carried out by a static condensation procedure as follows.

The lower half of equation (1) can be written as,

$$K_{\theta\theta} \theta + K_{x\theta} x = 0$$

from which the rotational displacements can be expressed in terms of the translation displacements as,

$$\theta = - K_{\theta\theta}^{-1} K_{x\theta} x \tag{2}$$

Substituting equation (2) into the upper half of equation (1), leads to

$$M\ddot{x} + C\dot{x} + Kx = 0 \tag{3}$$

where

$$K = K_{xx} - K_{x\theta} K_{\theta\theta}^{-1} K_{x\theta}$$

Equation (3) represents the n-degrees of freedom model of the upperstructure, with the rotational degrees of freedom being eliminated.

INTERACTION MODEL

The upperstructure foundation is assumed to be a rigid circular disk of mass m_F and centroidal moment of inertia I_F. The foundation is supported by the ground through soil medium exhibiting nonlinear behaviour. The soil nonlinearity is represented by nonlinear springs and dampers connecting the foundation to the ground. The equivalent stiffness and damping coefficients of the translational displacement are denoted, respectively, by k_x and c_x; and those of the rotational displacement by k_θ and c_θ. The degrees of freedom considered for the foundation movements are its lateral translation x_F, and rotation, θ_F.

STRUCTURE-INTERACTION MODEL

In order to simplify presenting the main ideas of this work, cubic-type non-linearity of the interaction will only be considered. A more general analysis can be found in (Moustafa and El-Gebeily 1989). For this case, we can write

$$k_x = k_{xo} (1 + f^2), \qquad c_x = c_{xo} (1+(3/8)f^2)$$

$$k_\theta = k_{\theta o} (1+f^2), \qquad c_\theta = c_{\theta o} (1+(3/8)f^2) \tag{4}$$

where

$$f = a_\theta \theta_F - a_x x_F$$

and k_{xo}, $k_{\theta o}$, c_{xo} and $c_{\theta o}$ represent the linear stiffness and damping properties of the interaction; a_x and a_θ are constants depending on the soil properties and foundation size.

Defining the displacement vector

$$y^T = [x_1, x_2, \ldots, x_n, x_F, \theta_F]$$

the equation of free vibrations can be derived as

$$M_o \ddot{y} + C_o \dot{y} + K_o y + (C_1 \dot{y} + K_1 y) = 0 \tag{5}$$

where M_o, C_o and K_o are the mass, damping and stiffness matrices of the entire structure-interaction model, respectively. They can be written as,

$$M_o = \left[\begin{array}{c|cc} M & & 0 \\ \hline & m_F & \\ 0 & & I_F \end{array} \right]$$

$$C_o = \left[\begin{array}{cc} C & C_{SF} \\ C_{SF}^T & C_{FF} \end{array} \right]$$

$$K_o = \left[\begin{array}{cc} K & K_{SF} \\ K_{SF}^T & K_{FF} \end{array} \right]$$

where C_{SF}, C_{FF}, K_{SF} and K_{FF} are matrices with appropriate dimensions. Matrices C_1 and K_1 represent the damping and stiffness effects due to soil nonlinearity. They can be derived as

$$K_1 = f^2 \{0, 0, \ldots, k_{xo}, k_{\theta o}\}$$

$$C_1 = (3/8)f^2 \{0, 0, \ldots, c_{xo}, c_{\theta o}\}$$

Let the modal matrix of the linear part of the structure-interaction system be denoted by U. Assuming classical damping and using the transformation

$$y = Uz$$

equation (5) can be written as

$$\ddot{z} + \{w_k\}^2 = -\{2\xi_k w_k\} \dot{z} - U^{-1} M_o^{-1} [C_1 U\dot{z} + K_1 Uz] \tag{6}$$

where ξ_k and w_k are the modal damping ratio and the natural frequency of the k th mode, respectively. Equation (6) can be written as the following set of nonlinear scalar equations (for $k = 1, 2, \ldots, n+2$)

$$\ddot{z}_k + w_k^2 z_k = -2\xi_k w_k \dot{z}_k - z^T Dz \phi_k^T z + f(z, \dot{z}) \tag{7}$$

The nonlinear function $f(z, \dot{z})$ can be shown to contain terms of order greater than 2 of the small parameter ε to be introduced later.

where z_k is the k th element of z, and

$$D = a_x^2 [u_{n+1} \ u_{n+1}^T] + a_\theta^2 [u_{n+2} \ u_{n+2}^T]$$

In the above, ϕ_k^T is the k th row of the matrix ϕ defined by

$$\phi = w_x^2 [\nu_{n+1} \ u_{n+1}^T] + w_\theta^2 [\nu_{n+2} \ u_{n+2}^T]$$

$$w_x^2 = k_{xo}/m_F, \ w_\theta^2 = k_{\theta o}/I_F$$

The above set of nonlinear equations (7) represents the nonlinear model of the structure-interaction system. These equations were analyzed by using perturbation theory (Nayfeh 1983, Nayfeh and Mook 1979). The structural natural frequencies were found to be affected as expected (Spidose and Skjastad 1986) by the model nonlinearity. The results will be presented in the next section.

NONLINEARITIES EFFECT ON FREQUENCIES

In applying the perturbation method of multiple-scales, the solution of equation (7) was represented in power series of ε as

$$z_k(t, \varepsilon) = \varepsilon z_{k1} + \varepsilon^2 z_{k2} + ... \tag{8}$$

where ε is a small positive parameter. Using equations (8) in equations (7), the solution up to a first order approximation of ε is

$$z_{k1} = g_k \exp(-\xi_k w_k t) \cos(w_k^* t + \phi_k) \tag{9}$$

where g_k and ϕ_k are constants of integrations depending on the initial conditions. The damped natural frequency of the k th mode was obtained as

$$w_{kd}^* = w_k [1 + \frac{\varepsilon^2}{w_k^2} \sum_j \frac{g_j^2 \beta_{kj} \exp(-2\xi_j w_j t)}{2 \xi_j w_j t}] \tag{10}$$

where

$$\beta_{kj} \begin{cases} (3/2) h_{kkkk} & j=k \\ h_{kjjk} + h_{kjkj} + h_{kkjj} & j \neq k \end{cases}$$

Note that the last term of equation (7) was expressed as

$$z^T Dz \ \phi_k^T z = \sum_{j,m,p} h_{kjmp} z_j z_m z_p$$

where the summation runs from 1 to n+2. Equations (9) and (10) reveal that the modal natural frequencies are dependent on time and amplitudes of oscillations. For small damping, $\xi_k \ll 1$, one can obtain from equation (10).

$$w_k^* \cong w_k [1 - \frac{\varepsilon^2}{w_k^2} \sum_j g_j^2 \beta_{kj}]$$

which gives the modified modal natural frequencies.

CONCLUSIONS

The modeling problem of a structure supported on a soil medium with non-linear stiffness and damping characteristics was reviewed. The upper-structure was modeled as a planer beam with lumped masses at a finite number of nodes. Static condensation was used to eliminate the rotational degrees of freedom. Cubic-type nonlinearity of the damping and stiffness properties of the interaction was considered. The entire structure-interaction model was analyzed to study the effect of nonlinearities on modal frequencies. A method was presented to compute the modified frequencies for lightly damped structures.

REFERENCES

Moustafa, K.A., and El-Gebeily, M. (1989). Free Vibration Characteristics of Structures with Nonlinear Underlying Soil, Submitted for publication to *Trans. of the* ASME *Journal of Vibrations, Acoustics, Stress, and Reliability in Design.*

Nayfeh, A.H. (1981). *Introduction to Perturbation Techniques.* John Wiley & Sons.

Nayfeh, A.H., and Mook., T.D. (1979). *Nonlinear Oscillations.* John Willey & Sons.

Spidsoe, N., and Skjastad, O. (1986). Measured Soil-Structure Interaction Properties of a Gravity Platform. *Proc. 18th Offshore Technology Conference,* Houston, Paper OTC 5222.

Derivation and Application of the Choleski Decomposition for the Positive Semi-definite Matrices used in Structural Dynamics Reanalysis

J. A. BRANDON

School of Engineering, University of Wales,
College of Cardiff, Newport Road, Cardiff CF2 1XH, UK

ABSTRACT

The paper describes the derivation and computation of a Choleski decomposition of a positive semi-definite matrix. The particular applications of the factorisation are described in receptance reanalysis.

KEYWORDS

Structural Dynamics; Choleski Decomposition; Semi-Definite Matrices.

INTRODUCTION

This algorithm was devised by the author approximately six years ago (Brandon (1983)), but was not published then in its own right. Subsequent events have suggested that the algorithm is more important than originally thought. For example, the reviewer of a recent paper commented (late 1987): "It appears that the singular Choleski algorithm is new and useful, but its coverage is not prominent in the paper".

The algorithm underlies the analysis presented in a number of papers published by the author and colleagues (Sadeghipour et al(1985), Brandon (1984a,b 1987) and Brandon et al (1988)). It extends the classic Choleski factorisation (see for example Wilkinson and Reinsch (1972)) from positive definite systems to positive semi-definite systems (ie with all eigenvalues positive or zero). These are particularly important in structural dynamics reanalysis, since they are essential to describe the modification matrices. A detailed numerical analysis of the Choleski decomposition for semi definite matrices has been given by Higham (1986, 1987, 1988).

The paper will outline the derivation of the factorisation and identify its importance in the area of reanalysis methods, both exact and differential (ie primarily sensitivity analysis).

225

THE MATRIX MODEL OF A STRUCTURE

The dynamic equations of a linear structure are usually expressed in the form

$$\underline{M}\,\underline{\ddot{x}} + \underline{C}\,\underline{\dot{x}} + \underline{K}\,\underline{x} = \underline{f}(t) \tag{1}$$

Where $\underline{M}$, $\underline{C}$ and $\underline{K}$ are the (nxn square, symmetric, constant) mass, viscous damping and stiffness matrices respectively; $\underline{x}$ is the vector of displacements at the n points representing the discrete structure; σ is the frequency; and $\underline{f}(t)$ the vector of external loads dependent on time t. For the vast majority of practical structures, where the damping is small, the replacement of hysteretic damping by a viscous damping approximation is widely regarded as being acceptable. A thorough treatment of the hysteretic damping model is provided by Bishop and Johnson (1960). Other additional terms of the system matrices are used in special applications, for example, to represent gyroscopic forces (an unsymmetric, frequency dependent matrix-see for example Meirovitch (1980)).

Modal Analysis is based on the treatment of the homogeneous form of (1) as an quadratic eigenvalue problem (using the trial solution $\underline{x} = \underline{\emptyset}\,e^{\sigma t}$):

$$(\sigma^2\,\underline{M} + \sigma\,\underline{C} + \underline{K})\,\underline{\emptyset} = \underline{0} \tag{2}$$

It is quite common (and extremely advantageous) to discount the effect of damping entirely (particularly in Finite Element Analysis) and solve the problem

$$(\sigma^2\,\underline{M} + \underline{K})\,\underline{\emptyset} = \underline{0} \tag{3}$$

If necessary damping is re-introduced using either proportional damping or a perturbation approach. Such an assumption enables analysis using only real arithmetic, whereas the inclusion of a general damping distribution not only demands the transition to complex arithmetic but also destroys advantageous numerical properties of the system matrices (in particular positive definiteness of the eigenvalue problem). The method described in the current paper is suitable for the exact incorporation of large localised damping distributions, for which proportional damping or perturbation approaches are unsuitable.

THE STANDARD CHOLESKI ALGORITHM

The eigenvalue problem equation (3) is a case of the generalised eigenvalue problem. The solution to generalised eigenvalue problems is most commonly approached by using a coordinate transformation to a standard eigenvalue problem ($\underline{X} - \alpha\,\underline{I})\,\underline{\theta} = \underline{0}$.

Unless the dynamic problem is poorly defined $\underline{M}$ will be positive definite, ie it is non-singular and more importantly all of its own eigenvalues are positive, and $\underline{K}$ will be at least positive semi-definite, ie, although it may be singular (if its coordinates are not sufficiently constrained), with at least one zero eigenvalue, all its non-zero eigenvalues are positive. In addition, for passive systems, $\underline{C}$ will also be (at least) positive semi-definite.

The generalised eigenvalue problem given by equation (3) can be transformed to a standard eigenvalue problem simply by pre-multiplying by the inverse of $\underline{M}$ to give

$$(\sigma^2 \ \underline{I} \ + \ \underline{M}^{-1} \ \underline{K}) \ \underline{\emptyset} \ = \ \underline{0} \qquad (4).$$

This transformation has the advantage that the eigenvectors are preserved, but the symmetry of the matrices is destroyed. It is desirable to retain symmetries of matrices if possible, not only to economise on storage (a symmetric matrix requires $n(n+1)/2$ rather than n^2) but also to take advantage of attractive numerical properties of real symmetric matrices. A common strategy uses the Choleski decomposition of the mass matrix.

A real, symmetric, positive-definite matrix can be factorised into the product of two triangular matrices (sometimes called Choleski square roots) which are transposes of each other (see Martin et al (1972))

$$\underline{M} \ = \ \underline{L} \ \underline{L}^T \qquad (5)$$

This factorisation is closely related to, and often used as a numerical improvement of, the common Gaussian elimination method for equation solution.

The transformation using the Choleski decomposition requires a change in the coordinate system. The coordinate system denoted by the vector $\underline{x}$ must be replaced by a system $\underline{y}$ such that

$$\underline{y} = \underline{L}^T \ \underline{x} \qquad (6)$$
or
$$\underline{x} = \underline{L}^{-T} \ \underline{y} \qquad (7)$$

The stiffness matrix becomes

$$\underline{K} = \underline{L} \ \underline{X} \ \underline{L}^T \qquad (8)$$
or
$$\underline{X} = \underline{L}^{-1} \ \underline{K} \ \underline{L}^{-T} \qquad (9)$$

with $\underline{X}$ symmetric for symmetric $\underline{K}$.

This allows the replacement of the generalised eigenvalue problem (3) with the standard symmetic eigenvalue problem (4) with the coordinate relationship between the eigenvectors

$$\underline{\emptyset} = \ \underline{L}^{-T} \ \underline{\theta}$$
or
$$\underline{\theta} = \ \underline{L}^T \ \underline{\emptyset}$$

THE CHOLESKI ALGORITHM FOR SEMI-DEFINITE MATRICES

Semi-definite matrices arise when there are insufficient constraints on a system. A typical case in structural dynamics is the common experimental configuration when a structure is suspended freely in space (practically by extremely compliant fixtures). In this case the stiffness matrix is singular and solution of the eigenvalue problem will give as many zero eigenvalues as there are insufficient constraints and corresponding eigenvectors, described as rigid body modes, which represent synchronous motion of the entire structure without deformation.

The second context where semi-definite matrices occur in structural dynamics is in the analysis of design modifications. The modification is commonly modelled as a "patch", a secondary structure connecting points on the original structure. For example a stiffness tie of stiffness k_m, joining points a and b, would be modelled as a sparse matrix of the same order as the primary structure, with only four non-zero elements: aa and bb elements of magnitude k_m and ab and ba elements $-k_m$. It is possible to exploit the sparseness by simply storing the locations and values rather than the complete matrix. Possibly of more use in this context would be storage of the patch matrix as a product of a nxr matrix with its transpose (since the patch matrix will be symmetric), a strategy which requires a singular Choleski factorisation.

Setting aside the management of the sparseness, which is primarily a "housekeeping" task, the singular Choleski method is best illustrated using an example. Consider a patch described by the matrix $\underline{P}$:

$$\underline{P} \quad = \quad \begin{vmatrix} 1 & 1 & 2 \\ 1 & 1 & 2 \\ 2 & 2 & 5 \end{vmatrix}$$

As described above factorise $\underline{P}$ into triangular factors $\underline{L}$ and $\underline{L}^T$ where

$$\underline{L} \quad = \quad \begin{vmatrix} a & 0 & 0 \\ b & c & 0 \\ d & e & f \end{vmatrix}$$

and solve successively for a ... f. This gives a = 1, b = 1, d = 2, c = 0 but e is indeterminate and f is constrained by the choice of e (f = $\sqrt{(1 - e^2)}$). It can be shown that where the procedure finds a zero in the diagonal all subsequent terms in that column are arbitrary. In this case there is a significant advantage in assigning these terms to zero. (In practical factorisations the zero detector must allow for computation using discrete arithmetic).

In the case of system modification using a patch, the rank of the modification matrix will be known, a priori, and the detection of zero terms in the factorisation will be unnecessary.

SOLVING THE QUADRATIC EIGENVALUE PROBLEM-
DUNCAN'S METHOD

This method was analysed by Frazer, Duncan and Collar (1938) and, as the state space method, forms the basis of much of modern control theory. The method is based on replacing the nxn quadratic eigenvalue problem with a 2nx2n linear eigenproblem.

By adjoining the identity

$$\underline{M}\,\underline{\dot{x}} - \underline{M}\,\underline{\dot{x}} \equiv \underline{0}$$

and using the coordinate transformation

$$\underline{y} = \begin{pmatrix} \underline{\dot{x}} \\ \underline{x} \end{pmatrix}$$

equation (3) can be re-written

$$\begin{pmatrix} \underline{0} & \underline{M} \\ \underline{M} & \underline{C} \end{pmatrix} \underline{\dot{y}} + \begin{pmatrix} -\underline{M} & \underline{0} \\ \underline{0} & \underline{K} \end{pmatrix} \underline{y} = \begin{pmatrix} \underline{0} \\ \underline{f}(t) \end{pmatrix} \qquad (10)$$

This form of the Duncan transformation, and its corresponding eigenvalue problem, have been widely used in structural dynamics. It is occasionally advantageous, (see Brandon(1984b)), to use an alternative form of the Duncan formulation based on the identity

$$\underline{K}\,\underline{\dot{x}} - \underline{K}\,\underline{\dot{x}} \equiv \underline{0}$$

with $\underline{y}$ defined as before, to give

$$\begin{pmatrix} \underline{M} & \underline{0} \\ \underline{0} & -\underline{K} \end{pmatrix} \underline{\dot{y}} + \begin{pmatrix} \underline{C} & \underline{K} \\ \underline{K} & \underline{0} \end{pmatrix} \underline{y} = \begin{pmatrix} \underline{f}(t) \\ \underline{0} \end{pmatrix} \qquad (11)$$

At first sight it would appear that the eigenvalue problem corresponding to this equation (or equation (10)) is no more arduous than that of equation (3). This is however far from the truth. The Duncan form, and indeed any method involving general distributions of viscous damping, requires the use of complex arithmetic (for both eigenvalues and eigenvectors). In addition the advantageous numerical properties of the undamped case can no longer be guaranteed and the Choleski decomposition no longer applied (see Martin and Wilkinson (1972)). This is a consequence of the loss of positive definiteness of the two matrices which form the generalised eigenvalue problem. Indeed the existence of a full set of eigenvectors for general eigenvalue problems cannot always be demonstrated.

ORTHOGONALITY PROPERTIES OF THE EIGENVALUE PROBLEM

A central feature of the solution and use of eigenvalue problems is the orthogonality property of the eigenvectors, leading to the reforming of sets of linear equations in terms of the spectral decomposition of the system.

Taking equation (3) and ordering the eigenvalues such that $\sigma_1{}^2 \leq \sigma_2{}^2 \ldots \leq \sigma_n{}^2$ the corresponding eigenvectors may be used to construct the modal matrix $\underline{\Phi}$

$$\underline{\Phi} = \{\ \underline{\emptyset}_1\ \underline{\emptyset}_2\ ..\ \underline{\emptyset}_n\ \}$$

The orthogonality conditions may be expressed in matrix form

$$\underline{\Phi}^T \underline{M} \underline{\Phi} = \text{diag}\{ m_i \} = \underline{m} \qquad (12)$$

$$\underline{\Phi}^T \underline{K} \underline{\Phi} = \text{diag}\{ k_i \} = \underline{k} \qquad (13)$$

These matrices are often referred as the modal mass and modal stiffness matrices, although their dimensions are energy/time squared and energy repectively.

CONSTRUCTING RECEPTANCE REPRESENTATIONS

The modal representation, in terms of the eigenvalues and eigenvectors, is not particularly useful in isolation. The structural properties most often required are the force-response characterisics of the structure. In the current work the emphasis will be on the force-displacement transfer function, the receptance. Other common tranfer functions based on force-velocity and force-acceleration are discussed by Ewins (1984).

Taking the Laplace transform of equation (3) and neglecting initial conditions gives

$$\underline{S} \underline{x} = (s^2 \underline{M} + s \underline{C} + \underline{K}) \underline{x} = \underline{f}(s) \qquad (14)$$

the term in brackets is referred to as the dynamic stiffness matrix. Provided that this term is non-singular then $\underline{x}$ is given by

$$\underline{x} = (s^2 \underline{M} + s\underline{C} + \underline{K})^{-1} \underline{f}(s) \qquad (15)$$

The condition for non-singularity is that $s = \sigma_i$, ie the solution cannot be evaluated when the structure is excited at a resonance. The matrix $\underline{R} = \underline{S}^{-1} = (s^2 \underline{M} + s\underline{C} + \underline{K})^{-1}$ is known as the receptance matrix. The inversion implied in equation (15) is not practicable for realistic engineering structures and alternative methods have been derived to generate a sufficiently accurate approximation to the receptance matrix for practical purposes. The most common form is the spectral decomposition of the receptance matix in terms of eigenvectors and modal masses.

Considering initially the undamped system, then $\underline{R}$ is given by

$$\underline{R}_c = (s^2 \underline{M} + \underline{K})^{-1} \qquad (16)$$

where the subscript c denotes the equivalent conservative system. Using the modal decomposition of the mass and stiffness matrices, equations (14) and (15) respectively gives

$$\underline{R}_c = (\underline{\Phi}^{-T} (s^2 \underline{m} + \underline{k}) \underline{\Phi}^{-1})^{-1} \qquad (17)$$

The inversion of this matrix is extremely straightforward, involving a trivial inversion, giving

$$\underline{R} = \underline{\Phi} \, \text{diag} \, \frac{1}{(s^2 m_i + k_i)} \, \underline{\Phi}^T \qquad (18)$$

This is more commonly expressed as the summation of a series of vector products

$$\underline{R} = \sum{}^n \frac{\underline{\emptyset}_i \underline{\emptyset}_i{}^T}{(s^2 m_i + k_i)} \tag{19}$$

In many practical applications it is not practicable, and fortunately not necessary, to evaluate all terms in this series, since many will be negligible within the frequency range of interest.

RECEPTANCE MATRICES FOR GENERAL DAMPING DISTRIBUTIONS

The receptance matrix for structures with general damping distributions can be constructed from the Duncan formulation, equation (10). As before the receptance matrix may be constructed using the spectral decomposition of the solution to the eigenvalue problem. Defining matrices $\underline{A}$ and $\underline{B}$

$$\underline{A} = \begin{vmatrix} \underline{0} & \underline{M} \\ \underline{M} & \underline{C} \end{vmatrix} \qquad \underline{B} = \begin{vmatrix} -\underline{M} & \underline{0} \\ \underline{0} & \underline{K} \end{vmatrix}$$

and 2n eigenvectors $\underline{\tau}_i$ and eigenvalues σ_i, allows the diagonalisation of $\underline{A}$ and $\underline{B}$ (subject to the proviso that the generalised eigenvalue problem has a full system of eigenvectors). These eigenvectors and eigenvalues are either real or occur in complex conjugate pairs.

As before the eigenvectors diagonalise the system matrices

$$a_i = \underline{\tau}_i{}^T \underline{A} \underline{\tau}_i, \qquad b_i = \underline{\tau}_i{}^T \underline{B} \underline{\tau}_i$$

considering the construction of $\underline{\tau}_i$

$$\underline{\tau}_i = \begin{vmatrix} \sigma_i \underline{\emptyset}_i \\ \underline{\emptyset}_i \end{vmatrix}$$

leads to the following expressions for a_i and b_i

$$a_i = 2 \sigma_i \underline{\emptyset}_i{}^T \underline{M} \underline{\emptyset}_i + \underline{\emptyset}_i{}^T \underline{C} \underline{\emptyset}_i$$

$$b_i = - \sigma^2 \underline{\emptyset}_i{}^T \underline{M} \underline{\emptyset}_i + \underline{\emptyset}_i{}^T \underline{K} \underline{\emptyset}_i$$

giving the state vector $\underline{y}$ in the form

$$\underline{y} = \sum{}^{2n} \frac{\underline{\tau}_i \underline{\tau}_i{}^T}{(s\, a_i + b_i)} \begin{vmatrix} 0 \\ \underline{f}(s) \end{vmatrix} \tag{20}$$

The receptance relates the lower half of $\underline{y}$ to the force vector $\underline{f}(s)$, which may be achieved by an appropriate partioning of the vectors

$$\underline{x}(s) = \sum{}^{2n} \frac{\underline{\emptyset}_i \underline{\emptyset}_i{}^T}{(s\, a_i + b_i)} \underline{f}(s) \tag{21}$$

232

USE OF EXPERIMENTALLY DERIVED RECEPTANCE MATRICES

As has been mentioned, the receptance matrix contains data about the force-response characteristics of the structure. This has been derived above from the system properties but it may be experimentally derived. Of interest in a number of applications (see Brandon (1984a, 1987), Brandon et al (1988)) is the effect of a design change on the observed receptance (or comparison of a number of candidate modifications).

The method is identified by Noble (1969) as a special case of Kron's method which has been described extensively by Simpson, (see for example the survey, Simpson (1980)). In the literature the application of the Kron methodology has been applied overwhelmingly to eigenvalue problems and in particular to the assembly of sub-structures. The appication to receptance methods is not well known. The method described by Sadeghipour et al (1985) Brandon (1987) and Brandon et al (1988) will therefore be described here.

A modification $\delta \underline{S}$, of rank r, to the dynamic stiffness matrix $\underline{S}$, (which may constitute a combination of mass, stiffness or damping, multiplied by the appropriate power of the forcing frequency) may be factorised into singular Choleski factors $\underline{D}$ and $\underline{E}$ of order nxr and rxn respectively, ie $\delta \underline{S} = \underline{DE}$, ie:

$$\underline{S}^* = \underline{S} + \delta \underline{S} \qquad (22)$$

where * denotes the property of the modified system.

The effect on the receptance matrix can be constructed by the formula, attributed to Kron by Noble (1969):

$$\underline{R}^* = \underline{S}^{*-1} = (\underline{S} + \delta \underline{S})^{-1} = \underline{R} - \underline{R}\,\underline{D}\,\underline{G}\,\underline{E}\,\underline{R} \qquad (23)$$
where
$$\underline{G} = (\underline{I} + \underline{ERD})^{-1} \qquad (24)$$

In contrast to earlier methods, often described as pseudoforce methods, surveyed by Brandon et al (1988), this formulation allows computation using matrices of low order, with storage minimised, since $(\underline{I} + \underline{ERD})$ is symmetric and hence may be expressed in a Choleski form, although without the numerical benefits of the standard Choleski decomposition, since $\underline{R}$ is in general symmetric but complex. Sadeghipour et al (1985) discuss methods for manipulating the matrices using real arithmetic (based on separating $\underline{R}$ into real and imaginary parts).

CONCLUSIONS

The derivation and application of the Choleski decomposition for semi definite matrices has been described. The storage and algebraic requirements have significant advantages over pseudoforce techniques.

REFERENCES

Bishop, R.E.D. and Johnson, D.C. (1960). The Mechanics of Vibration, Cambridge University Press.

Brandon, J. A., (1983). Reanalysis methods for design and experimental models in structural dynamics, PhD Thesis, U Manchester

Brandon, J. A., (1984a). Limitations of an Exact Formula for Receptance Reanalysis, International Journal Numerical Methods in Engineering, 20, pp1575-80.

Brandon, J. A., (1984b). Discussion of Alternative Duncan Formulations of the Eigenproblem for the Solution of Nonclassically, Viscously Damped Linear Systems, Transactions of the American Society of Mechanical Engineers, Journal of Applied Mechanics, 51, pp904-6.

Brandon, J. A., (1987). Eliminating indirect analysis- the potential for receptance sensitivities, International Journal of Analytical and Experimental Modal Analysis, 2, pp73-75.

Brandon, J. A. (1988). On the robustness of algorithms for the computation of the pseudo-inverse for modal analysis, 6th International Modal Analysis Conference, Orlando, Fa

Brandon, J. A., Sadeghipour, K. and Cowley, A., (1988). Exact reanalysis techniques for predicting the effects of modification on the dynamic behaviour of structures, their potential and limitations, International Journal Machine Tool Design and Manufacture, 28, pp351-7

Ewins, D.J. (1984). Modal Analysis: Theory and Applications, Research Studies Press.

Frazer, R.A., Duncan, W.J. and Collar, A.R. (1938). Elementary Matrices, Cambridge University Press.

Higham, N. J., (1986), Computing a nearest symmetric positive semi definite matrix, Numerical Analysis Report 126, Department of Mathematics, University of Manchester

Higham, N. J., (1987), Analysis of the Choleski decomposition of a semi-definite matrix, Numerical Analysis Report 128, Department of Mathematics, University of Manchester

Higham, N. J., (1988), Matrix nearness problems and applications, Numerical Analysis Report 161, Department of Mathematics, University of Manchester

Martin, R.S., Peters, G. and Wilkinson, J.H., (1972). Symmetric Decomposition of a Positive Definite Matrix, in Wilkinson, J. H. and Reinsch, C. (Editors), Handbook for Automatic Computation, Vol 2: Linear Algebra, Springer-Verlag, pp9-30

Martin, R.S., and Wilkinson, J.H., (1972). Reduction of the Symmetric Eigenproblem, $A x = \lambda B x$ and Related Problems to Standard Form, in Wilkinson, J. H. and Reinsch, C. (Editors), Handbook for Automatic Computation, Vol 2: Linear Algebra, Springer-Verlag, pp303-314

Meirovitch, L. (1967). Analytical Methods in Vibrations, Macmillan

Meirovitch, L. (1980). Computational Methods in Structural Dynamics, Sijthoff & Noordhoff

Meirovitch, L. (1986). Elements of Vibration Analysis, McGraw-Hill.

Noble, B. (1969). Applied Linear Algebra, Prentice-Hall

Sadeghipour, K., Brandon, J. A. and Cowley, A., (1985). The receptance modification strategy of a complex vibrating system, International Journal of the Mechanical Sciences, 27, pp841-6.

Simpson, A. (1980). The Kron methodology and practical algorithms for eigenvalue, sensitivity and response analyses of large scale structural systems, Aeronautical Journal, 84, pp417-433.

Old Lamps for New!
A Photoelastic Design Tool for Weight
and Cost Saving on Aircraft Structures

E. W. O'BRIEN

*Structures Department, British Aerospace Commercial Aircraft Ltd.,
Airbus Division, Bristol BS99 7AR, UK*

ABSTRACT

Advances in the 'state of the art' in aircraft structures design
demand higher levels of confidence in the analysis techniques
employed by the designer to minimise weight and costs in order to
achieve the required performance .
In order to meet the challenge of this objective by determining
an 'all over stress distribution' the old technique of
Photoelasticity has been re-introduced and adapted as an addition
to the existing array of tools available to the designer .

KEYWORDS

Photoelasticity , Aircraft Design , Stress , Weight and Cost
saving .

PHOTOELASTIC DESIGN TOOL FOR COST & WEIGHT SAVING

The performance of commercial aircraft structures is very much
dependent on the weight of the structure required to meet the
exacting strength and stiffness criteria . The competitive nature
of the airline industry means that structural weight and hence
performance can be calibrated against costs , the current premium
is estimated at between $400 and $800 per kilo saved per aircraft
, depending on project and application .

Every effort must be made by the designer to use every available
means in achieving the industry objective of designing and
building successful aircraft that sell in sufficient numbers to
be profitable .

Sophisticated computer techniques such as finite element models
and boundary elements have permitted great advances in the design
of more efficient structures . However there are two significant
factors that have a bearing on component design :-

1) Fatigue life is often governed by local features which are difficult to model precisely by computer .

2) Weight saving by computer requires many expensive and time consuming iterations .

The object of our exercise is to overcome these design difficulties , each of which can be solved providing there is sufficient stress information to work on .

To achieve an 'all over' stress distribution in order to make the improved engineering judgements , we have re-introduced the old technique of Photoelasticity (Old Lamps for New !) and adapted the approach to use as a design tool .

Traditionally Photoelasticity gives a 'feed back' from the test phase of a project , here it is used in a 'feed forward' mode by bringing the analysis right into the design/stress department at an early stage in the design process . This provides sufficient stress data including fine local detail in a form that can easily be modified for strength and weight optimisation .

SHORT REVIEW OF PHOTOELASTICITY - an optical strain measurement technique .

Sir Isaac Newton (1642 - 1722) **Fig.1**

Newtons experiments in the late 1600's were so fundamental that they greatly extended the frontiers of scientific knowledge - this was particularly true in the understanding of the theory of light and optics .
Many university professors and teachers actually repeated Newtons experiments as teaching aids for their students . In 1816 at St Andrews , Scotland Sir David Brewster was doing that very thing when he discovered the Photoelastic Phenomenon using polarised light from Iceland Spar .

FIG.1

Bi-refringence.
Iceland Spar crystals demonstrate very powerfully the physical property of bi-refringence or double refraction . A double image is clearly seen through a large crystal , first explained by Erasmus Bartholinus (1669). **Fig. 2** .

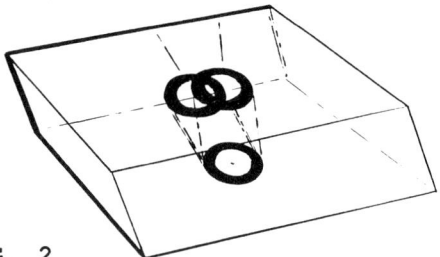

ICELAND SPAR - CALCITE

SHOWING DOUBLE REFRACTION EFFECT

FIG. 2

Certain clear crystals have this property of double refractive indices , such as calcite (CACO$_3$) and Quartz (SIO$_2$) . These are known as uniaxial crystals since they have a single optical axis , Two different sets of secondary wavelets are propagated within the the crystal for one incident wavefront . The relevence to Photoelasticity is that the extraordinary emergent ray from such a crystal is almost completely 'plain polarised' light .

The molecules of matter have an electromagnetic field around their centre of gravity , the more unsymmetrical this field is then the less uniform are its physical properties . This is known as anisotropy , it manifests itself in various ways . The one of interest to us is the optical anisotropy in which an incident ray of light is refracted on two discrete fronts as in the Iceland Spar . Some materials exhibit the bi-refringence only when under strain , the refractive indices being directly related to the magnitude of the strain . This is often known as temporary bi-refringence as it disappears when the load is removed - this was the basis of Brewsters discovery .

Using two Iceland Spar crystals it can be observed that certain objects or models placed between the crystals demonstrate stress patterns when under load . Incidently this is not confined to the visible range , but also occurs in the infra-red and X-ray range (noted for further research one day !).

PRACTICAL PHOTOELASTICITY

The Plane Polariscope .

If polarised light is passed through a bi-refringent stress model , then the light beam splits into two component waves coinciding with the major and minor principal planes , the magnitude of these waves is in exact proportion to the principal stress levels . The emergent image is made visible by passing the light through a further polarised filter , the optical signal at this stage however carries directional information known as ISOCLINICS indicating the principal plane direction . This optical system is a PLANE POLARISCOPE as shown in **Fig.3** .

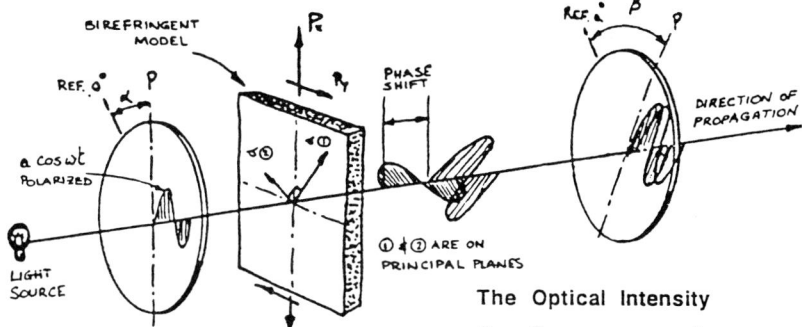

Fig.3 PLANE POLARISCOPE

The Optical Intensity

$$I = a^2\sin^2 2(\beta - \alpha) \sin^2 \pi \delta / \lambda.$$

The Circular Polariscope .

If quarter wave plates are introduced into the plain polariscope either side of the bi-refringent stress model then the directional effects disappear leaving only the strain fringes known as ISOCHROMATICS . These are lines of equal value of principal strain difference (ε_1 - ε_2) , this optical system is a CIRCULAR POLARISCOPE as shown in **Fig.4** .

The Optical Intensity $I = a^2 \sin^2 \pi \delta / \lambda$.

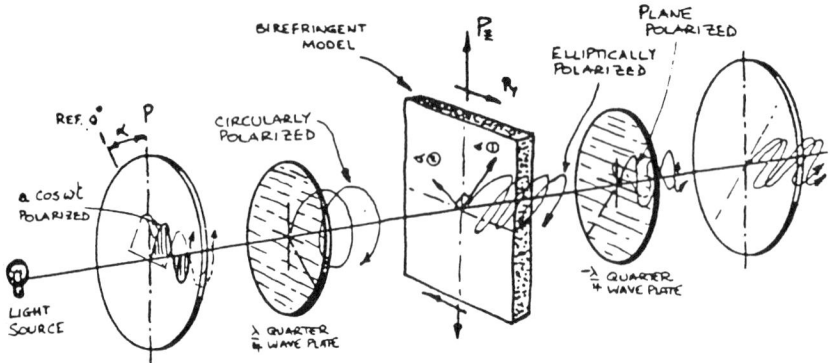

FIG.4   **DIAGRAMATIC CIRCULAR POLARISCOPE**

It is of immediate interest to remember that at a free edge of a stress model ε_2 is zero , leaving ε_1 as the observed signal . With a calibrated polariscope the value of the isochromatic fringe is measured by detecting the phase retardation between the two beams split by the principal strain in the stress model . The Photoelastic effect in the stress model is exactly related to the full elastic behaviour of the real component with no assumptions or terms missing etc .

Photoelastic Models

1)   Two dimensional models are usually made from bi-refringent plastic which is easy to profile and modify in a progressive design approach as in this Airbus Slat Track stress model. **Fig 5** .

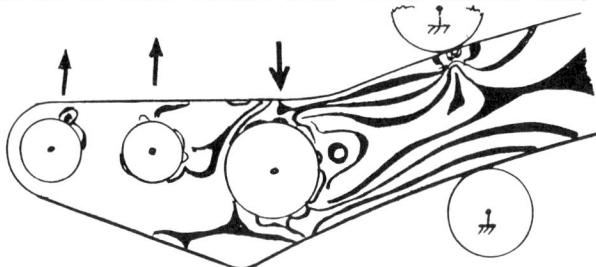

FIG.5

2)   Thin coatings of bi-refringent plastic bonded to the surface of a component or even an epoxy model like an all over strain gauge . In the examples quoted elsewhere , the Concorde Rudder Bracket is the actual aluminium component but the Airbus Landing Gears and Pintle Fitting are models made from hot cure epoxy . In this type of analysis the coating is bonded with a reflective adhesive which acts as a mirror , thus effectively folding the optical system in half and doubling the sensitivity because the light passes through the coating twice – on the way in and on the way out .

3) Stress freezing – solid models are made from special plastics that are heat treated while under load , the strain pattern is retained after cooling . The model is then sliced for analysis .

Fringe Patterns

The optical strain signals through the circular polariscope appear as fringes , they are very much like contour lines on a map starting with black at the zero stress value and counted in ascending order every time the red/blue is crossed . The fringe is identified as the tint of passage when the colour yellow is missing from the spectrum , the wave length being 22.7×10^{-6} ins. or its multiples . **Fig 6.**

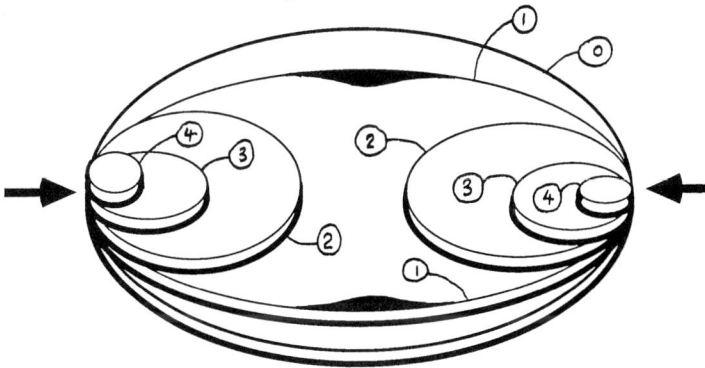

FIG 6   COUNTING FRINGE VALUES

Photoelastic Equations .   For transmission polariscope .

Principal strain difference $\varepsilon_1 - \varepsilon_2 = Nf/t$

Principal stress difference $\sigma_1 - \sigma_2 = NfE/t(1+\nu)$

N = number of fringes .   t = model thickness .
f = stress optic factor . ν = poissons ratio .
E = Youngs modulus .

Summary – Photoelasticity may be regarded as an all over strain gauge with zero gauge length or a finite element model with an infinitely small mesh size and no assumptions in the stiffness definition .

VERIFICATION OF NUMERICAL METHODS

First encounters with Photoelasticity a number of years ago was in the context of backing up my finite element results with a practical technique , however it was soon found to be such a powerful method that it became a prime technique in its own right . It has been useful on a number of occasions - adjudicating between numerical models that should have given similar results !

We have two interesting examples of correlation between Photoelastic models and numerical analysis results .

Example (1)
Pin Loaded Lug .
FEM - 8 noded quadrilateral plate elements with extremely fine mesh .
Photoelastic model - polyurethane sheet 3 mm thick .
The correlation was very good mainly due to the fine mesh of the finite element model .
The polyurethane model took about 30 minutes to make and analyse , the FEM took somewhat longer ! The pin in the Photoelastic model was thought to be more realistic and the model was easy to modify subsequently to achieve reduction in geometric concentrations . **Fig. 7**

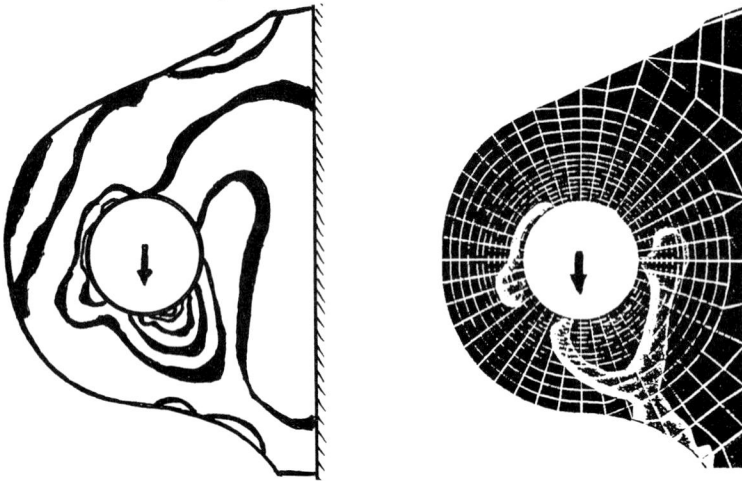

FIG.7   PE - FEM CORRELATION

Example (2)
Pin loaded lug/bracket.   Concorde Rudder Cradle Fitting
Boundary Element - 2-D analysis with medium sized discretisation.
Photoelastic model - 3 mm thick polyurethane sheet .
Both methods were very quick and gave good correlaton with each other , it was immediately clear with the Photoelastic model that the exact boundary fixations were critical . However neither model represented the actual practical problem very well as the cause in this case proved to be due to a 3-dimensional effect described in the next section .

PROBLEMS POSITIONING STRAIN GAUGES

In the early days , the Concorde Rudder Cradle Fitting gave a very revealing example of problems with strain gauge location . A mystery persisted after 2-D photoelastic and boundary element studies indicated fairly low stress levels at a position where fatigue cracking had been experienced on a similar component . Strain gauges were applied to an uncracked test component using the usual calibrated engineering judgement regards positioning - the stresses measured by these gauges was also relatively low . It was then decided to bring in Photoelasticity and cost the critical zone to try to understand the nature of the stress distribution . The results showed that most of the gauges had been placed at least 100 mm either side of the peak stress which was consequently underestimated by over 100% . Now the mystery was unravelled - the high fatigue stress was due to the 3-dimensional 'dog leg' effect causing an additional geometric concentration .With this additional strain information it was possible to progressively remove a small amount of material and reduce the stress raiser to about half its previous level . The Photoelastic strain patterns were recorded during the fatigue test against a time base for further analysis . The results were extremely good such that even the vibration due to stiction in the loading jacks was detectable . **Fig. 8**

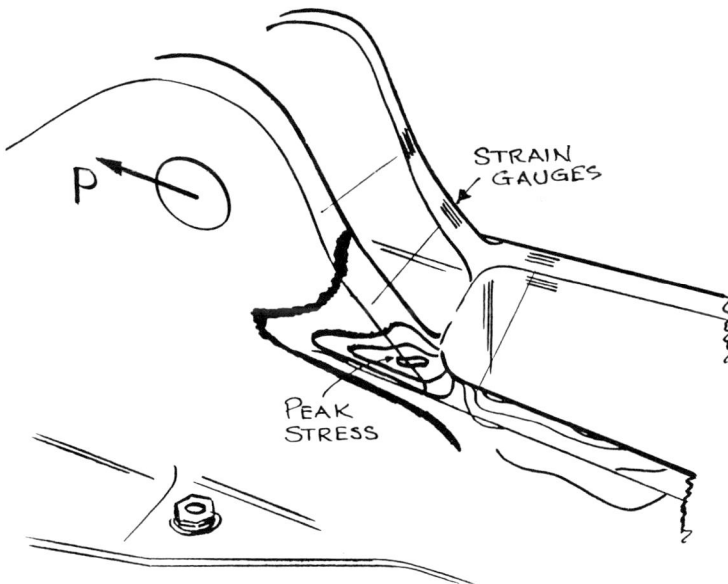

FIG. 8

WEIGHT AND COST SAVING

The Concorde Rudder was a trivial example of weight saving however the cost saving was significant as it was possible to easily modify the fleet of aircraft using the existing very expensive component .

Some examples of weight saving at the design stage are :-

Spoiler Bracket .
Note the progressive or iterative approach each time removing low stress material to 'tune up' the complete component to a more even stress distribution . **Fig. 9**

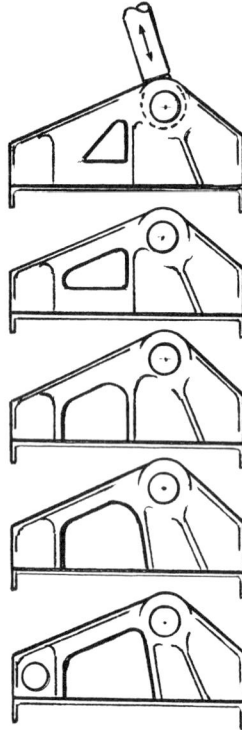

1) INITIAL DESIGN PROPOSAL .

2) ENLARGED APERTURE TO PERMIT ROUTING OF SYSTEMS .

3) WEB REMOVED IN LOW STRESS AREA FOR WEIGHT REDUCTION .

FIG. 9

A340 SPOILER ACTUATOR MOUNTING .

4) MORE WEB REMOVED FOR WEIGHT SAVING .

5) LIGHTENING HOLE ADDED AND RADIUS REDUCED .

Landing Gear Pintle Fitting.
The stress model is made from hot cure epoxy CT200 which has proven to be best material so far because of its high strength and zero creep characteristics . Normally very large components cannot be made in this material since the exothermic reaction during mix is too violent with big volumes .

Stress Engineering Services who manufacture the A340 landing gear and pintle models for us have overcome this problem by modifying the constituents to their own formula .

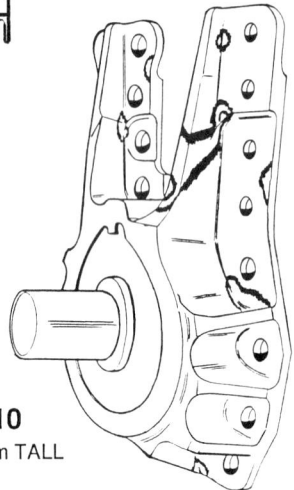

FIG.10
700 mm TALL

The material is cast then machined precisely to the aircraft drawings then coated with a Photoelastic coating . The low modulus of the model allows for relatively low loads to be applied in light weight test rigs . It is possible with programme management to obtain the stress results and conduct weight saving studies quick enough to be incorporated in the production drawing issue , it also possible to avoid expensive mistakes at the forging drawing stage by this technique .

With the stress envelope now analyzed the low stress areas are clearly identified , these areas can be progressively removed and retested on an iterative basis . During this process it must be understood that the functional requirements of the component must be maintained .

Thus Photoelasticity used in this way not only saves weight and costs but also provides a quality assurance on the design which is not normally possible .

An additional feature of Photoelasticity for design is that sometimes a component that is expensive for stress reasons may be simplified as a result of the additional strain information , thus saving costs .

The process of design using all the available techniques including Photoelasticity clearly need to be managed in order to maximise the principle of 'fitness for purpose' . It may need a change of emphasis from the old idea of applying a strain gauge when the design is already fixed to make measurements - hopefully to find the maximum stress .

The true design ethos should be :-
1) Determine the stress distribution within a structure to ascertain fitness for purpose .
2) Make measurements to show how failure can be avoided .
3) Ascertain a realistic service load history - from continuous onboard recording if possible .
4) Obtain strain information to allow full weight saving procedures .
5) Involve and co-ordinate strain measurement and design engineers to impact the product .

Typical progressive design approach for a Landing Gear

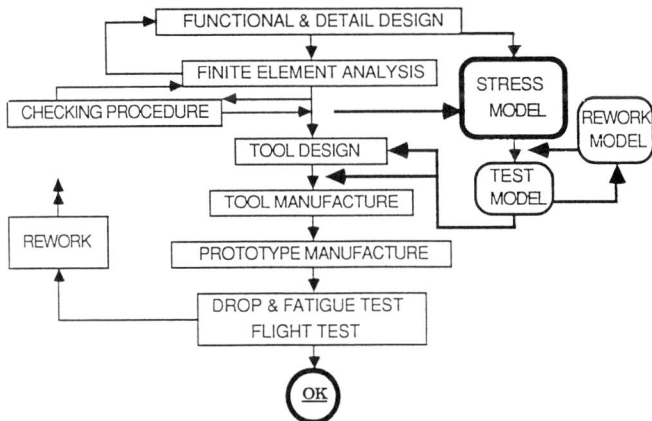

Fig 11

CONCLUSION

Like all measurement techniques this process will initially cost money , but the overall pay off for aircraft structures is identifiable and clearly worth going for with the help of *old lamps for new !* .

An Improved Method for the Determination of Photoelastic Stress Intensity Factors using the Westergaard Stress Function

T. H. HYDE and N. A. WARRIOR

*Department of Mechanical Engineering, Nottingham University,
Nottingham NG7 2RD, UK*

ABSTRACT

Methods for photoelastic prediction of stress intensity factors traditionally compare the maximum in-plane shear stress obtained from the observed isochromatic fringe pattern ahead of a manufactured crack with the crack-tip stresses predicted by a truncated series derived from the Westergaard stress function. These truncated equations can significantly misrepresent the stress field, even close to the crack tip, when an appreciable mode 2, or far-field stress component is present. For practical 3d photoelastic models where the crack tip position is ill-defined, it is shown that an optimum position exists where the origin of the experimental crack-tip stress field coincides with the origin of the Westergaard stress function. The conditions, and the change of optimum position with varying mode mixity is demonstrated for practical crack-like flaws. A new method, for prediction of mixed-mode stress intensity factors, based on Sanford and Dally's MPODM, and incorporating a full expansion of the crack-tip stress series due to Westergaard and an optimum crack-tip location routine is introduced, and accuracies for practical notch type geometries are presented. Stress intensity factors for semi-circular surface cracks in plates and blocks under tension and bending are calculated using the new method.

KEYWORDS

Stress intensity factor; Photoelasticity; Westergaard stress function.

INTRODUCTION

It is important that designers have data relating mixed-mode stress intensity factors to crack and component dimensions. However, analytical solutions exist only for relatively simple components and loadings, and for more complex problems experimental or numerical methods must be used to determine K_I and K_{II}. By careful experimental technique it is possible to introduce flaws into photoelastic models, and to produce flaw-tip isochromatic fringe patterns which characterise the stress fields near natural cracks. The maximum shear stresses determined from the isochromatic fringes can be related to K_I and K_{II} by using the crack tip stress functions developed by Westergaard (1939), and many practitioners have proposed photoelastic stress intensity factor calculation methods based on these relationships. Most of these methods can be shown to give accurate predictions using large scale two-dimensional photoelastic models loaded to within prescribed ranges of K_I and far-field stress (the difference between the remote stresses applied perdendicular and parallel to the crack). However, limitations to the practical maximum size of three-dimensional photoelastic models make a nominal crack length of around 5mm suitable, and difficulties in growing a natural crack to the size and shape required make it neccessary to use a manufactured crack-like flaw to simulate the actual crack. These practical considerations limit the suitability of the existing methods.

An existing algorithm has been modified to give a method suited to analysis of three-dimensional cracked photoelastic models under general loading conditions, and the new method has been used to predict stress intensity factors for a range of practical geometries.

NOTATION

a	crack radius or crack half-length
f_1 to f_6	functions of r, Θ and a (see equation 2)
f	material fringe value
g, g_1 to g_5	functions of f_1 to f_6 and r (see equation 4)
k	factor of load biaxiality (Fig.1)
K_I, K_{II}	mode I and mode II stress intensity factors
m	number of data points for a slice
N	fringe value
(r,Θ)	coordinates relative to the crack tip
R	coordinate relative to maximum depth of notch
S	non-dimensional stress
t	slice thickness
T	photoelastic model thickness
u	displacement
Y_I, Y_{II}	non-dimensional mode I and mode II stress intensity factors
β	crack inclination
σ	direct stress
τ	shear stress
ϕ	angular position along crack tip
ρ	notch root radius
Z'	dZ/dz (in table 1)

Subscripts

c.t.	for optimum crack tip position (Fig.4)
k	refers to a particular data point
max	maximum (shear stress, equation 3)
ox	far-field stress ($\sigma_{ox}=(1-k)\sigma$)
x,y,z	coordinate directions

THE WESTERGAARD STRESS FUNCTION

Westergaard (1939) applied an analytical stress function, Z, of the complex variable z ($= x+iy$) to an infinite plate under uniform biaxial tension at infinity:

$$Z = Z(z) = Z(x+iy) = Re[Z] + iIm[Z]$$

Sih (1966) expanded upon Westergaard's work to show that, where $z = (a+re^{i\Theta})$, the crack tip stresses for the symmetric (Fig. 1(a)) and skew-symmetric (Fig. 1(b)) load cases can be expressed in terms of the stress function $Z(z)$ as in table 1:

Table 1: Expressions for crack-tip stresses under mode I and II.

	symmetric (mode I)	skew-symmetric (mode II)
σ_x	$Re\ [Z(z)] - y\ Im\ [Z'(z)] - (1 - k)\sigma/2$	$2\ Im\ [Z(z)] + y\ Re\ [Z'(z)]$
σ_y	$Re\ [Z(z)] + y\ Im\ [Z'(z)] + (1 - k)\sigma/2$	$-y\ Re\ [Z'(z)]$
τ_{xy}	$- y\ Re\ [Z'(z)]$	$Re\ [Z(z)] - y\ Im\ [Z'(z)]$
$Z(z)$	$\dfrac{\sigma(a + re^{i\Theta})\ (1 + re^{i\Theta}/2a)^{-1/2}}{(2\ a\ re^{i\Theta})^{1/2}} - \dfrac{(1-k)\sigma}{2}$	$\dfrac{\tau(a + re^{i\Theta})\ (1 + re^{i\Theta}/2a)^{-1/2}}{(2\ a\ re^{i\Theta}\ )^{1/2}}$

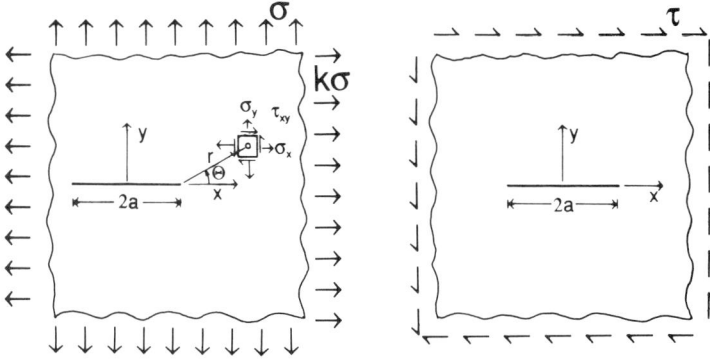

Fig. 1: Cracks in infinite plates subjected to (a) uniform biaxial tension and (b) uniform shear remote stress conditions

The $(1 + re^{i\Theta}/2a)^{-1/2}$ term can be expanded as a binomial series, and the symmetric and skew-symmetric cases combined by superposition to give the in-plane stresses in the vicinity of a crack tip:

$$\sigma_x = \frac{K_I}{\sqrt{2\pi r}}\ f_1\ (r,\ \Theta,\ a) \quad + \quad \frac{K_{II}}{\sqrt{2\pi r}}\ f_2\ (r,\ \Theta,\ a) - \sigma_{ox}$$

$$\sigma_y = \frac{K_I}{\sqrt{2\pi r}}\ f_3\ (r,\ \Theta,\ a) \quad + \quad \frac{K_{II}}{\sqrt{2\pi r}}\ f_4\ (r,\ \Theta,\ a) \tag{1}$$

$$\tau_{xy} = \frac{K_I}{\sqrt{2\pi r}}\ f_5\ (r,\ \Theta,\ a) \quad + \quad \frac{K_{II}}{\sqrt{2\pi r}}\ f_6\ (r,\ \Theta,\ a)$$

where f_1 to f_6 are infinite series in r (Hyde and Warrior, 1989).

The equations in table 1 are only strictly applicable to an internally cracked infinite plate with remotely applied uniform stresses. With other types of stress fields alternative forms of these equations should be used. Often, for simplicity, the case where $r<<a$ is considered, and $(1 + re^{i\Theta}/2a)^{-1/2}$ series are truncated after the $r^{-1/2}$ term. Since the $r^{-1/2}$ terms are independent of crack length, a, then for the truncated case, in equation 1, $f_1 = f_1\ (r, \Theta)$ (i.e. independent of a) etc., and the region described by the resulting crack-tip stress equations referred to as the "singularity dominated zone". The extent of this zone for various mode I dominant geometries has been estimated by many authors, and their methods for prediction of stress intensity factors from selected data in this region have

been used with some success (Hyde and Warrior 1987). However, Theocaris and Gdoutos (1978) showed that the truncated crack-tip stress equations can significantly misrepresent the stress field, even close to the crack tip, when an appreciable mode II, or far-field stress component is present. Their results illustrate these singularity dominated zone methods' ability to produce accurate, or inaccurate solutions, for stress intensity factor, depending on the mode mixity, the regions chosen for data collection, and the user's interpretation of the extent of the singularity dominated zone. It can be concluded then, that for accurate prediction of mixed-mode stress intensity factors the reference to which the experimental stress field should be compared is the full series expansion of the Westergaard stress function.

THE EXPANDED-MPODM

Using the stress optic law, to relate τ_{max} to isochromatic fringe order, N and then to the Cartesian stresses, σ_x, σ_y and τ_{xy} gives:

$$\tau_{max}^2 = \left(\frac{N f}{2 t}\right)^2 = \frac{(\sigma_x - \sigma_y)^2 + \tau_{xy}^2}{4} \tag{2}$$

Expressions for σ_x, σ_y and τ_{xy} can be substituted into equation 2 resulting in :

$$
\begin{aligned}
g(K_I, K_{II}, \sigma_{ox}) = \quad & K_I^2 g_I(r, \Theta, a) \quad + \quad K_{II}^2 g_2(r, \Theta, a) \\
+ \quad & \sigma_{ox}^2 + K_I K_{II} g_3(r, \Theta, a) \quad + \quad K_I \sigma_{ox} g_4(r, \Theta, a) \\
+ \quad & K_{II} \sigma_{ox} g_5(r, \Theta, a) \quad - \quad \left(\frac{N f}{t}\right)^2 = \quad 0
\end{aligned} \tag{3}
$$

where

$$
\begin{aligned}
g_I &= [(f_1 - f_3)^2 + 2f_5^2] \,/\, 2\pi r & g_4 &= (f_3 - f_1)/\, \pi r \\
g_2 &= [(f_2 - f_4)^2 + 2f_6^2] \,/\, 2\pi r & g_5 &= (f_4 - f_2)/\, \pi r \\
g_3 &= [(f_1 - f_3)(f_2 - f_4) + 2f_5f_6] \,/\, 4\pi r &
\end{aligned}
$$

Work (Hyde and Warrior, 1987) has shown that the most reliable method of photoelastic stress intensity factor prediction is the full field method, or MPODM, due to Sanford and Dally (1979). The fringe value N_k $(k = 1,2...m)$ at m positions in the vicinity of the crack tip is measured, and each $(r, \Theta, N)_k$ data substituted into equation 3. The resulting over-determined set of equations is reduced to one equation in three variables by the least-squares minimisation process, then approximate solutions for K_I, K_{II} and σ_{ox} are improved by using the Newton-Raphson method. For the work presented here, the method has been extended to incorporate the full series expansion of the stress equations; eleven terms were found to be sufficient. It will be seen that, unless the K_I, K_{II}, σ_{ox} solution is an exact fit to the data, $g(K_I, K_{II}, \sigma_{ox}) \neq 0$ in equation 3, and a residual error will result. The magnitude of this residual is an indication of the accuracy of fit of the solution to the experimental data.

ON THE VALIDITY OF REPRESENTING CRACKS BY NOTCHES

The isochromatic fringe pattern in the close vicinity of the tip of a photoelastic crack is not representative of the τ_{max} field near an actual crack tip. This "none crack-like" near-field can be attributed to the physical effect of the very high stresses at the crack tip, and the artificial crack - or sharp notch effect. The rapid variation of stress gradient and the existence of the triaxial stress field lead to blunting, finite rotation and deformation at the crack tip. Smith et al. (1975,1978) predicted that non-singular behaviour would be confined to within a radius of $3 \times 10^{-3}a$, and concluded that these effects are too near the crack tip to be picked up in the photoelastic data. In a photoelastic model, the 'crack' will generally be cut, or cast into the specimen at the manufacturing stage. The 'crack', then will be a sharp notch, and the stress field at the tip of that notch will not be the same as that ahead of a crack of equivalent length. Many researchers have proposed crack tip correction terms. Creager and Paris (1967) expanded the Westergaard mode I stress solutions by adding a single term dependent on ρ, the notch tip radius, at the crack tip. Phang and Ruiz (1984) presented results using a tip radius correction factor based on Creager and Paris's work, but showed that any improvement was insignificant, and it must be concluded that this approach is unsatisfactory.

The Extent of the Notch Tip Influence

Schroedl et al. (1972) compared the Kossoff-Inglis solution for an elliptical hole with the crack solution to investigate the influence of notch root radius on the τ_{max} distribution. The results showed the notch effects extend for five to ten notch root radii into the material. Schroedl and Smith, (1975) demonstrated the near field effects of various crack shapes using a conformal mapping approach. Their results show that the shape of the notch tip is of little significance to the extent of the notch effect, and that the notch width is the important parameter.

The near-field effects can be studied using the finite element method. In the present work, a range of notch sizes were modelled in 'infinite' plates, using eight-noded isoparametric elements (Fig. 2). In order to represent the high stress gradient present in the vicinity of the notch tip, an extremely fine mesh was used there. Two solutions with the boundary conditions shown in Table 2., were obtained for each notch radius.

Table 2. Boundary conditions for infinite plates

plane	ab	bc	cd	de
Mode I	$u_y = 0$	---	$\sigma_x = const.$	$u_x = 0$
Mode II	$u_x = \sigma_y = 0$	$\tau_{xy} = const.$	$\tau_{yx} (=\tau_{xy}) = const.$	$u_y = \sigma_x = 0$

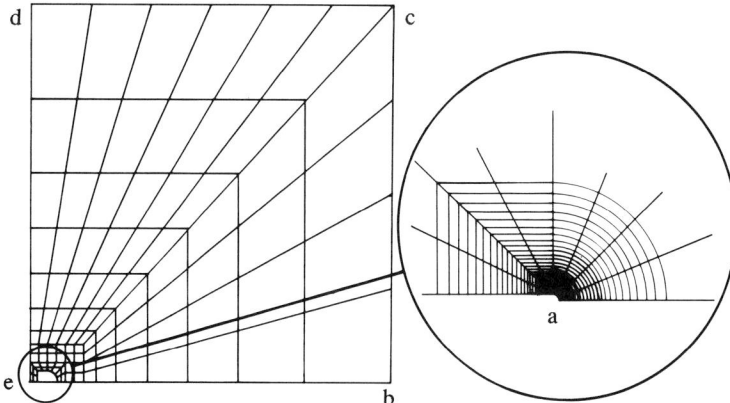

Fig. 2: Typical finite element mesh

By adding various proportions of the solutions for σ_x, σ_y and τ_{xy} and using these to obtain the appropriate τ_{max} values, results for a range of mode mixities were obtained. These very accurate solutions were then used to study the extent to which the near-field (for various notch radii) affects the accuracy of stress intensity factors determined using τ_{max} results from photoelastic models. In Fig. 3, the variation of τ_{max}/σ, obtained from the finite element results, with distance from the centre of the notch radius (for $\rho/a = 0$, 5×10^{-2}, 5×10^{-3} and 5×10^{-4}) along $\Theta = 90°$, under pure mode I loading is compared with the corresponding exact solution for a crack, (—). It can be seen that for $\rho/a = 5\times10^{-3}$ (a practical photoelastic value) the near-field effects are negligible when $r/a > 0.04$. Results from other Θ values for mixed-mode situations, with various far-field stress magnitudes, also show that near-field effects are negligible for $r/a > 0.04$.

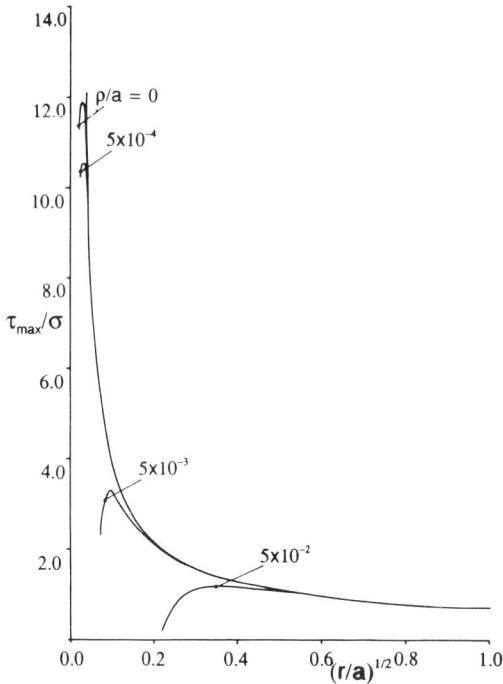

Fig. 3: The variation of τ_{max}/σ with $(r/a)^{1/2}$ for various notch tip radii

The Effect of the Notch Width on the Prediction of Stress Intensity Factor

From the preceeding, it can be seen that if the stresses near to the notch tip are excluded, the rest of the stress field for a sharp notch is very similar to that which would be obtained with an actual crack. It could be argued that the position of the equivalent crack tip should be at the centre of the notch radius, or at the maximum depth of the notch. Choosing either of these positions would result in two different sets of r and Θ values corresponding to the measurement positions in the vicinity of the notch. Significant differences in the values of the stress intensity factors, obtained using the expanded-MPODM, would result from assuming the origins of the equivalent crack to be at these positions. In the present work, in order to allow for any possible errors in the choice (or setting up) of the equivalent crack tip position, the position within the vicinity of the crack tip which resulted in the lowest residual from the Newton-Raphson method was taken to be the optimum equivalent crack tip position. Using the expanded-MPODM with the τ_{max} values obtained from the finite element solutions, the variation in optimum crack tip position varies with the relative magnitudes of mode I and mode II loading was studied. For each loading mode combination, a field of τ_{max} was selected from the region $0.06 < r/a < 0.6$; thus avoiding the near-field effects, but including the data further away from the notch tip than would usually be possible if only the singular terms were included. For a range of crack tip positions, the expanded MPODM was used to evaluate K_I, K_{II}, σ_{ox}, and residual. By fitting a second order polynomial to the (coordinate, residual) data, a position of crack-tip for best fit (ie. minimum residual) was obtained.

Fig. (4a) shows the extent of the variation of optimum crack-tip position for ρ/a values of 0 (theoretical crack), 5×10^{-4} and 5×10^{-3}. Figs. (4b-c) show the accuracy of the stress intensity factor solution at that crack tip position, using the expanded-MPODM. It will be seen that an accurate K_I and K_{II} solution is possible from these notch geometries, using the Westergaard stress function, but that the position of origin of the stress function axes is dependent on remote stress conditions and notch root radius.

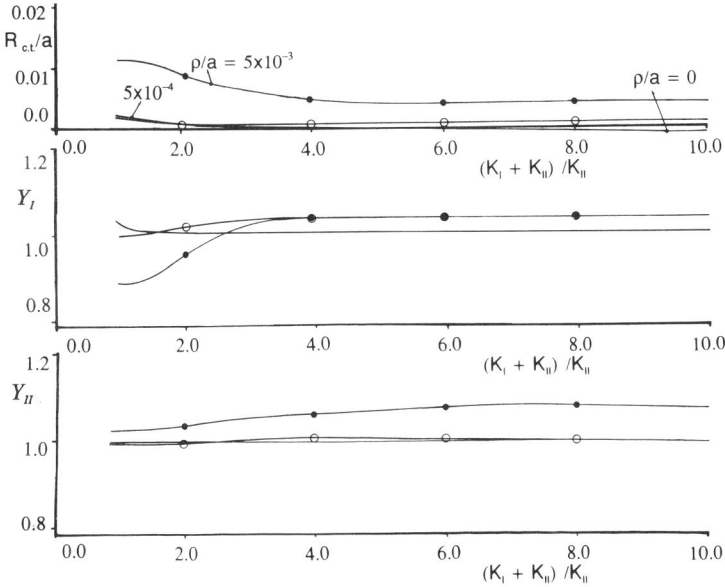

Fig. 4: The variation of solution parameters, (a) optimum crack-tip position, (b) Y_I, (c) Y_{II}, with mode mixity.

It is evident that if the stress intensity factor solution from the comparison between the predicted crack tip stresses and the actual stress field due to the notch is to be accurate, then the origins of the respective stress system axes must be coincident. Since the magnitude of mode mixity will be unknown, it is required to let the crack tip "float" to the position of minimum residual within the approximate bounds defined by the curves in Fig. 4a for the relevant ρ/a ratio.

TYPICAL RESULTS AND DISCUSSION

In this section, results for plates and blocks, subjected to tension and bending, are used to illustrate the likely accuracy of the proposed new method. The photoelastic models were cast (using Araldite CT200 with HT907 hardener) in moulds which incorporated 0.05mm thick semi-circular shims. After loading and stress freezing, slices were machined at various angular positions, ϕ, from the crack-like features. Stress intensity factors, K_I, K_{II}, and far-field stress σ_{ox}, were determined from the isochromatic fringes near to the crack tip using the expanded-MPODM, letting the crack tip float to the position of minimum residual within the vicinity of the notch tip.

For $\beta=90°$, the photoelastically determined results, (—), can be compared with finite element results, (---), obtained by Newman and Raju (1981). The K_I, K_{II} and σ_{ox} results have been normalised to give the non-dimensional stress intensity factors, Y_I and Y_{II}, (where $Y_i=K_i/(\sigma\sqrt{\pi a})$) and the non-dimensional far-field stress, S_{ox}, (where $S_{ox}=\sigma_{ox}/\sigma$ and σ is either the nominal tensile stress or nominal surface bending stress).

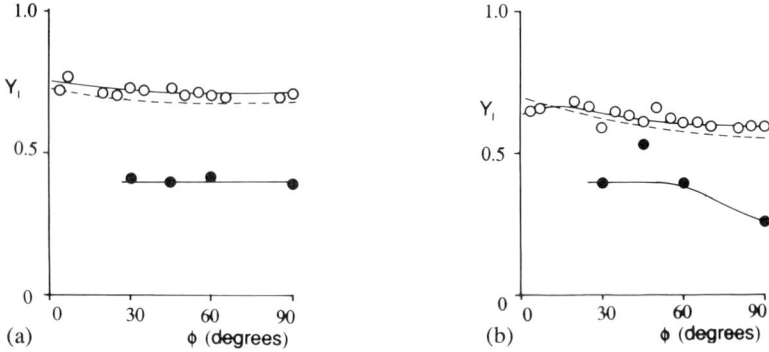

Fig. 5: Variation of Y_I values along crack front under (a) tension and (b) bending loading conditions.

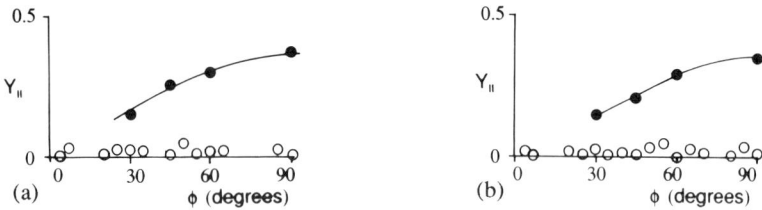

Fig. 6: Variation of Y_{II} values along crack front under (a) tension and (b) bending loading conditions.

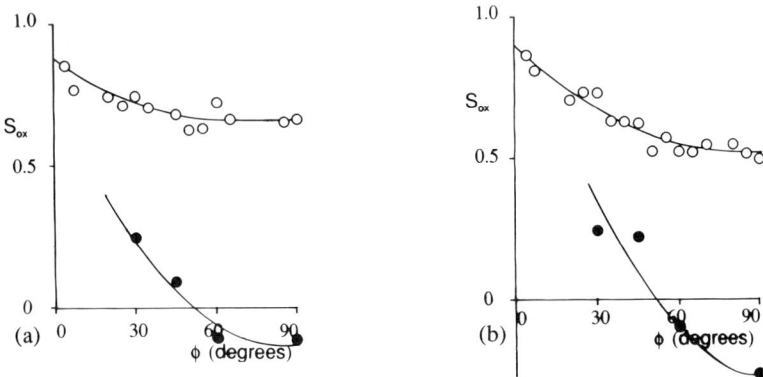

Fig. 7: Variation of S_{ox} values along crack front under (a) tension and (b) bending loading conditions.

Y_I values obtained for blocks (a/T=0.125), with β=45° (●) and 90° (○), under tensile and bending conditions are given in Figs. 5. The finite element results for ß=90°, (Newman and Raju, 1981), are superimposed on Figs. 5. It can be seen that the present results are within 5% of the finite element results. The corresponding Y_{II}, and S_{ox} values are given in Figs. 6 and 7. From Figs. 6, it can be seen that small Y_{II} values were obtained with β=90°; these should be zero. The magnitudes of these values give an indication of the accuracy of the Y-values (both Y_I and Y_{II}), ie. Y-values are likely to be accurate to within about ± 0.03. The individual S_{ox} values, Figs. 7, exhibit relatively little deviation from the 'average' curves drawn through all of the results, giving a further indication of the consistency and likely accuracy of the method, both in tension and bending situations.

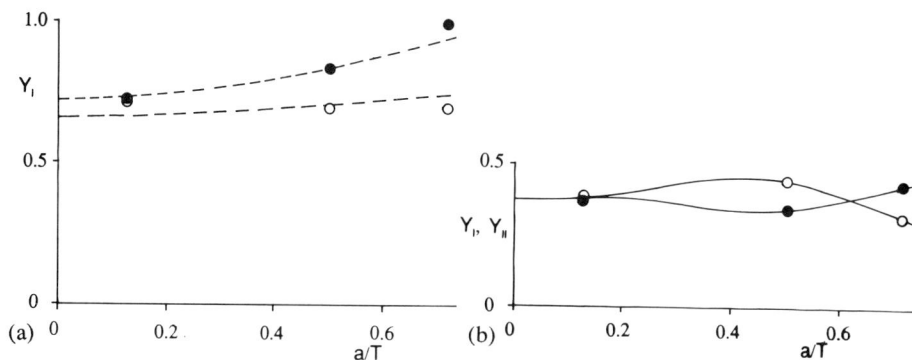

Fig. 8: Variation of Y values with a/T ratio under (a) tension and (b) bending loading conditions.

By reducing the thickness of the blocks, results for plate-type components (ie. with high a/T values) were obtained. Typical results are presented in Figs. 8(a) and 8(b) in which the variations of Y_I with a/T, for β=90°, and Y_I and Y_{II} with a/T for β=45° are presented respectively. The pure mode-I results, ie β=90° are compared with the finite element results in Fig. 8(a). From Fig. 8(a), it can be seen that the accuracy which is evident from the detailed comparisons shown in Figs. 5-7, with a/T=0.125, is typical of the accuracy attainable for all a/T values.

CONCLUSIONS

By incorporating eleven terms in the crack-tip stress field equations, the necessity of arbitrarily selecting data within a so-called "singularity dominated" region has been eliminated. The extent of the notch tip influence has been investigated and quantified for practical photoelastic model geometries; for $\rho/a < 5\times10^{-3}$, data for $r/a < 0.05$ should be neglected.

Accurate stress intensity factor solutions can be obtained from the isochromatic fringe patterns measured in the vicinity of crack-like flaws created by incorporating 0.05mm thick shims in the moulds used to manufacture the models.

254

ACKNOWLEDGEMENTS

The authors would like to thank the technicians of the Department of Mechanical Engineering, Nottingham University for their skilled assistance. N.A. Warrior held an SERC studentship for the duration of this work.

REFERENCES

Creager M. and Paris P.C. (1967). Elastic Field Equations for Blunt Cracks with Reference to Stress Corrosion Cracking, *Int. J. Fracture.*, 3, 247-252

Gdoutos E.E. and Theocaris P.S. (1978). A Photoelastic Determination of Mixed-Mode Stress Intensity Factors, *Exp. Mech.*, 18, 87-96

Hyde T.H. and Warrior N.A. (1987). A Critical Assessment of Photoelastic Methods of Determining Stress Intensity Factors, *App. Solid Mech. 2* (A.S. Tooth and J. Spence, Ed.), Elsevier Science Publishers, 23-40

Hyde T. H. and Warrior N.A. (1989). Photoelastic Determination of Mixed-Mode Stress Intensity Factors for Inclined Semi-Circular Cracks in Thin Tubes, To be published *J. Strain Anal.*

McGowan J.J. and Smith C.W. (1975). A Finite Deformation Analysis of the Near Field Surrounding the Tip of Crack-Like Elliptical Perforations, *Int. J. of Fracture*, 11, 977-987

Newman J.C. and Raju I.S. (1981). An Empirical Stress-Intensity Factor Equation For The Surface Crack. *Engng. Fracture Mech.*, 15 , 185-192.

Phang Y. and Ruiz C. (1984).Photoelastic Determination of Stress Intensity Factors for Single and Interacting Cracks and Comparison with Calculated Results (Parts I & II), *J. Strain Anal.*, 19, 23-41

Sanford R.J. and Dally J.W. (1979). A General Method for Determining Mixed-Mode Stress Intensity Factors for Isochromatic Fringe Patterns, *Engng. Fracture Mech.*, 11, 621-633

Schroedl M.A., Mcgowan J.J. and Smith C.W. (1972). An Assessment of Factors Influencing Data Obtained by the Photoelastic Stress Freezing Technique for Stress Fields Near Crack Tips, *Engng. Fracture Mech.*, 4, 801-809

Schroedl M.A. and Smith C. W. (1975). A Study of Near and Far-Field Effects in Photoelastic Stress Intensity Determination, *Engng. Fracture Mech.*, 7, 341-355

Sih G.C. (1966). On the Westergaard Method of Crack Analysis, *J. Fracture. Mech.*, 2, 628-630

Smith C.W. McGowan J.J. and Peters W.H. (1978). A Study of Crack-Tip Nonlinearities in Frozen Stress Fields, *Exp. Mech.*, 18, 309-315

Westergaard H.M (1939). Bearing Pressures and Cracks, *J. Appl. Mech.*, 61, A49

Modelling and Testing of Aluminium Bonded Structural Components

I. J. McGREGOR and D. NARDINI

*Alcan International Limited, Banbury Laboratories,
Oxon OX16 7SP, UK*

ABSTRACT

This paper describes some of the work conducted by Alcan on the modelling of bonded aluminium structural components. The different ways in which the bondline can be represented in a component are described and discussed. As an example, the modelling and experimental results of a design exercise on a vehicle front cross-member are presented. The use of detailed joint modelling in determining the influence of local joint geometry on both adhesive and metal stresses is also demonstrated.

KEYWORDS

Bondline; modelling; stresses; aluminium; design.

INTRODUCTION

In recent years, adhesive bonding has increased in application in many areas of engineering. One very versatile application is in joining parts made of aluminium, and this was first used in the aircraft industry. A unique property of a pretreated aluminium surface is that it is extremely resistant to environmental influences, thereby providing a durable and reliable bond. The technique of adhesive bonding has further been developed by Alcan in connection with the design of aluminium structured vehicles, and some of this work is described elsewhere (Marwick and Sheasby, 1987; Nardini and Seeds, 1989; Seeds *et al.*, 1989; Warren *et al.*, 1989; Wheeler *et al.*,1987).

This paper is concerned with some of the detailed studies of bonded joint behaviour in terms of its stiffness and strength, as well as its influence on properties of a component that it is a part of.

When modelling bonded structural components and whole structures, it is often difficult to include the bondline in the model because of the very small detail required to represent the adhesive. It is often the case, therefore, that structural components are modelled without the adhesive. This approach can be used to model the behaviour of the metal in the component, but gives no information on the behaviour of the bondline.

In order to obtain the full potential of a structure, it is important to consider the design of the joints. By modelling the adhesive in a component, it is possible to assess the stress condition of the joint for any given load on the component. This approach can indicate the areas in the component where high stresses are concentrated in the bondline. However, it is not possible with this size of model to investigate the influence of local structural details, such as a forming radius, bondline thickness and adhesive fillet size on the stress level in the adhesive. To investigate this influence, it is necessary to produce detailed joint models, where the joint alone is modelled under a variety of loading conditions.

This paper presents some of the work Alcan has conducted on the modelling of bonded structural components. The different ways in which the bondline can be represented in a component are presented, and a discussion is made on the influence of each representation on joint stiffness and its ability to produce adhesive stress information.

To illustrate the advantages and limitations of the different approaches of modelling bonded components, the results of a design exercise on a vehicle front cross-member are presented. The results of detailed modelling of joints in peel are also presented, showing the influence of a forming radius, bondline thickness and adhesive fillet size on the stresses in the adhesive and metal.

BONDLINE MODELLING TECHNIQUES

There are several different ways to model the geometry of two sheets of metal joined by adhesive. The complexity and refinement will depend on the required precision of the solution. This is especially true when representing details such as corner radii and the adhesive fillet, and their influence will be discussed later on. The main objective in modelling the bondline is to compute the stress state of the adhesive and to determine its safety against failure under the applied loading. Namely, a good structural design should always favour that the failure first occurs in the metal, which by its nature has large reserves of plastic behaviour, and, therefore, the opportunity to redistribute the stresses.

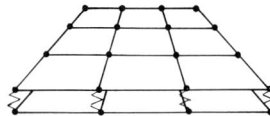

Fig. 1 - Thin Plates and Springs Model

In any case, reliable strength criteria should be derived to predict the onset of failure in the adhesive, and it can be stress or strain related with fracture mechanics foundation. The resolution of the model will however also have a large influence on the application of such a criteria, i.e. fine models will always give higher maximum stress/strain levels than the courser ones. One of the commonly used methods to model the influence of the adhesive between two plates, is to connect them with spring elements with stiffness in three coordinate directons [Fig. 1]. With adequate spring rates, this method, in most cases, yields a good representation of stiffness behaviour of the component. However, there is very little chance of finding the adhesive stresses.

Keeping to the plate/shell representation of the metal parts, one can use solid elements to model the adhesive [Fig. 2a]. This approach, although quite attractive, has some drawbacks. Namely, if one wants to preserve the distances between the metal plates in order to correctly represent their joint behaviour, the modelled adhesive thickness will be much greater than in reality [Fig. 2b]. This will have consequences in unrealistic stiffness representation of the joint. One way to overcome this is to modify the material constants of the adhesive, thereby arriving at an orthotropic description (requirements are different for shear, bending and tension). However, even with the orthotropic model, the joint will still be inconsistent in some deformation modes.

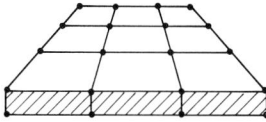

Fig. 2a - Thin Plates and Solids Model   (b) Real Bondline

The most accurate way to model the joint will be by use of a thick plate element or a solid element, that possess a geometric thickness [Fig. 3]. In that case there are no geometrical approximations involved, and the accuracy of the solution is most satisfactory. The only drawback of such modelling is the increase in model preparation effort. It is interesting to note that computer solution times need not be significantly greater than in the thin plate/shell model since the total number of degrees of freedom (dof) will be the same (thin shell has 6 dof per node, thick has 3 dof per node, but twice as many nodes).

The following design example demonstrates the use of solid elements in modelling the bondline in a structural component.

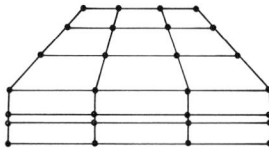

Fig. 3 - Thick Plates and Solids Model

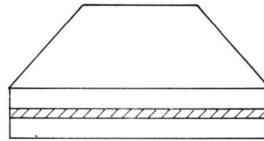

258

DESIGN EXAMPLE

To illustrate the advantages and limitations of the different approaches of modelling bonded components, the results of a design exercise on a vehicle front cross-member are presented. The design requirement of the cross-member was to carry a 10kN load (the case of jacking-up a fully loaded vehicle) with a permanent deformation of less than 1mm on unloading. The geometry of the component is shown in Fig. 4. The cross-member was loaded in three point bending, and consisted of a top-hat section with bonded closure and reinforcement plates.

In the first model of this component, only the aluminium was modelled. A quarter of the component was meshed with three dimensional eight noded isoparametric solid elements. This model was used for parametric studies on the influence of aluminium gauge and reinforcement length on the stresses in the aluminium.

From early experimental tests, it was found that the original form did not satisfy the design requirement, due to the formation of plastic hinges at point A in Fig. 4. It was considered that by suitable change in the reinforcement length, it would be possible to move the critical section away from the end of the reinforcement to position B in Fig. 4, and thereby produce a more stable yield performance.

By conducting a parametric study on the influence of the reinforcement length on the ratio of stress at point B to point A, the length of reinforcement necessary to produce initial yield at point B could be determined. Figure 5 shows the change in the ratio of stress at point B to point A as a function of reinforcement length. When the stress ratio is greater than 1, yield occurs at point B before point A and improved performance of the cross-member should be achieved. The analysis indicated that a reinforcement length of around 400mm would be sufficient.

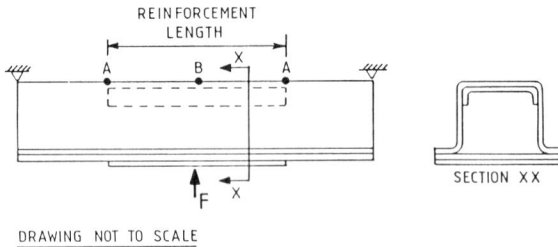

Fig. 4 - Drawing of vehicle cross-member

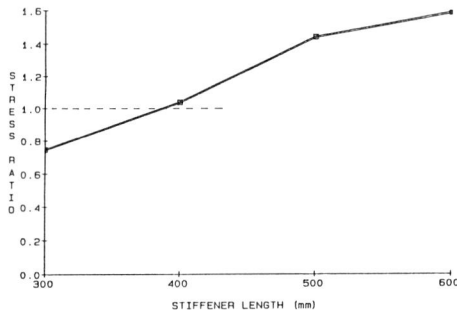

Fig. 5 - Modelled results for influence of stiffener length on stress ratio

It was not possible with this model, however, to obtain any information regarding the stress levels within the adhesive bondline. A new model was, therefore, developed which included the bondline by representing the adhesive with solid elements. Figure 6a shows the complete aluminium and adhesive model, and Fig. 6b shows the adhesive elements only.

The top row of adhesive elements represents the adhesive between the top-hat section and the inner reinforcement, and the bottom two layers of elements the adhesive between the top-hat and the closure plate and the closure plate and the bottom reinforcement.

A non-linear analysis was conducted including material non-linearity for both the aluminium and the adhesive. Figure 7 shows the correlation between the experimentally measured and the predicted load displacement curve for the component. It can be seen that the modified component design satisfies the design requirement of carrying a 10kN load with less than 1mm permanent deformation on unloading. It can also be seen that the model gives good correlation with the experimental curve over the full loading range. This is especially so in view of the localized non-linear behaviour of the cross-member in bending.

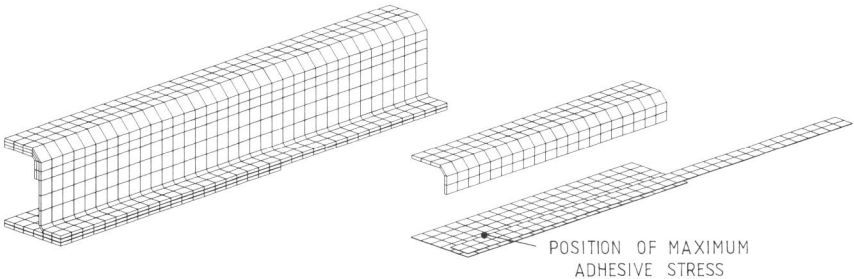

Fig. 6(a) - Aluminium and Adhesive Model   *(b) Adhesive elements*

The maximum predicted stress in the adhesive was 33N/mm^2. Other work suggests that this is well within the allowable stress for the adhesive. The position of maximum stress is indicated in Fig. 6b. This position is close to the forming radius of the top-hat section. It is often found that the highest stresses in the bondline are in this area of the joint, and it is known that changes in the joint design, such as spew fillet size, can significantly affect the stresses in the bondline. It is very difficult and inefficient to model small detailed changes in the joint design on a model of a component such as that presented here. The alternative is to make a detailed model of the joint and investigate the influence of design parameters on the stresses in the bondline under a variety of loading conditions.

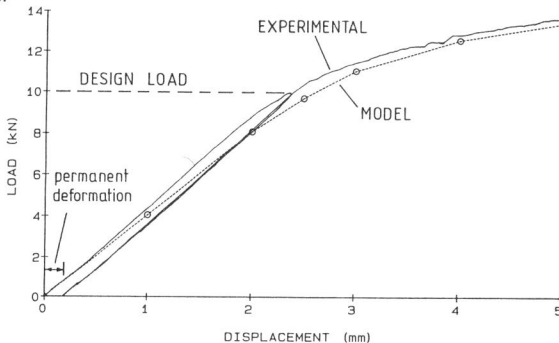

Fig. 7 - Comparison between Experimental and Theoretical load displacment curve

DETAILED JOINT MODELLING

One of the most severe loading conditions for a joint is perpendicular to the plane of the bondline. The local joint design parameters which influence the stresses both in the adhesive and the aluminium in this type of joint, have been modelled. The objective was to obtain an understanding of how these parameters affect the stresses in the joint, and to use this knowledge to reduce stress levels. A model of a typical tension specimen is shown in Fig. 8. Two dimensional eight noded isoparametric solid elements are used to model the specimen, only half of which is modelled. Figure 9 shows the bondline in more detail. Severe peel stresses are generated in the bondline, with the maximum principal stress in the adhesive occurring at point A in Fig. 9. The maximum stresses in the aluminium occur in the inside radius of the aluminium at position B in Fig. 9.

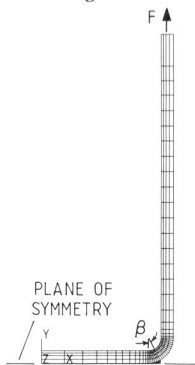

Fig. 8 - Model of typical tension specimen   *Fig. 9 - Model of tension specimen showing bondline detail*

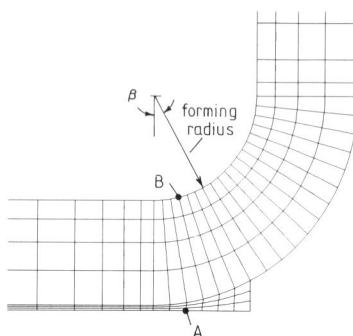

Figures 10, 11 and 12 present the results of an elastic analysis, undertaken to look at the influence of detailed joint design parameters on the stresses in the tension specimen. The results are presented in terms of the stress ratio, which is the maximum stress in either the adhesive or the aluminium per unit stress applied to the end of the specimen.

Figure 10 presents the results for the influence of the forming radius on the stresses in the adhesive and aluminium. It can be seen that as forming radius increases there is little influence on the stresses in the bondline, however, the stresses in the aluminium increase significantly. This is a result of the significant local bending that takes place in the specimen around the forming radius. As the forming radius increases, the bending effects increase, producing higher stresses in the adherend. Therefore, by reducing the forming radius, the bending stress in the aluminium can be kept to a minimum.

Figure 11 shows the influence of bondline thickness on the stresses in the adhesive. The stress ratio drops to a constant value as bondline thickness increases. For this particular sample geometry, the stress ratio levels out at a bondline thickness of 1mm. This result suggests that there will be an advantage in increasing the bondline thickness up to a certain value.

Figure 12 shows perhaps the most important result for this joint. The parameter β in Fig. 12 is a measure of the spew fillet size, and is shown in Fig. 9. It is seen that as fillet size increases, the stresses in both the adhesive and aluminium reduce significantly. This is a result of the adhesive stresses in a tension joint being concentrated in the spew fillet. Figure 13 shows the distribution of the normal stresses (σ_y) along the mid-line of the bondline. It is seen that most of the load transfer of the specimen takes place through the spew fillet. Therefore, with a large fillet, the stress will be less concentrated in the adhesive, resulting in greater joint strength. It is

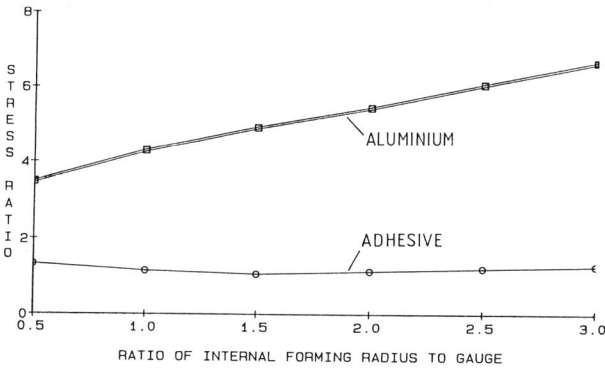

Fig. 10 - Influence of forming radius on tension specimen stresses

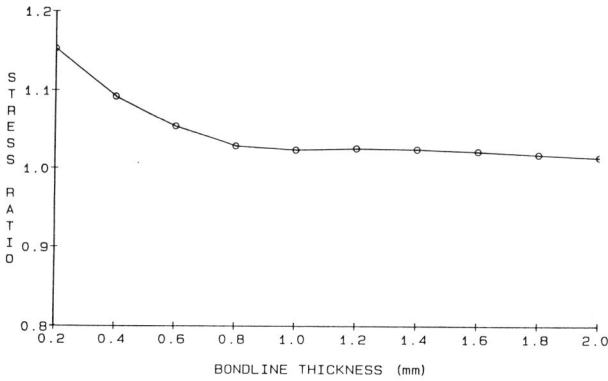

Fig. 11 - Influence of bondline thickness on tension specimen stresses

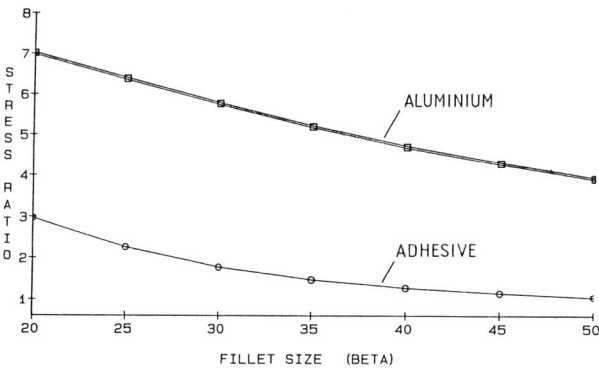

Fig. 12 - Influence of adhesive fillet size on tension specimen stresses

262

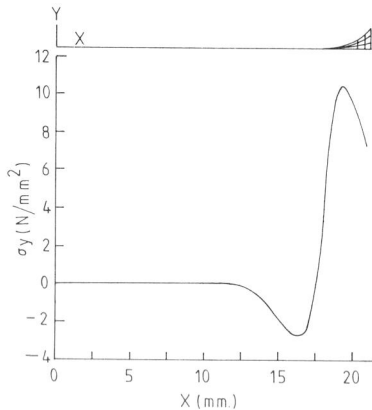

Fig. 13 - Distribution of normal stresses along the mid-line of the bondline

important, therefore, that for consistent joint strength, the fillet size is carefully controlled during joint manufacture.

CONCLUSIONS

In order to assess the strength and stiffness of joints and bonded components, models of varying complexity may be employed. The accuracy of the modelled stresses and stiffness will depend greatly on the type of bondline representation used. By using detailed modelling, it is possible to analyse the influence of joining details on stresses in the metal and the adhesive thereby determining the prefered joint geometry and configuration.

As an example, a front cross-member of a vehicle frame has been tested and analysed, in the non-linear range, and the results have shown a very good correlation. It is the authors' opinion that analyses of this kind will contribute to the development of detailed codes for adhesively bonded aluminium structures, which will simplify their design.

REFERENCES

Marwick, W.F. and P.G. Sheasby (1987). Evaluation of Adhesives for Aluminium Structured Vehicles. SAE Technical Paper Series, 870151.
Nardini, D and A. Seeds (1989). Structural Design Considerations for Bonded Aluminium Structural Vehicles. SAE Technical Paper Series, 890716.
Seeds, A., D. Nardini and F. Cassese (1989). The Development of Centre Cell Structure in Bonded Aluminium for the Ferrari 408 Research Vehicle. SAE Technical Paper Series, 890717.
Warren, A.S., J.E. Wheatley, W.F. Marwick and D.J. Meadows (1989). The Building and Test-Track Evaluation of an Aluminium Structured Bertone X1/9 Replica Vehicle. SAE Technical Paper Series, 890718.
Wheeler, M.J., P.G. Sheasby and D. Kewley (1987). Aluminium Structured Vehicle Technology - A Comprehensive Approach to Vehicle Design and Manufacturing in Aluminium. SAE Technical Paper Series, 870146.

Characterisation and Measurement of the Dynamic Initiation Fracture Toughness of Metallic Materials: A Review

R. A. W. MINES

*Impact Research Centre, Faculty of Engineering,
The University of Liverpool, P.O. Box 147,
Liverpool L69 3BX, UK*

ABSTRACT

Dynamic initiation of cracks in metals occurs as a result of impact loading. For the case of small scale yielding, the fracture toughness of the material reduces with elevated strain rate, whereas for large scale yielding fracture toughness increases with strain rate. The review concentrates on the modelling of the crack tip stress field and the micromechanical causes of the rate dependency of dynamic initiation fracture toughness. The validation of developed fracture criteria requires careful experimental testing, and problems associated with instrumented impact testing are discussed. Some results are given for 150M12 and 817M40/En24 structural steels.

KEYWORDS

Dynamic, impact, fracture toughness, strain rate, measurement, characterisation, metals.

INTRODUCTION

The effect of elevated strain rate on the flow of metallic materials is to increase the yield stress and work hardening index and to reduce the fracture strain. Body centred cubic structures, e.g., ferritic steels, show the greatest strain rate sensitivity in yield stress. The possibility of cleavage, for example in structural steels, means a reduction in cleavage stress with strain rate and the possibility of transition from fibrous to cleavage crack initiation, as a result of elevated strain rate. For ferritic steels, a thirty per cent increase in yield stress occurs for an increase in strain rate of five orders of magnitude (Harding, 1987). For structural steels, a reduction in fracture toughness of fifty per cent for an increase in strain rate of five orders of magnitude has been found (Klepaczko, 1982). Hence, the study of material behaviour at high rates of strain has wide technological implications.

Material failure can be a result of overstraining a homogeneous continuum. In this case micromechanical models have to be developed which characterise void formation and coalescence to form macrocracks (Curran et al., 1987). Only when such macrocracks are formed, or are pre-existing, can the macromechanical concepts of fracture mechanics be employed. This review will address this latter subject and will cover characterisation of material behaviour at the crack tip, experimental derivation of fracture parameters and, briefly, the application of developed criteria to the assessment of the performance of machine and structural elements. The review will be restricted to mode I, tensile, loading. Also crack initiation only will be discussed. The subject of dynamic crack propagation and arrest is a large subject and is reviewed by Kanninen and Popelar (1985).

LINEAR ELASTICITY AND SMALL SCALE YIELDING (SSY)

Characterisation for Small Scale Yielding (SSY)

Under static conditions, the constraint conditions for valid plane strain fracture toughness (K_{IC}) measurement is given by Rolfe and Barsom (1977):

$$B, \ (W-a) \geq 2.5 \ \left(\frac{K_{IC}}{\sigma_y}\right)^2 \tag{1}$$

where B is the section thickness and (W-a) the ligament size. Table 1 gives constraint values for two materials currently being studied by the author.

It can be assumed that a relation, similar to that given in eq. (1), will hold for dynamic conditions (Rolfe and Barsom, 1977) and hence with the decrease in K_{IC} and increase in σ_y with increasing strain rate, the section size required for K dominance will reduce. Table 1 gives the effect on constraint limits for an increased strain rate.

Nilsson (1984) has stated that for a stationary crack loaded dynamically the spatial distribution of stress around the crack tip is the same as for the static case, i.e., an inverse square root singularity and Westergaard's solution are applicable. The strength of the K singularity is dependent on remotely applied loads, which are time dependent and which may be a result of stress wave or structural inertia effects. Analytic solutions for dynamic fracture problems deal with fundamental but idealised problems, e.g., a sub surface semi-infinite crack in an semi-infinite plane (Brock et al., 1985), or with the strength of material models of finite geometries, e.g., the three point bend specimen (Nash, 1969). Numerical solutions use singularity elements which assume the crack tip solution but which are numerically efficient (Kanninen, 1978). Experimental methods, such as dynamic photoelasticity, for transparent (model) materials, and dynamic caustics, for opaque materials, are well established (Burger 1987, Kalthoff 1987). Hence dynamic analyses can either be applied to laboratory specimens for deriving dynamic fracture parameters or to engineering components for the application of parameters. In the latter case numerical methods are the most usual. Dynamic fracture problems are usually posed in terms of plane stress or strain, although the stress intensity factor can vary along the crack front. Nakamura et al. (1986b) have published some three dimensional numerical work which deals with both elastic, and elasto-plastic behaviour. Such analyses are computationally demanding

Table 1. Mechanical Properties for 150M12 and 817M40/En 24 Steels (RT - Room Temperature)

	Temperature (° C)	150M12	En24/ 817M40[*]
σ_{ys}: Static (MPa)	RT	400	1450
	-80	$(440)^{\Delta}$	$(1640)^{+}$
σ_{yd}: Dynamic (MPa) @ $\dot{\varepsilon}$ = 100 s⁻¹ (150M12) $\dot{\varepsilon}$ = 300 s⁻¹ (817M40)	RT	500	$(1560)^{\phi}$
	-80	$(640)^{\nabla}$	$(1852)^{\phi}$
K_{IC}: Static (MPa √m)	RT	-	$(70 \times 10^{6})^{\phi}$
	-80	-	$(54 \times 10^{6})^{\phi}$
K_{Id}: Dynamic (MPa √m) @ $\dot{K}_I$ = ~ 10^6 MPa √m s⁻¹ ($\dot{\varepsilon}$ = 102 s⁻¹ from eq. 3)	-80	43×10^{6}	48×10^{6}
J_{IC}: Static (Jm⁻²)	RT	$(20.5. \times 10^{3})^{\Delta}$	(21.2×10^{3})
$2.5 \left(\frac{K_{IC}}{\sigma_{ys}}\right)^2$ (mm)	RT	-	(6)
$2.5 \left(\frac{K_{Id}}{\sigma_{yd}}\right)^2$ (mm) @ $\dot{\varepsilon}$ = 100-300 s⁻¹	-80	(14)	(2)
$25 \left(\frac{J_{IC}}{\sigma_y}\right)$ (mm)	RT	(1.25)	(0.4)

( ) derived quantities

[*] tempered at 375° C
[∇] Symonds, 1967
[+] Lee and Kang, 1986
[Δ] Bann and Geary, 1987
[ϕ] Tanimura and Duffy, 1986

requiring approximately 100 hours CPU time on a Cray 2 to calculate the real time response of 2 milli-seconds in an impacted three point bend specimen, for a rate dependent material. A similar two-dimensional analysis would take 1 hour of CPU time.

Given the ability to calculate or measure a dynamic stress intensity factor ($K_I(t)$), the next step is to define the criterion for fracture. The usual criterion is:

$$K_I(t) = K_{Id} \text{ at fracture} \qquad (2)$$

which is a similar methodology to the static case. K_{Id} is the dynamic initiation fracture toughness and is assumed to be a material property.

Such a criterion is simplistic as it takes no account of history effects and the requirement of a finite material damage time, which could be of importance at very high loading rates (Ravi-Chandar and Knauss, 1984). Also possible crack tip heating (Rice and Levy, 1969) and crack blunting, i.e., loss of singularity, are not taken into account. Such a simplistic criterion is acceptable for the rates met at intermediate impact rates, e.g., $\dot{K}_I < 10^6$ MPa $\sqrt{m}$ s^{-1} where $\dot{K}_I = K_{Id}/t_f$, and t_f is the time to fracture. Klepaczko (1984) has related the rate of change of stress intensity factor, $\dot{K}_I$, to an average strain rate ($\dot{\varepsilon}$) from the size of the plastic zone:

$$\dot{\varepsilon} = \frac{\sigma_y}{E} \frac{\dot{K}_I}{K_{Ic}} \qquad (3)$$

Table 2 relates $\dot{K}_I$ and $\dot{\varepsilon}$ values for various notched geometries and impact conditions.

Table 2. Some Common Dynamic Fracture Tests with Associated Strain Rates

	Ranges		
	$\dot{\varepsilon}$ (s^{-1})	$\dot{K}_I$ (MPa $\sqrt{m}$ s^{-1})	V_{impact} (ms^{-1})
Circularly grooved bar (Wilson et al., 1980)	102-1020	10^6-10^7	(30 μs stress ramp)
Compact tension in high rate servohydraulic machine (Shum et al., 1985)	0-1	Static-10^4	0-15
Compact tension in split Hopkinson Bar (Klepaczko, 1982)	0-102	Static-10^6	0-30
Instrumented Charpy (Wullaert, 1980)	10.2-102	10^5-10^6	2-5
Three point bend (span ~ 120 mm) (Kanninen et al., 1979)	0-10.2	Static-10^5	0-15
Single point bend (Kalthoff, 1985)	102-1020	10^6-10^7	30-100

N.B. 1. $\dot{\varepsilon}$ related to $\dot{K}_I$ using equation 3

N.B. 2. Data given for En24/817M40 (see Table 1)
i.e., $\dot{\varepsilon} = 0.102 \times 10^{-9}\ \dot{K}_I$

Another aspect of importance is defining the zone of K dominance for a real material, i.e., the zone in which K can be said to control the stress field. Figure 1 shows the sizes of the process zone and singular field for 817M40 steel for static conditions and room temperature.

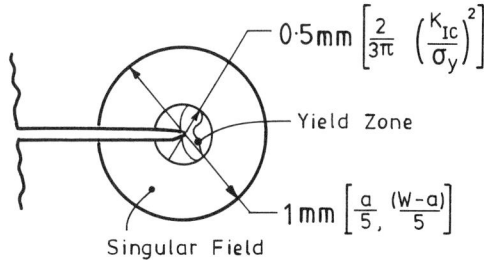

$$0.5\,\text{mm} \left[\frac{2}{3\pi} \left(\frac{K_{Ic}}{\sigma_y}\right)^2\right]$$

Yield Zone

$$1\,\text{mm} \left[\frac{a}{5},\ \frac{(W-a)}{5}\right]$$

Singular Field

Fig. 1.   Zone of K Dominance for 817M40 Steel (Room
temperature, static, fatigue precracked Charpy)

An alternative, and widely used, approach to linear elastic fracture uses the methodology of energy. In the case of the dynamic initiation of stationary cracks, an energy balance is only possible for the case of a quasi static stress field at any given time (Nakamura et al., 1985). Such an approach has been used for the dynamic fracture of polymers (Williams, 1987), but the approach is not discussed further here given that global energy balances give little information on the behaviour of the material at the crack tip.

Measurement of Dynamic Initiation Fracture Toughness (K_{Id}) Under Small Scale Yielding

The task of laboratory instrumented impact tests is to derive a fracture toughness surface, K_{Id} ($\dot{K}_I$, T), where T is the temperature. Figure 2 shows a schematic surface for a structural steel derived from data given in Klepaczko (1985a). The achievement of such a surface is both complex and expensive given the fact that no single fracture toughness test can cover all strain rates. Also understanding, and hence quantifying, the inertial effects of the impact machine on specimen behaviour is often difficult. Table 2 gives the most widely used dynamic fracture toughness tests with their range of strain rates. An up to date review has been provided by the American Society of Metals (Newby, 1985). Loading mechanisms can be by drop weights, drop weight pendulum, gas guns, high rate servohydraulic machines and stress wave bars.

As an example of the problems associated with the accurate derivation of dynamic initiation fracture toughness values, a technique developed by the author, known as the Hopkinson Pressure Bar Loaded Instrumented Charpy Test (Mines and Ruiz, 1985, Ruiz and Mines, 1985) will be briefly described. Figure 3 shows the physical system and inertial model, whilst Fig. 4 shows a typical force-time plot.   Stress wave bars and one

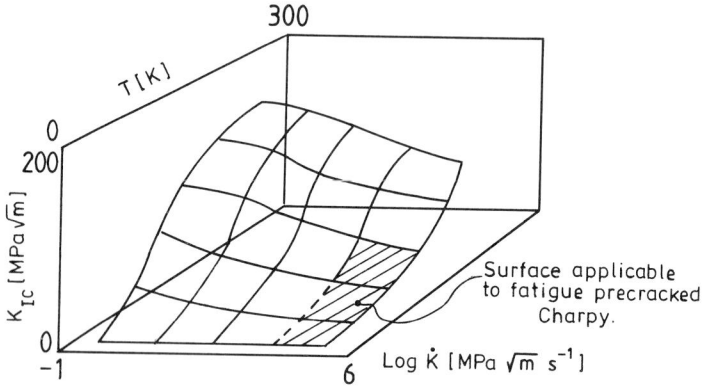

Fig. 2. Fracture Toughness Surface for Fe E460
Steel (German Standard)

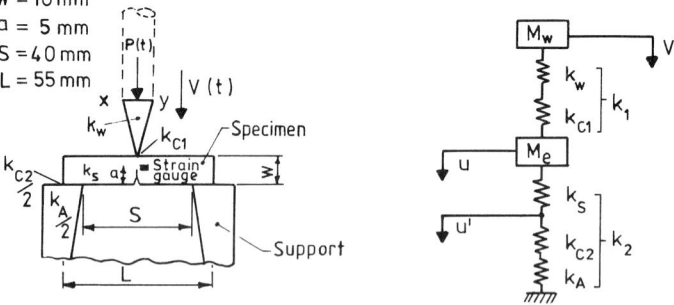

Fig. 3. Physical System and Inertial Model for Hopkinson
Pressure Bar Loaded Instrumented Charpy Test

dimensional stress wave theory are used to derive accurate values of load and velocity at section XY in Fig. 3 (Ruiz and Mines, 1985). The impact test can be modelled as a wedge stiffness (k_w), contact stiffness (k_{C1}), specimen effective mass (M_e), wedge mass (M_w), specimen stiffness (k_s), support contact stiffness (k_{C2}) and support stiffness (k_A). These component stiffnesses and masses can be calculated and input into the inertial model. The forcing function is as measured in a test using one dimensional stress wave theory, and numerical results are derived using a Runge-Kutta algorithm. Numerical and experimental results are compared in Fig. 4. The specimen applied force at failure is used to calculate a K_I value using a calibration function where the time of crack initiation (t_f) is measured using a crack tip strain gauge (see Fig. 3). From Fig. 4 it

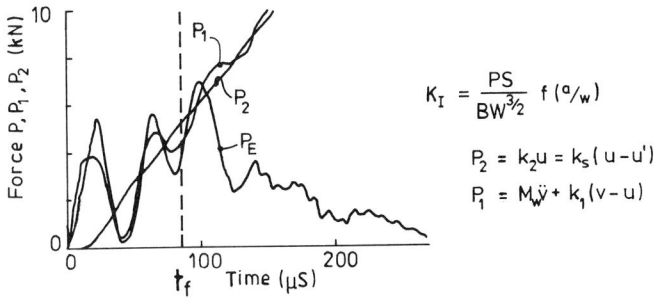

$$K_I = \frac{PS}{BW^{3/2}} f(a/w)$$

$$P_2 = k_2 u = k_s(u-u')$$

$$P_1 = M_w \ddot{w} + k_1(v-u)$$

Fig. 4. Comparison of Experimental and Inertial Model Forces for Instrumented Charpy Test for 817M40/En24 Steel at T = -20° C and Velocity of Impact = 2.86 ms^{-1}

can be seen that there is a difference between measured load, i.e., P_1 from inertial model or P (experimental), and the force experienced by the specimen (P_2). This difference is due to contact stiffness effects. Also the remotely measured load does not instantaneously reflect crack initiation. Inaccuracies in K_{Id} determination can be due to (1) the measurement of total not specimen force, (2) the inaccurate measurement of crack initiation and (3) the use of a quasi static calibration function. Such problems are typical of instrumented impact tests. Also of importance is the applicability of the K_I parameter. Table 1 shows that for structural steels small fracture specimens, such as the fatigue pre-cracked Charpy specimen, can give invalid results. The fatigue pre-cracked Charpy specimen is also restricted in the strain rates it can cover (see Table 2 and Fig. 2).

Models for the Strain Rate Sensitivity of K_{Id}

The development of models is dependent on a knowledge of the state of stress at the crack tip at failure and of the failure mode. Development of models also requires the relating of macroscopic parameters, e.g., crack tip stress fields, to micromechanical properties, e.g, critical stress or strain. Two major contributions to the solution of this problem have been given by Klepaczko (1985b) and Nilsson and Stahle (1989). From Fig. 2 it can be seen that fracture toughness under small scale yielding for structural steels reduces with increasing strain rate and decreasing temperature. Both the flow and fracture of the material are dependent on environmental conditions, e.g., constraint, temperature and strain rate.

Klepaczko (1985b) combines the local conditions for unstable cleavage, viz. the RKR criterion (Ritchie et al., 1973), with the static crack tip stress field, viz. the HRR criterion (Hutchinson, 1968, Rice and Rosengren, 1968) to give:

$$K_{IC} = A_L \sigma_y \left(\frac{\sigma_f}{\sigma_y}\right)^{\lambda(n)} \tag{4}$$

where A_L is a constant, σ_f the fracture stress, σ_y the yield stress and $\lambda(n)$ a function of the work hardening index for the material. The effect of elevated strain rate is to increase σ_y and n. The rate dependency of σ_f is less well defined, and Klepaczko (1985b) cites an increase of 15.2 per cent for a SA 302B steel from static to $\dot{K}_I$ of 10^5 MPa $\sqrt{m}$ s^{-1}. Given this model, the reduction in fracture toughness due to strain rate effects in both flow and fracture can be studied.

Nilsson et al. (1989) propose a strip yield model as being simpler than full singularity modelling and as a first step to modelling rate dependent crack tip problems. They investigate the relation between the intensity of their crack tip model and critical stress, for cleavage fracture and critical strain, for fibrous fracture. For cleavage fracture they show a reducing fracture toughness with increasing strain rate in line with Klepaczko's model (1985b) and experimental results (Klepaczko, 1982).

LARGE SCALE YIELDING

Characterisation for Large Scale Yielding

Table 1 shows the constraint values for K dominance. Hence for small sections and ductile materials, K dominance is lost. The major problem then is the characterisation parameter. Fracture parameters such as the J integral, Crack Opening Displacement, Crack Tip Opening Displacement and Crack Tip Opening Angle can be developed from static fracture mechanics (Kanninen and Popelar, 1985). A wide class of materials can be modelled as power law hardening (Ramberg and Osgood, 1943) and hence the HRR singularity (Hutchinson, 1968, Rice and Rosengren, 1968) can be said to dominate. The singularity has the form:

$$\sigma_{yy} = \sigma_O \, f(n) \, \left(\frac{J}{\varepsilon_O \, Ix}\right)^{1/n+1} \tag{5}$$

where σ_{yy} is the normal stress at a distance x directly ahead of the crack tip, σ_O is the flow stress, J is a measure of geometry and applied load, ε_O is the yield strain, n is the Ramberg-Osgood hardening exponent, I is a numerical constant weakly dependent on n and f(n) the value of the normalised angular distribution of σ_{yy} (Ritchie et al., 1973). Figure 5 shows a comparison of the power law hardening material model for 150M12 and 817M40/En24 steels. In the case of the former material, dynamic values are shown. The J coefficient in eq. (5) is equivalent to the J contour integral and is valid for proportional loading, monotonic loading and deformation theory. J dominance includes linear elastic behaviour and can be applicable up to uncontained yielding. J values can be influenced by whether tension or bending loading dominates. The constraint conditions for J dominance are given by:

$$B, \, W-a \geq 25 \left(\frac{J_{IC}}{\sigma_y}\right) \tag{6}$$

(Rolfe and Barsom, 1977). As with K dominance, the zone of J dominance has to be defined. Figure 6 shows the relative sizes of the process zone and singularity zone for 817M40/En24 at room temperature. The relative size of the singular region to the process zone is larger than in the case of K

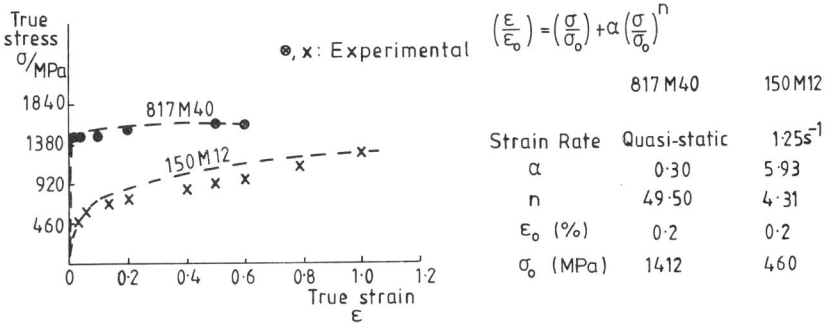

$$\left(\frac{\varepsilon}{\varepsilon_0}\right) = \left(\frac{\sigma}{\sigma_0}\right) + \alpha \left(\frac{\sigma}{\sigma_0}\right)^n$$

⊛, x: Experimental

	817M40	150M12
Strain Rate	Quasi-static	$1.25s^{-1}$
α	0.30	5.93
n	49.50	4.31
ε_0 (%)	0.2	0.2
σ_0 (MPa)	1412	460

Fig. 5.   Comparison of Power-Law Hardening Model with
Data for 150M12 and 817M40/En 24 Steels at
room temperature.   817M40 tempered at 427° C

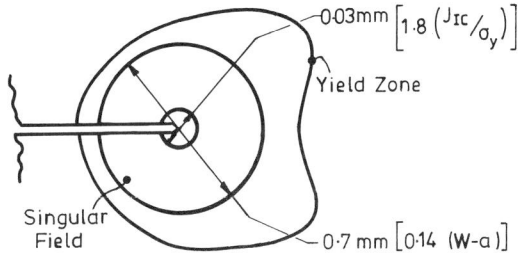

Fig. 6   Zone of J Dominance for 817M40 Steel (Room
temperature, static, fatigue pre-cracked Charpy)

dominance (see Fig. 1).   The size of the J dominant region depends on the
yield stress.   For an increase in σ_y, the process zone size would reduce.

Under dynamic loading conditions the crack tip stress field will be
modified by local inertia effects and the strain rate dependence of
material flow.   It is easier to develop Rate Independent (RI) analyses,
i.e., inertia only, before studying Rate Dependent (RD) effects.   No
formulation of the effect of inertia on the HRR singularity has yet been
developed (Hoff et al., 1985).   The J integral formulation is similar to
the static case:

$$J = \lim_{\Gamma \to 0} \int_\Gamma \left(Un_1 - \sigma_{ij} n_j \frac{\partial u_i}{\partial x_1}\right) d\Gamma \tag{7}$$

where Γ is defined in Fig. 7(a), U is the elastic strain energy density, n is the unit outward normal to the contour and σ the stress component (see Fig. 7(a)). In the dynamic case the J integral is not path independent and hence the material inertia in the vicinity of the crack tip has to be taken into account (Mall, 1982):

$$ J = \int_{\Gamma_0} (Un_1 - \sigma_{ij} \, n_j \, \frac{\partial u_i}{\partial x_1}) \, d\Gamma + \int_{A_0} \rho \, \frac{\partial u}{\partial t} \cdot \frac{\partial u_i}{\partial x_1} \, dA \qquad (8) $$

where Γ_0 and A_0 are defined in Fig. 7(b), ρ is the material density and $\partial u/\partial t$ is the material velocity. Contour Γ' is shrunk onto the crack tip. This equation can be solved numerically using specialised computer codes such as ABAQUS (Nakamura et al., 1986a). A rate independent (RI) two dimensional analysis with inertia is discussed by Nakamura et al. (1986a) and they show that at low loads there can be a 20 per cent difference between a static and dynamic analysis. To investigate the effect of rate on material flow, Hoff et al. (1985) consider an inertialess material. For a fifty per cent increase in flow stress with strain rate, the J value increases by seven per cent. Another possible source of inaccuracy in J determination is the neglecting of three dimensional effects. J can vary by a factor of two across a crack front in a three point bend specimen (Nakamura et al., 1986b).

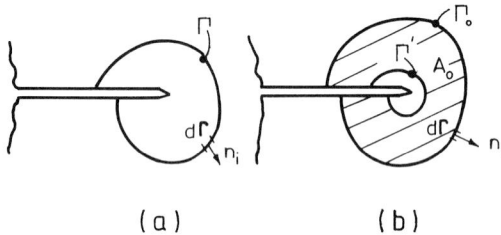

(a)   (b)

Fig. 7   (a)   Contour for Quasistatic J
         (b)   Contours for Dynamic J

As in the linear elastic case, the criterion for failure can be given by:

$$ J_I(t) = J_{Id} \text{ at fracture} \qquad (9) $$

where $J_I(t)$ is the time dependent J integral and J_{Id} the dynamic initiation fracture toughness under large scale yielding. It should be noted that if the material departs from power law hardening, then the order of the crack singularity changes (Nilsson and Stahle, 1989). In this case the intensity of the singularity is given by a conservation integral, I.

The relating of the strain rate at the crack tip $(\dot{J})$ with the strain rate of the material from a uniaxial dynamic tensile test $(\dot{\varepsilon})$ is problematic for the case of J dominance. Hoff et al. (1983) investigate a number of derivations, including the assumption of a linear increase of strain in the process zone and of an HRR singularity. For these calculations:

$$\dot{\varepsilon} = C \; \frac{\dot{J}_I}{J_I} \tag{10}$$

where

$$0.00024 < C < 0.0025 \quad \text{in the plastic zone}$$
$$C \sim 0.35 \quad \text{in the process zone}$$

The relative size of process to plastic zone for a structural steel is typically 0.03 (Hoff et al., 1983). Hoff et al. (1983) conclude that rate effects depend on the behaviour of both the process and plastic zone and that relationships given in eq. (10) depend on the properties of the material.

It has been shown that the Crack Opening Displacement (COD) is consistent with the J integral for materials obeying the constitutive behaviour for a valid J (Kanninen and Popelar, 1985):

$$\delta = d_n \frac{J}{\sigma_y} \tag{11}$$

where δ is the COD and d_n is a function of material properties. An advantage of the COD is that it is less rate sensitive than J (Hoff et al. 1983), the rate dependency being dependent on the behaviour of the process zone. Disadvantages of COD are that the parameter is difficult to measure and calculate (Kanninen and Popelar, 1985).

As in the small scale yielding case, the energetic interpretation of J is not discussed here given the limited significance of global energy balances for the dynamic initiation of stationary cracks (Nakamura et al., 1985).

Measurement of Dynamic Initiation Fracture Toughness (J_{Id}) Under Large Scale Yielding

The usual method of determining J values from experimental tests is by the use of deep crack formulations (Rice et al., 1973). For the three point bend specimen:

$$J_I = \frac{2}{b} \int_0^\Delta P d\Delta \tag{12}$$

where Δ is the displacement of the central loading point, P is the applied load and b the section thickness. It should be noted that care should be taken for the assessment of applied load. Contact stiffness effects should be taken into account (see Fig. 4). Equation 12 assumes a quasi-static stress distribution in the specimen - a reasonable assumption for the longer fracture times in elasto-plastic problems, but the upper limit of strain rate needs to be defined. Nakamura et al. (1986a) define a transition time, which is similar to empirically derived quasi-static limits (Server, 1978). In many instrumented impact tests these quasi-static limits are not met. An added complication in the use of deep crack formulations is the assessment of the various contributions of elastic and plastic effects (Sumpter and Turner, 1976). The incorporation of dynamic and elasto-plastic effects into deep crack formulations is dependent on the geometry and on the ductility of the material. The conduct of an instrumented impact test is similar to the elasto-dynamic case. The major problem is the identification of the point of crack initiation. Crack tunnelling in the specimen may occur, making specimen surface strain measurements invalid. No satisfactory method has yet been found for measurement of crack initiation in the elasto-plastic regime (Nilsson, 1984).

The Strain Rate Sensitivity of J_{Id}

The measurement of fracture toughness surfaces and the development of strain rate sensitive models is less well developed compared with the small scale yielding case. In the elasto-plastic regime physical phenomena become more complex viz. crack tunnelling, ductile-brittle transitions, large scale yielding and crack tip heating. The crack tip field for a power law hardening material can be characterised by extending the static concepts of the J integral and the HRR singularity. At a critical J value the crack can initiate by the mechanism of strain induced linking of voids (Ritchie et al., 1979). Hence a rate dependent model could be developed by linking the strain in the vicinity of the crack tip to a critical strain, ε_f. This methodology has been pursued for the application of the J integral to the case of small scale yielding with fibrous fracture. Ritchie et al. (1979) and Hutchinson (1987) state:

$$(J_{IC})_{SSY} = \text{Constant}.\varepsilon_f.\ell_o.\sigma_y \qquad (13)$$

where ℓ_o is a characteristic distance over which the critical plastic strain is exceeded. Klepaczko (1985b) shows that ε_f reduces with increasing strain rate. The variation of J_{Id} with strain rate is therefore dependent on two competing mechanisms, described by ε_f and σ_y. It is proposed that J_{Id} will increase with increasing strain rate for most metals. No systematic study of the J_{Id} variation with strain rate has yet been published.

A major role of the, rather complex, calculation of J_{Id} is to quantify the transition from fibrous to cleavage dominated failure. This has been observed in instrumented Charpy tests (Server, 1985). Tvergaard (1988) has noted the necessity of developing dynamic viscoplastic material models to study ductile brittle transitions. An approach, favoured in France, is the use of probabilistic methods, e.g., Pluvinage, 1987. In this a statistically determined array of microcracks in front of the crack tip is said to influence the mode and initiation of fracture. Also the transition from J dominated to K dominated fracture, as a result of elevated strain rate, needs to be quantified. This will be influenced by the strain rate sensitivity of yield stress and work hardening index and the density of the material.

The development of rate sensitive J_{Id} analyses should also be regarded as a step to a more general, conservation integral, analysis for materials with constitutive behaviours that depart from power law hardening (Nilsson and Stahle, 1989). In this case crack tip stress fields will change as will the micromechanics of failure. The aim is to develop a completely general analysis for the dynamic initiation of cracks.

APPLICATIONS

The reduction of K_{Id} with elevated strain rate is of importance, for example, in the rail industry (Cannon and Sharpe, 1980), the nuclear industry, where an added hazard is embrittlement due to irradiation (Little, 1986) and in the gas industry where crack propagation in pipelines has to be guarded against (Fearnehough, 1972). A major problem in locating examples of the application of dynamic fracture methodologies, is the association of the technique with accident investigations making developed reports confidential. An indication of the current level of interest in K_{Id} testing, in Europe, can be found in Baker, 1988. Developed methodologies may also be important for machine elements, e.g., impact hammers and mining equipment.

The limitation of K dominance for many engineering applications requires the development of a parameter that can deal with elasto-plastic material behaviour. An increase of J_{Id} with strain rate for many materials gives a static analysis as a lower bound, but the possible transition from fibrous to cleavage fracture needs to be quantified (Sumpter et al., 1988). An associated problem is crack 'pop-in', which can occur under static loading, but which is a dynamic event (Satoh and Toyada, 1986).

CONCLUSIONS

The review has been mainly concerned with steels, but developed methodologies could be of use to non-ferrous and polymeric materials. Also the review has been concerned with a specific type of material behaviour, viz. material failure in the intermediate strain rate regime in a singularity dominated stress field.

In the case of small-scale yielding, dynamic fracture toughness surfaces need to be developed for common materials. Such an undertaking requires well defined experimental procedures and data reduction techniques. It is proposed that K_{Id} can be regarded as a material property for intermediate strain rates, although no systematic study of the independence of geometry on K_{Id} has yet been undertaken. Application of K_{Id} values to structural problems requires simplistic failure criteria (see eq. 2). In the case of large-sale yielding, the development and experimental validation of fracture characterisation parameters need to be progressed further. The development of the J integral would seem to be the most promising, although the use of the more general conservation integral will provide a basis for a completely general treatment of the dynamic initiation of cracks. The basis for the strain rate dependence of macroscopic parameters needs to be sought at the micromechanical level. Again, application of J_{Id} values to structural problems currently requires simplistic failure criteria (see eq. 8).

ACKNOWLEDGMENTS

The work was supported by the Science and Engineering Research Council Grant GR/E/00570 and the Health and Safety Executive Contract No.

2163/R31.08. Dr. S. McParland and Mr. G. Dutton developed the materials data. Mr. G. Dutton developed the inertial model of the Hopkinson bar Loaded Instrumented Charpy Test. Mrs. M. White typed the manuscript and Mrs. A. Green traced the figures.

REFERENCES

Baker, R. G. (1988). Report of a study on metrological requirements in fracture mechanics testing. Report No. X11/353/88, Commission of the European Communities, Brussels.

Bann, P. J. and Geary, W. (1987). Fracture toughness of 150M12 mining steel to BS 2772 under static loading conditions. Report No. R31.07.MQM.01, Health and Safety Executive, Sheffield.

Brock, L. M., Jolles, M. and Schroedl, M. (1985). Dynamic impact over a sub surface crack: applications to the dynamic tear test. J. Appl. Mech., ASME 52, 287-290.

Burger, C. P. (1987). Photoelasticity. In: Handbook on Experimental Mechanics (A. S. Kobayashi, ed.), Chap. 5, 162-281, Prentice Hall, New Jersey.

Cannon, D. and Sharpe, K. A. (1980). Wheel rail impact loads: a contributory cause of rail fracture. In: Analytical and Experimental Fracture Mechanics (G. C. Sih and M. Mirabile, eds.), 643-656, Sijhoff and Noordhoff, Netherlands.

Curran, D. R., Seaman, L. and Shockey, D. A. (1987). Dynamic failure of solids. Physics Reports, 5 and 6, 253-388.

Fearnehough, G. D. (1972). The small scale test and its application to fracture propagation problems. In: Dynamic Crack Propagation (G. C. Sih, ed.), 77-101, Noordhoff, Netherlands.

Harding, J. (1987). Effect of strain rate on material properties. In: Materials at High Strain Rates (T. Z. Blazynski, ed.), Chap.4, 133-186, Elsevier Applied Science, London.

Hoff, R., Rubin, C. A. and Hahn, G. T. (1983). High rate deformation in the field of crack. In: Material Behaviour Under High Stress and Ultrahigh Loading Rates (J. Mescall and V. Weiss, eds.), 223-240, Plenum Press, New York.

Hoff, R., Rubin, C. A. and Hahn, G. T. (1985). Strain rate dependence of the deformation at the tip of a stationary crack. In: Fracture Mechanics: Sixteenth Symposium, ASTM, STP 868 (M. F. Kanninen and A. T. Hopper, eds), 409-430, ASTM, Philadelphia.

Hutchinson, J. W. (1968). Singularity behaviour at the end of a tensile crack in a hardening material. J. Mech. Phy. Solids, 16, 13-31.

Hutchinson, J. W. (1987). Micromechanics of damage in deformation and fracture. Technical University of Denmark, Lyngby.

Kalthoff, J. F. (1985). On the measurement of dynamic fracture toughness - A review of recent work. Int. Journ. Fract., 27, 277-298.

Kalthoff, J. F.(1987). Shadow optical method of caustics. In: Handbook on Experimental Mechanics (A. S. Kobayashi, ed.), Chap. 9, 430-500, Prentice Hall, New Jersey.

Kanninen, M. F. (1978). A critical appraisal of solution techniques in dynamic fracture mechanics. In: Numerical Methods in Fracture Mechanics (A. R. Luxmoore, ed.), 612-633, Pineridge Press, Swansea.

Kanninen, M. F., Gehlen, P. C., Barnes, C. R., Hoagland, R. G., Hahn, G. T. and Popelar, C. H. (1979). Dynamic crack propagation under impact loading. In: Nonlinear and Dynamic Fracture Mechanics (N. Perrone, S. N. Atluri, eds.), ASME AMD, Vol. 35, 185-200.

Kanninen, M. F. and Popelar, C. H. (1985). Advanced Fracture Mechanics. Oxford University Press, New York.

Klepaczko, J. (1982). Discussion of a new experimental method in measuring fracture toughness initiation at high loading rates by stress waves. J. Eng. Mat and Technology, ASME, 104, 29-35.

Klepaczko, J. R. (1984). Load rate spectra for fracture initiation in metals. Theor. and Appl. Fract. Mech., 1, 181-191.

Klepaczko, J. R. (1985a). Fracture initiation under impact. Int. Journ. Impact Eng., 3, 191-210.

Klepaczko, J. R. (1985b). Initiation of fracture at different loading rates: An attempt of modelling based on dynamic plasticity. In: Journal de Physique, Colloque C5, Supplement au no. 8, C5245-C5250, Les Editions de Physique, France.

Lee, W. S. and Kang, I. C. (1986). Tensile properties and fracture behaviour of AISI 4340 steel as a function of test temperature (in Korean), Journal of Korean Institute of Metals, 24, 15-23.

Little, E. A. (1986). Fracture mechanics evaluations of neutron irradiated type 321 austenitic steel. Atomic Energy Research Establishment Report No. R11913, Harwell, Oxfordshire.

Mall, S. (1982). A finite element analysis of transient crack problems with a path independent integral. In: Advances in Fracture Research (Fracture 81) (D. Francois, ed.), Vol. 5, 2171-2178, Pergamon Press, Oxford.

Mines, R. A. W. and Ruiz, C. (1985). The dynamic behaviour of the instrumented Charpy test. In: Journal de Physique, Colloque C5, Supplement au no. 8, C5187-C5196, Les Editions de Physique, France.

Nakamura, T., Shih, C. F. and Freund, L. B. (1985). Computational methods based on an energy integral in dynamic fracture. Int. Journ. Fract., 27, 229-243.

Nakamura, T., Shih, C. F. and Freund, L. B. (1986a). Analysis of a dynamically loaded three point bend ductile fracture specimen. Eng. Fract. Mech., 25, 323-339.

Nakamura, T., Shih, C. F. and Freund, L. B. (1986b). Three dimensional transient analysis of a dynamically loaded three point bend specimen. Office of Naval Research Report No. ONR 0365/3.

Nash, G. E. (1969). An analysis of the forces and bending moments generated during the notched beam impact test. Int. Journ. Fract. Mech., 5, 269-285.

Newby, J. R. (1985). American Society of Metals Handbook (9th Edition), Vol. 8 - Mechanical Testing, ASM, Ohio.

Nilsson, F. (1984). Crack growth initiation and propagation under dynamic loading. In: Mechanical Properties at High Rates of Strain, 1984 (J. Harding, ed.), Chap. 2, 185-204, Institute of Physics, London.

Nilsson, F., Ohlsson, P., Sjoberg, F. and Stahle, P.(1989). A strip yield rate dependent model for rapid loading conditions, to be published - Engineering Fracture Mechanics.

Nilsson, F. and Stahle, P. (1989). Crack growth criteria and crack tip models. To be published - S. M. Archives.

Pluvinage, G. (1987). Probabilistic aspects of dynamic fracture mechanics. In: Impact Loading and Dynamic Behaviour of Metals (Impact '87) (C. Y. Chiem, ed.). Paper DF9, Deutsch Gesellschaft fur Metallkunde, Koln.

Ramberg, W. and Osgood, W. R. (1943). Description of stress-strain curves by three parameters. NACA Technical Note No. 902.

Ravi-Chandar, K. and Knauss, W. G. (1984). An experimental investigation into dynamic fracture: I Crack initiation and arrest. Int. J. Fract., 25, 247-262.

278

Rice, J. R. and Rosengren, G. F. (1968). Plane strain deformation near a crack tip in a power law hardening material. J. Mech. Phys. Solids, 16, 1-12.

Rice, J. R. and Levy, N. (1969). Local heating by plastic deformation at a crack tip. In: Physics of Strength and Plasticity (A. S. Argon, ed.), Chap. 20, 277-293, MIT Press, Massachusetts.

Rice, J. R., Paris, P. C. and Merkle, J. G. (1973). Some further results of J-integral analysis and estimates. In: Progress in Flaw Growth and Fracture Toughness Testing. ASTM STP 536 (J. F. Kaufman, ed.), 231-245, ASTM, Philadelphia.

Ritchie, R. O., Knott, J. F. and Rice, J. R. (1973). On the relationship between critical tensile stress and fracture toughness in mild steel. J. Mech. Phys. Solids, 21, 395-410.

Ritchie, R. O., Server, W. L. and Wullaert, R. A. (1979). Critical fracture stress and fracture strain models for the prediction of lower and upper shelf toughness in nuclear pressure vessel steels. Met. Trans., ASM, 10A, 1557-1570.

Rolfe, S. T. and Barsom, J. M. (1977). Fracture and Fatigue Control in Structures. Prentice Hall, New Jersey.

Ruiz, C. and Mines, R. A. W.(1985). The Hopkinson pressure bar: An alternative to the instrumented pendulum for charpy tests. Int. Journ. Fract., 29, 101-109.

Satoh, K. and Toyoda, M. (1986). Guideline for fracture mechanics testing of weld metal in the heat affected zone. International Institute of Welding Document X-1113-86.

Server, W. L. (1978). Impact three point bend testing for notched and precracked specimens. Journ. of Test. and Eval., 6, 29-34.

Server, W. L. (1985). Charpy impact testing. In: Metals Handbook 9th Edition, Vol. 8, Mechanical Testing (J. R. Newby, ed.), 261-268, ASM, Ohio.

Shum, D., Bassim, M. N. and Bayoumi, M. R. (1985). Dynamic tensile fracture toughness evaluation using compact tension specimens. Int. Journ. Fract., 29, R3-R10.

Sumpter, J. D. G. and Turner, C. E. (1976). Method for laboratory determination of J_C. In: Cracks and Fracture, ASTM STP 601, 3-18. ASTM, Philadelphia.

Sumpter, J. D. G., Bird, J., Clarke, J. D. and Caudrey, A. J. (1988). Fracture toughness of ship steels. Paper No. 2, Royal Institution of Naval Architects Spring Meetings, London.

Symonds, P. S. (1967). Survey of methods of analysis for plastic deformation of structures under dynamic loading. Report BU/NSRDC/1-67, Brown University, U.S.A.

Tanimura, S. and Duffy, T. (1986). Strain rate effects and temperature history effects for three different tempers of 4340 VAR steel. Int. Journ of Plast., 2, 21-35.

Tvergaard, V. (1988). Material failure by void growth to coalescence. Technical University of Denmark, Lyngby.

Williams, J. G. (1987). The analysis of dynamic fracture using lumped mass spring models. Int. Journ. Fract., 33, 47-59.

Wilson, M. L., Hawley, R. H. and Duffy, J. (1980). The effect of loading rate and temperature on fracture initiation in 1020 hot rolled steel. Eng. Fract. Mech., 13, 371-385.

Wullaert, R. A. (1980). CSNI specialist meeting on instrumented precracked Charpy testing. Electric Power Research Institute Report No. NP-2102-LD, Palo Alto, California.

A Finite Element Technique for the Investigation of the Shape Development of Planar Cracks with Initially Irregular Profiles

M. I. CHIPALO*, M. D. GILCHRIST** and R. A. SMITH**†

*Department of Engineering, Cambridge University,
Trumpington Street, Cambridge CB2 1PZ, UK
**Department of Mechanical and Process Engineering,
Sheffield University, Mappin Street, Sheffield S1 3JD, UK
†Author to whom correspondence should be addressed

ABSTRACT

The introduction to this paper describes a general finite element model used to calculate opening mode stress intensity factors (K_I) along the front of an irregular planar crack. Crack advance is calculated as a function of K_I and a new profile is thus defined. Automatic reconfiguration of the finite element mesh enables the crack development to be followed. The use of this technique to investigate various problems of practical interest is then described. These examples include the effect of constrained and unconstrained corners on developing cracks, the interaction of two adjacent thumbnail cracks, the breakout of an internally initiated crack and the effect of a surface crack in a leak before break situation.

KEYWORDS

Finite element method; linear elastic fracture mechanics; crack growth; fatigue; opening mode stress intensity factor.

INTRODUCTION

This work contributes to the analysis of the stability and growth of irregular planar defects normal to the remote applied stress by utilising the concepts of linear elastic fracture mechanics. K_I, the opening mode stress intensity factor is used to characterise such problems. A general finite element model used to calculate K_I along the crack front is described. Local crack growth is calculated as a function of K_I along the crack front and a new profile is consequently defined. The finite element mesh is automatically reconfigured to the new profile thus allowing the model to follow the development of the crack shape by repeated iterations.

SOLUTION PROCEDURE

At any point on the crack front, a section plane can be defined to exist orthogonal to the plane of the crack and normal to the tangent to the crack front at the point, as shown in Fig. 1. The displacements of the crack surface are calculated by a finite element analysis and subsequently used in the near crack tip stress field equations to obtain the local K_I value. This procedure is repeated for other points on the crack front and a local normal increment of growth is calculated for each of the points choosen. For the work described here the Paris growth law equation is used to relate local K_I to crack advance. Figure 1 details how the advance of these points establishes a new location of the crack front as a result of fatigue crack growth. Redefinition of the finite element mesh to assume the position of the new crack front, and subsequent iteration of the analysis procedure permits the actual progression of the crack front growth to be followed (Smith and Cooper, 1989). While this investigation examines fatigue crack growth it is equally feasible to analyse any other process that is dependent on a local stress/strain field parameter calculated by the finite element method.

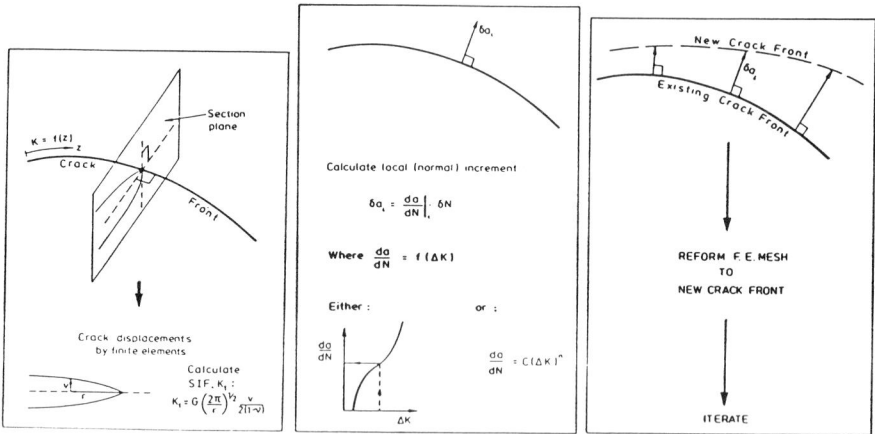

Fig. 1.   Modelling the growth of crack fronts with varying stress intensity factors, K_I, by finite elements and linear elastic fracture mechanics.

The advantages of this particular analysis are twofold; it facilitates the automatic remeshing of the finite element model from one crack front position to the next and it considerable reduces the complexity of the finite element mesh thereby allowing detailed examination of the volume around the crack.

The schematic of a typical problem is shown in Fig. 2. Two blocks are used for the analysis; the larger, block 1, contains the loading surface and the smaller, block 2, contains a definition of the crack shape on its base, face B. The bulk of the uncracked material is represented by block 1 and remains unchanged throughout the analysis. Block 2, modelled with a finer mesh, follows the growth of the crack and is reassembled after each growth step. Faces B and A_1 in this instance define two planes of symmetry while faces A_1 and A_2 are used to apply loading and/or boundary conditions (planes of symmetry or free surfaces etc.) appropriate to the problem under consideration.

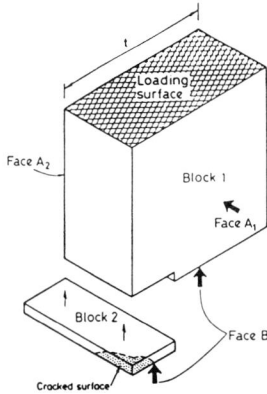

Fig. 2.   Schematic of formation of computational model.

The base of block 2 contains the cracked surface. This plane is defined by the 2-D mesh of Fig. 3 by means of MENTAT, the preprocessing program associated with the MARC finite element solver. The cracked surface is represented as shaded for clarity and the 2-D elements are eight-noded, quadrilateral and isoparametric. However the elements abutting the crack front have their midside nodes located at the quarter points closest to the front, following the work by Barsoum (1976), which avoids the need for using special crack tip elements.

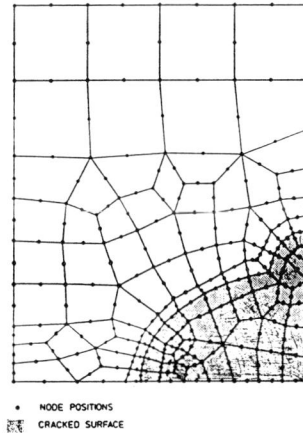

Fig. 3. 2-D representation of cracked surface on face B of block 2.

A purpose written program expands the 2-D mesh of Fig. 3 into the 3-D mesh of block 2 (Smith and Cooper, 1989). Away from the crack front the elements are 20-noded isoparametric bricks. 2-D Elements abutting the crack front are expanded into four 20-noded bricks, two of which are degenerated into 15-noded wedges, as shown in Fig. 4. The midside nodes of elements surrounding the crack tip are automatically set at the quarter points.

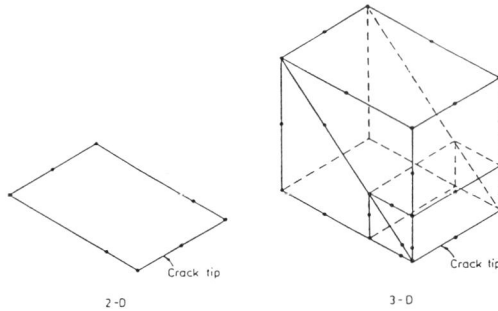

Fig. 4. Expansion of 2-D crack front element into four 3-D elements.
Only visible nodes are shown.

A further purpose written program joins blocks 1 and 2 togther and ensures complete nodal continuity across the interface by interpolation. Coincident nodes on the two surfaces are merged whilst those nodes which are not, are tied to the interface by means of constraint equations (Smith and Cooper, 1989).

It is important to orientate elements so that their edges intersect the crack front orthogonally in order that the true magnitude of the stress intensity factor is calculated for the corner point nodes on the crack tip. This is not possible where the crack makes a sharp change of direction or where the crack tip approaches a free surface. However it was found that sharp corners rapidly developed into smooth curves as the crack grew and consequently a normal mesh layout was possible. The problems of a crack approaching a free surface are due to the stress condition changing from plane strain within the material to plane stress on the free surface. However, the associated potential error, some 5% (Smith and Cooper, 1989), of assuming plane strain conditions throughout is an acceptable limitation of the present method especially since good comparisons with experimental results have been obtained.

Quality checks on the accuracy of the K_I computations have been made against both those computed by Newman and Raju (1981), for the case of an elliptical surface 'thumbnail' crack and by Murakami and Nemat-Nasser (1982, 1983) for a slot with a square protrusion defect. These comparisons are detailed by Smith and Cooper (1989) and are found to be satisfactory.

PRACTICAL APPLICATIONS

A wide variety of practical problems have been analysed; these include such irregular planar defects normal to the remote applied stresses as semi-elliptical, slot type, deep or shallow cracks as may exist either on or beneath the surfaces of a component. It has generally been found that if a crack is permitted to grow indefinitely it will tend towards an iso-K_I configuration, that is K_I round the boundary of the crack becomes constant. In other words, the ratio of the maximum to minimum stress intensity factor for a particular crack profile, K_{Imax}/K_{Imin}, tends towards unity as the crack develops.

The slot with a square protrusion, shown in Fig. 5, illustrates the effect of both constrained and unconstrained internal corners subject to a uniform remote stress. The greatest initial fatigue growth associated with this shape occurs at the unconstrained (or re-entrant) corner, C, whilst the least occurs at the two constrained corners, B and D. The tendency of the crack to assume an iso-K configuration is clearly shown in Fig. 6. Essentially, therefore, constrained corners have low local K's while re-entrant corners have high local K's (Smith and Cooper, 1989). Consequently the growth of arbitrary irregular cracks can be qualitatively predicted.

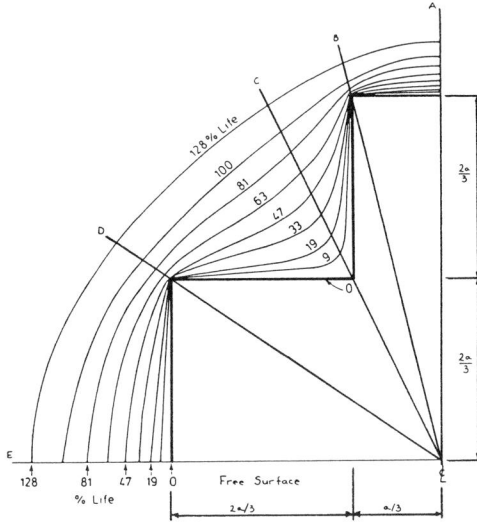

Fig. 5.   Fatigue shape development of a slot defect with a protrusion.

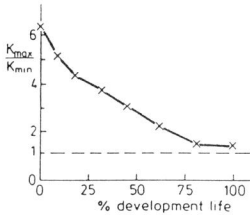

Fig. 6.   Variation of K_{max}/K_{min} along crack front as the effect of sharp corners diminishes.

The interaction and coalescence of two semi-elliptical cracks are investigated by Kishimoto et al. (1989) and Soboyejo et al. (1989) and detailed below in Figs. 7 and 8. The two cracks grow almost independently until their adjacent crack tips coalesce. There appears to be virtually no fatigue crack interaction before coalescence. After coalescence the crack front forms a single cusp shape and grows as a single larger semi-ellipse. The highest crack growth rates occur during coalescence and this causes a rise in the ratio of K_{max}/K_{min} with the growth of the cracks and is shown in Fig. 8. Consequently the K_{max} to K_{min} ratio does not decrease monotonically; however it does eventually tend to unity after the two cracks form a single larger semi-elliptical crack.

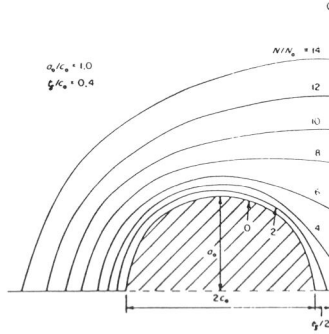

Fig. 7.   Shape development of two semi-elliptical surface cracks by fatigue. (initial aspect ratio $a_0/c_0 = 1$)

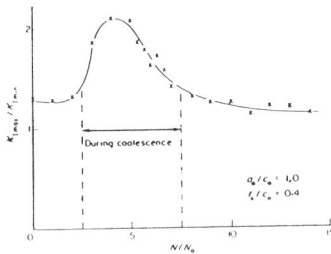

Fig. 8.   Variation of K_{max}/K_{min} along crack front during shape development.

Having established these general characteristics associated with the growth of cracks, the model is being developed to analyse further problems and boundary conditions. Surface cracks in components with one external boundary, (semi-infinite plate), have been discussed above. Internal cracks which break out to become visible on the surface, and cracks in components with two or more external boundaries, such as leak before break problems, can now be examined by this method.

Figure 9 shows the variation of the computed crack profiles for the growth of a subsurface elliptical defect due to a remote applied tensile stress. The ratio N/N_0 for these curves represents a dimensionless number of loading cycles. The profiles are such that the same number of loading cycles is taken to develop from one contour to the next. The defect is seen to grow uniformly, generally maintaining its elliptical shape, until it approaches the free surface. The highest values of K_I then occur at the points on the crack front that intersect the material free surface. Consequently the area of most rapid growth occurs along the free surface whilst the growth inwards into the material is virtually negligible. This pattern of growth continues until the crack profile attains a smooth thumbnail appearance, at which stage an iso-K configuration is established. Figure 10 shows the variation of maximum to minimum stress intensity factor as the crack develops. This decreases towards unity but only after rising to its highest value when the crack interacts with the free surface. Indeed, both Figs. 9 and 10 are very similar to Figs. 7 and 8, respectively, which define the growth and interaction of two semi-elliptical surface cracks.

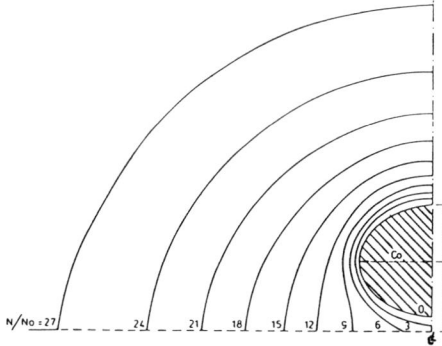

Fig. 9. Fatigue shape development of subsurface elliptical defect.

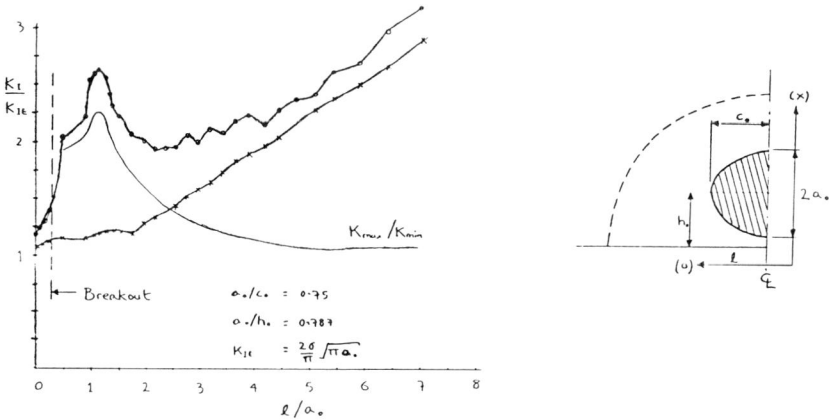

Fig. 10. Variation of K_{max}/K_{min} along crack front during shape development.

An interesting result of this particular analysis concerns the observed surface growth rates. An observer examining the surface of this component would see the sudden appearance of a surface crack having a high growth rate which would decrease before accelerating again. From assumption of a semi-elliptical surface crack, the resultant calculated growth rates against stress intensity factor curve would actually be different from the usual Paris law curve. It is suggested, therefore, that such errors in da/dN Vs ΔK curves may well be due to incorrectly assuming an internal defect to be a surface defect.

The growth of two extreme flaws (short deep crack and long shallow crack) in a leak before break analysis is detailed in Figs. 11 and 12. The flaws are normal to the remote applied tensile stress; the same number of loading cycles is taken to develop the crack from one profile to the next; the ratio N/N_0 represents a dimensionless number of loading cycles. In both cases the cracks grow into smooth thumbnail configurations: growth for the short deep crack is initially highest on the free surface of the plate whilst it is highest inwards into the plate for the long shallow crack. Checks have been made on the stress intensity factors at the begining and end of the analyses against theoretical values and errors of up to a few percent were found in all cases.

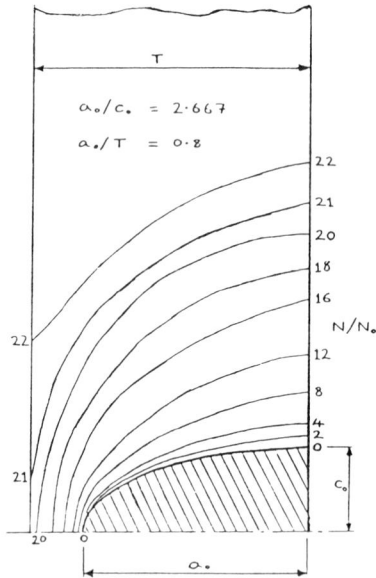

Fig. 11.   Shape development of short deep crack in leak before break analysis.

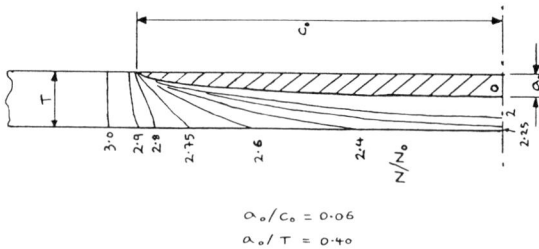

Fig. 12.   Shape development of long narrow crack in leak before break analysis.

CONCLUSIONS

The importance of this particular finite element model lies in its versatility and adaptability in analysing so many different classes of planar cracks. The automatic remeshing of the model for following the development of a crack makes such analyses realistic. The trend of different crack shapes to develop into an iso-K configuration, or smooth thumbnail shape, has been established. It has also been shown that the ratio of maximum to minimum stress intensity factor for a crack profile tends towards unity as the crack develops and that the decrease of this ratio is not always in monotonic fashion.

ACKNOWLEDGEMENTS

Thanks are recorded to MARC for the use of their finite element programs and to both the Marine Technology Directorate of the SERC and the UKAEA for financial support.

REFERENCES

Barsoum, R. S. (1976). On the use of isoparametric finite elements in linear fracture mechanics. *Int. J. Num. Methods in Engng.*, **10**, 25-37.

Kishimoto, K., W. O. Soboyejo, R. A. Smith and J. F. Knott (1989). A numerical investigation of the interaction and coalescence of twin coplanar semi-elliptical fatigue cracks. *Int. J. Fatigue*, **11**, 91-96.

Murakami, Y. and S. Nemat-Nasser (1982). Interacting dissimilar semi-elliptical surface flaws under tension and bending. *Engng. Fract. Mech.*, **16**, 373-386.

Murakami, Y. and S. Nemat-Nasser (1983). Growth and stability of interacting surface flaws of arbitrary shape. *Engng. Fract. Mech.*, **17**, 193-210.

Newman, Jr., J. C. and I. S. Raju (1981). An empirical stress intensity factor equation for the surface crack. *Engng. Fract. Mech.*, **15**, 185-192.

Smith, R. A. and J. F. Cooper (1989). A finite element model for the shape development of irregular planar cracks. *Int. J. Pres. Ves. & Piping*, **36**, 315-326.

Soboyejo, W. O., K. Kishimoto, R. A. Smith and J. F. Knott (1989). A study of the interaction and coalescence of two coplanar fatigue cracks in bending. To appear in *Fatigue Fract. Engng. Mater. Struct.*

A Photoelastic Technique to Predict the Direction of Edge Crack Extension using Blunt Cracks

A. D. NURSE and E. A. PATTERSON

Department of Mechanical and Process Engineering,
University of Sheffield, Sheffield S1 3JD, UK

ABSTRACT

It has been shown that for low values of the ratio of the stress intensity factors (K_{II}/K_I) the axis of symmetry of the isochromatic fringe loops in the neighbourhood of the crack tip is almost equal to the direction of the maximum circumferential stress. Thus, measurement of the former angle enables an accurate prediction of the crack propagation angle to be made.

KEYWORDS

Photoelasticity; crack propagation angle; blunt cracks; stress field equations; fracture tests; edge cracks.

INTRODUCTION

In a mixed-mode fracture test the results will include the failure stress and the direction in which propagation initiates.

The first theory for the direction of crack propagation postulated that the crack will open in a direction, θ_ρ, in which the circumferential stress, σ_θ, is a maximum (Erdogan and Sih, 1963), Fig. 1. This is known as the maximum circumferential stress theory (MCST) and is equivalent to predicting that fracture occurs along the line of zero circumferential or radial shear stress, i.e.

$$\tau_{r\theta} = 0, \quad \sigma_\theta > 0, \quad \frac{\partial^2 \sigma_\theta}{\partial \theta^2} < 0 . \tag{1}$$

The MCST was then extended to incorporate the second order term, σ_{ox}, to include the effects of loads in the direction of the crack (Williams and Ewing, 1972; Finnie and Saith, 1973).

Restricting the use of the MCST to σ_θ has been questioned (Chrysakis, 1986) and it was suggested that the other stresses, σ_r and $\tau_{r\theta}$, can act in an opening mode, Fig. 1 .

289

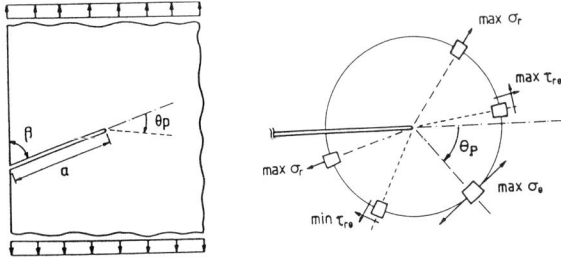

Fig. 1. The direction of maximum σ_θ and the
"opening" action of σ_r and $\tau_{r\theta}$.

An attempt to predict θ_p in the direction of minimum strain energy density
was initiated (Sih, 1973). The minimum strain energy density theory
(MSET) assumes that the direction of propagation is material dependent,
i.e. on Poisson's ratio (ν). This is significantly different from the
MCST. The MSET is reported to give results close to those of the MCST and
experimental data. The second order term effects on the MSET have also
been studied (Liebowitz et al, 1978).

The distribution of strain energy density, S, at the crack tip is given
by

$$2ES = (\sigma_x^2 + \sigma_y^2)(1-\nu^2) - 2\nu(1+\nu)\,\sigma_x\,\sigma_y + 2(1+\nu)\,\tau^2_{xy} \tag{2}$$

where E is the Young's modulus. This is equivalent to the distribution of
principal shear stress, τ^2_m, given by

$$\tau^2_m = \left\{\frac{\sigma_y - \sigma_x}{2}\right\}^2 + \tau^2_{xy} \tag{3}$$

if the assumptions that plane strain conditions exist at the crack tip and
$\nu = 0.5$ are made (Rouhi et al, 1977). With respect to θ τ^2_m is a minimum
along the direction θ_p and this direction can be conveniently found using
the isochromatic fringe pattern of a photoelastic model, Fig. 2 . The
concept of employing the minimum τ^2_m value to determine θ_p has been termed
the minimum principal shear stress theory (MPST). This idea was extended
to incorporate different values of ν for a more appreciative use of the
MSET (Gdoutos, 1980).

A hitherto apparently unnoticed observation is that the direction of
minimum τ^2_m is theoretically close to the value of θ_p found using the MCST
as well as the MSET. The photoelastic fringe pattern is not dependent on
material properties (except for the material fringe constant) and so it
would seem more plausible to compare it with the MCST which is also
independent of material properties.

The authors have investigated the use of photoelasticity to predict θ_p.
Experimental data for the direction of initial crack extension of edge
cracks has been obtained.

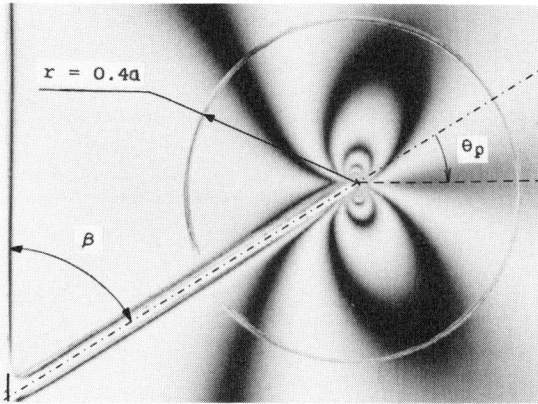

Fig. 2. The isochromatic fringe pattern for an edge crack, $\beta = 60°$, and the direction of the MPST (θ_p).

THE MINIMUM PRINCIPAL SHEAR STRESS THEORY (MPST)

Using the singular solution of the stress field (Theocaris and Spyropoulos, 1983) the principal shear stress can be expressed as

$$\tau^2_m = \frac{1}{2\pi r} \left[\frac{1}{4} \sin^2\theta + K^2 (1 - \frac{3}{4} \sin^2\theta) + K\sin\theta\cos\theta \right] \tag{4}$$

where K is the ratio of the stress intensity factors (K_{II}/K_I). The direction of the minimum is found using

$$\frac{\partial \tau^2_m}{\partial \theta} = 0, \quad \frac{\partial^2 \tau^2_m}{\partial \theta^2} > 0 \tag{5}$$

and

$$\frac{\partial \tau^2_m}{\partial \theta} = \frac{1}{2\pi r} \left[\frac{1}{4} (1 - 3K^2) \sin 2\theta + K\cos 2\theta \right] = 0 \tag{6}$$

The direction of propagation is found by combining (5) and (6) to give

$$\tan 2\theta_p = \frac{4K}{3K^2 - 1}, \quad \frac{\partial^2 \tau^2_m}{\partial \theta^2} > 0 \quad . \tag{7}$$

Comparison of the MPST with the MCST

The direction of the maximum circumferential stress is found by equating $\tau_{r\theta}$ to zero. For the singular solution of the stress field this becomes

$$\sin\theta_p + K (3 \cos\theta_p - 1) = 0, \quad \sigma_\theta > 0. \tag{8}$$

The graph in Fig. 3 compares the MPST, eq . (7), and the MCST, eq . (8). The difference in θ_p between the two solutions is less than 5% for K<0.7.

The quality of agreement between the MPST and the MCST for low values of K

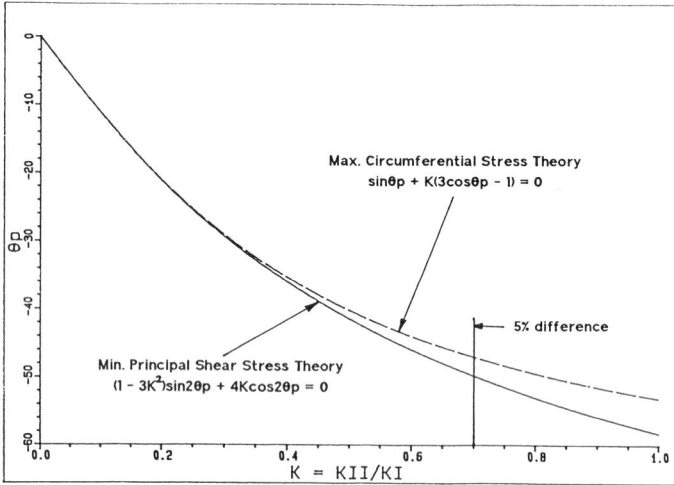

Fig. 3. Comparison of the MPST and MCST
using the singular stress field.

can be ascertained using the following mathematical treatment. It is
necessary to use the Taylor approximation for sine and cosine which are:

$$\cos\theta = 1 - \frac{\theta^2}{2} \quad , \quad \sin\theta = \theta - \frac{\theta^3}{6} \tag{9}$$

and ignoring terms of the fourth derivative and higher. Eq . (8) becomes

$$\theta - \frac{\theta^3}{6} + K \left(2 - 3 \frac{\theta^2}{2}\right) = 0 . \tag{10}$$

From this it can be written for the MCST

$$K = \frac{\theta^3 - 6\theta}{12 - 9\theta^2} . \tag{11}$$

Similarly from using eq . (9) in eq . (6) the MPST can be expressed as

$$(1 - 3K^2) \left(2\theta - \frac{4\theta^3}{3}\right) + 4K (1 - 2\theta^2) = 0 . \tag{12}$$

It can be shown that eq. (12) remains valid when K is set to the value in
eq. (11) and terms of θ^4 and higher are ignored. This demonstrates that
the two theories are quantitatively identical for low values of θ.

DETERMINATION OF THE STRESS FIELD AROUND THE CRACK TIP

The Stress Field Equations

The theoretical distribution of stress in the neighbourhood of the crack
tip (Theocaris and Spyropoulos, 1983) is given as

$$\sigma_x = \frac{K_I}{\sqrt{2\pi r}} \left\{ \left[\cos\frac{\theta}{2} (1 - \sin\frac{\theta}{2} \sin\frac{3\theta}{2}) \right] + C_n[\cos(n+\tfrac{1}{2})\theta - (n+\tfrac{1}{2}) \sin\theta\sin(n-\tfrac{1}{2})\theta] \right\}$$

$$+ \frac{K_{II}}{\sqrt{2\pi r}} \left\{ \left[-\sin\frac{\theta}{2} (2 + \cos\frac{\theta}{2} \cos\frac{3\theta}{2}) \right] + C_n[2\sin(n+\tfrac{1}{2})\theta + (n+\tfrac{1}{2})\sin\theta\cos(n-\tfrac{1}{2})\theta] \right\}$$

$$\sigma_y = \frac{K_I}{\sqrt{2\pi r}} \left\{ \left[\cos\frac{\theta}{2} (1 + \sin\frac{\theta}{2} \sin\frac{3\theta}{2}) \right] + C_n[\cos(n+\tfrac{1}{2})\theta + (n+\tfrac{1}{2})\sin\theta\sin(n-\tfrac{1}{2})\theta] \right\}$$

$$+ \frac{K_{II}}{\sqrt{2\pi r}} \left\{ \left[\sin\frac{\theta}{2} \cos\frac{\theta}{2} \cos\frac{3\theta}{2} \right] - C_n [(n+\tfrac{1}{2})\sin\theta\cos(n-\tfrac{1}{2})\sin\theta\cos(n-\tfrac{1}{2})\theta] \right\}$$

$$\tau_{xy} = \frac{K_I}{\sqrt{2\pi r}} \left\{ \left[\sin\frac{\theta}{2} \cos\frac{\theta}{2} \cos\frac{3\theta}{2} \right] - C_n[(n+\tfrac{1}{2}) \sin\theta\cos (n-\tfrac{1}{2})\theta] \right\} \tag{13}$$

$$+ \frac{K_{II}}{\sqrt{2\pi r}} \left\{ \left[\cos\frac{\theta}{2} (1 - \sin\frac{\theta}{2} \sin\frac{3\theta}{2}) \right] + C_n[\cos(n+\tfrac{1}{2})\theta - (n+\tfrac{1}{2}) \sin\theta\sin(n-\tfrac{1}{2})\theta] \right\}$$

where $n = 0,1,2 \ldots$ and

$$C_n = (-1)^n \cdot \left(\frac{r}{2a}\right)^{n+1} \frac{1.3\ldots(2n-1)}{2.4\ldots(2n)} \cdot \tag{14}$$

The higher order terms account for changes in the singular solution as r increases to a size comparable with the crack length, a, where the value of integer n determines the limit of r to which the stress field can be studied accurately (Theocaris and Spyropoulos, 1983).

The second order terms, σ_{ox} and τ_{ox} (Doyle et al, 1981), are added as follows

$$\sigma_x = \ldots -\sigma_{ox} , \qquad \tau_{xy} = \ldots -\tau_{ox} . \tag{15}$$

These account for the far-field loads and, hence, are constants added to the stress field.

If the crack tip is blunt and has a radius ρ the following correction terms are added to the singular solution (Creager and Paris, 1967).

$$\sigma_x = -\frac{K_I}{\sqrt{2\pi r}} \frac{\rho}{2r} \cos\frac{3\theta}{2} + \frac{K_{II}}{\sqrt{2\pi r}} \frac{\rho}{2r} \sin\frac{3\theta}{2} + \ldots$$

$$\sigma_y = \frac{K_I}{\sqrt{2\pi r}} \frac{\rho}{2r} \cos\frac{3\theta}{2} - \frac{K_{II}}{\sqrt{2\pi r}} \frac{\rho}{2r} \sin\frac{3\theta}{2} + \ldots \tag{16}$$

$$\tau_{xy} = -\frac{K_I}{\sqrt{2\pi r}} \frac{\rho}{2r} \sin\frac{3\theta}{2} - \frac{K_{II}}{\sqrt{2\pi r}} \frac{\rho}{2r} \cos\frac{3\theta}{2} + \ldots \cdot$$

The use of boundary parameters, $B^{(n)}{}_{I,II}$ is recommended to correct the higher order terms of the stress field if the geometry of the cracked body is significant. It is therefore unnecessary to assume that the crack is contained in an infinite plate and in the case of the edge crack, for instance, the effect of the finite boundaries can be incorporated, so that eq.(14) now becomes (Etheridge et al, 1978)

$$C_n = B^{(n)}_{I,II} (-1)^n \left(\frac{r}{2\alpha}\right)^{n+1} \frac{1.3 \dots (2n-1)}{2.4 \dots (2n)}$$ (17)

where the subscript refers to mode I and mode II and $B_{I,II}$ (e.g. $B^{(1)}_{I}$, $B^{(1)}_{II}$, $B^{(2)}_{I}$, $B^{(2)}_{II}$ ...) is a parameter for each order n.

A Photoelastic Technique to Determine the Stress Field

A photoelastic technique has been developed by Sanford and Dally which solves an over-determined set of equations incorporating a Newton-Raphson iteration scheme with a least squares approach (Sanford and Dally, 1978). The stress-optic law in photoelasticity which relates the fringe order N to the in-plane principal shear stress is

$$\tau^2_m = \left(\frac{Nf_\sigma}{2h}\right)^2$$ (18)

where f_σ is the material fringe value and h is the model thickness. Combining eqs. (3) and (18) it can be written

$$\left(\frac{Nf_\sigma}{2h}\right)^2 = \left(\frac{\sigma_y - \sigma_x}{2}\right)^2 + \tau^2_{xy} .$$ (19)

For a point (r,θ) on an isochromatic fringe of order N an equation can be obtained of the form of eq. (19) with K_I, K_{II}, σ_{ox}, τ_{ox} and $B^{(n)}_{I,II}$ the unknowns. If a set of arbitrary points, equal to the number of variables, is selected then the Newton-Raphson scheme will solve the set of simultaneous non-linear equations. The over-deterministic approach is to select an over specified number of data points and to use a least squares minimization process with Newton-Raphson to solve for the variables. The over-determined method tends to minimize errors incurred in the measurement of r and θ.

The use of boundary parameters and higher order terms extends the region in which the data points can be collected. However, as $B^{(n)}_{I,II}$ is a parameter and no longer a predetermined constant the freedom to choose the limit of r and thereby specify n for a given accuracy is not valid. This problem is solved by performing the analysis for a set of data points using a series of values of n. The optimum n will be found when the nth order terms make a negligible contribution to the stress field.

EXPERIMENTAL PROCEDURE

In this study on edge crack propagation it was necessary to obtain isochromatic fringe patterns of the stress field and to fracture edge-cracked specimens to measure the direction of initial extension (θ_p).

Preparation and Testing of Specimens

Specimens were made out of cast epoxy resin (CT-200, $\nu=0.35$) with dimensions of length 225mm, width 80mm and thickness 6.35mm. An initial crack was machined into each specimen such that the crack tip was 112.5mm from each end. Various angles and lengths were used for the initial crack in each specimen. Initially, a circular saw blade coated in diamond crystals

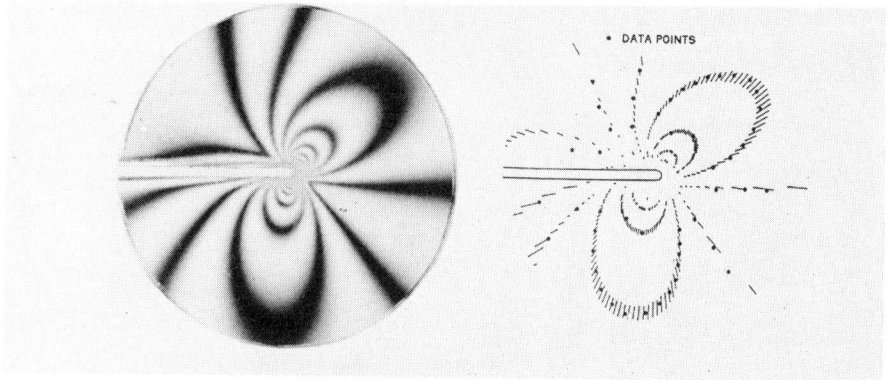

Fig. 4.   Comparison of a theoretical and actual
isochromatic fringe pattern, $\beta = 60°$.

was used to machine a slit of width 0.4mm with a square tip. Care was taken to ensure that the crack tip was straight through the thickness of the specimen. To produce a radius in the tip a razor saw with rounded teeth was used in the final stages . To provide a reference field for study a circle of radius 0.4 x crack length (0.4a) centred on the tip was neatly scribed on the surface of each specimen (Fig.2).

A JJ Instruments T22K tensile testing machine fitted with a polariscope and quarter-wave plates was used to analyse the specimens. Photographs were taken of the light-field isochromatics within the reference circle. Specimens with cracks of a = 0.25 x width of specimen (0.25w) or less were loaded until the isochromatic loop of order 2.5 was as large as the reference circle (Fig. 2). For longer cracks the specimens were loaded until the 1.5 fringe order loop was as large as the circle. To use the MPST to predict θ_p another photograph of the fringe pattern was taken when a fringe formed a line of symmetry between the loops (Fig.2). This enabled θ_p to be measured directly from the photograph. Finally, each specimen was loaded to failure at a displacement rate of 1mm/min.

A clean, plane strain, brittle fracture occurred in all the specimens. Using a Societe Genevoise microscopic instrument it could be seen that the line of failure was straight for the first 2mm (10ρ) of crack growth in each specimen. The instrument was used to measure the angle of fracture from both sides of the specimen and θ_p was determined to be the average .

The over-determined method involving the Newton-Raphson and least squares techniques was used to find the stress field around the crack tip. A set of 50 data points (r,θ ,N) was taken from each photograph of the iso-chromatic fringe pattern with the aid of a digitizing pad attached to a computer. These were filed on the computer and a Fortran program was written to solve K_I, K_{II}, σ_{ox}, τ_{ox} and $B^{(n)}_{I,II}$. The program was run for different values of n between 0 nd 8 and if it was found that there was a value of n such that the nth order terms contributed less than 2% to the stress field at a radius r = 0.4a then the data set was accepted. Other-wise, the data set was discarded and another one was tried. After successfully finding a data set for each specimen it was found that for cracks of length 0.25w or less n=6 and for longer cracks n=5. As a final

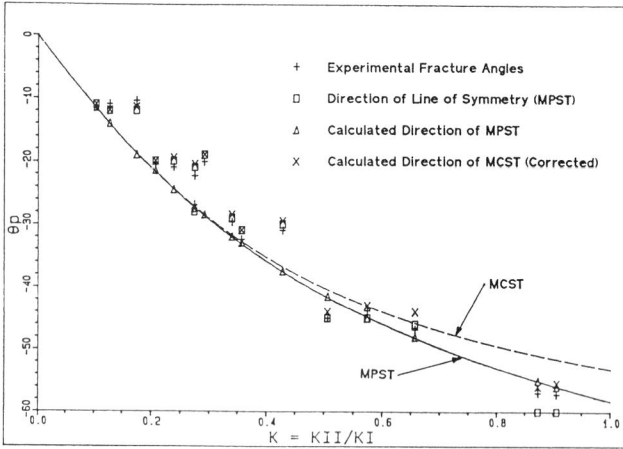

Fig. 5.   Experimental fracture angles and
           theoretical results.

test the computer program plotted a corresponding theoretical fringe
pattern to compare it with the original photograph. An example of this is
shown in Fig.4.

The theoretical values of θ_ρ were found using the MPST and MCST by
processing the stress field on a computer. Only the singular solution of
the stress field was used for both theories (eqs. (7) and (8)), otherwise,
the results would be dependent upon r. For this reason it was also
necessary to exclude the crack tip radius terms. Finally, predictions of
the MPST were also obtained directly from the photographs especially taken
for the purpose (e.g. Fig.2). For each specimen the centre-line of the
fringe which forms a line of symmetry between the isochromatic fringe
loops was observed to be straight between $\rho < r < 20\rho$. The angle of this
line was measured to find the prediction of the MPST for the fracture
angle.

RESULTS AND DISCUSSION

The results of the fracture angles and the MPST and MCST predictions are
shown in Fig.5. In each case the ratio $K = K_{II}/K_I$ has been calculated
using the stress field solutions and, therefore, the points on the graph
vertical to one another correspond to the same specimen.

There is good agreement between the experimental fracture angles and the
MPST results using the line of symmetry from the photographs.
Significantly, large differences occur for the two cases where $K \simeq 0.9$ as
expected. These correspond to initial crack angles of $\beta = 30°$.

A comparison of the measured and calculated predictions of the MPST show
large discrepancies in some cases. This is almost certainly due to the
fact that only the singular solutions of the stress field equations were
included in the calculations. The curve for MPST plotted in Fig.3 fits

the calculated results as shown in Fig.5. Therefore, one would also expect the results of the MCST to reproduce the equivalent curve in Fig.3. Hence, the large discrepancies also occur between the MCST results and the experimental fracture angles. However, the MCST results can be corrected to give accurate predictions of the fracture angles. To do this it is assumed the angular difference, ϕ, between the actual values and calculated values of the MCST predictions is the same as the difference between the measured values and calculated values of the MPST. Fig.5 shows that in each case by subtracting the appropriate angle ϕ from the original MCST results improved predictions can be obtained.

Conclusions

It has been shown that the MPST accurately predicts the direction of initial crack extension if the ratio of the stress intensity factors, K, is less than 0.7.

The direction of the MPST can be found immediately from photoelastic isochromatics if a fringe forms a line of symmetry between the loops around the crack tip.

REFERENCES

Chrysakis, A.C. (1986). Dependence of mixed-mode crack propagation on the variation of σ_r and $\tau_{r\theta}$. Engng. Fract. Mechs., 24, 351–360.
Creager, M. and P.C. Paris (1967). Elastic field equations for blunt cracks with reference to stress corrosion cracking. Int. J. Fract., 3, 247–251.
Doyle, J.F., S. Kamle and J. Takesaki (1981). Error analysis of photo-elasticity in fracture mechanics. Expt. Mechs., 21, 429–435.
Ergodan, F. and G.S. Sih (1963). On the crack extension in plates under plane loading and transverse shear. J.Basic Engng. Trans. ASME, 85D, 519–525.
Etheridge, J.M., J.W. Dally and T. Kobayashi (1978). A new method of determining the stress intensity factor K from isochromatic fringe loops. Engng. Fract. Mechs., 10, 81–93.
Finnie, I. and A. Saith (1973). A note on the angled crack problem and the directional stability of cracks. Int. J. Fract., 9, 484–486.
Gdoutos, E.E. (1980). A photelastic prediction of the crack propagation angle. J. Phys. E: Sci.Instrum., 13, 776–777.
Liebowitz, H., J.D. Lee and J. Eftis (1978). Biaxial load effects in fracture mechanics. Engng.Fract. Mechs., 10, 315–335.
Rouhi, M.R., W.T. Evans and B.I.G. Barr (1977). A photoelastic approach to fracture path prediction. Int.J. Fract., 13, 370–375.
Sanford, R.J. and J.W. Dally (1978). A general method for determining mixed-mode stress intensity factors from isochromatic fringe patterns. Engng. Fract. Mechs., 11, 621–633.
Sih, G.C. (1973). A special theory of crack propagation. In: Methods of Analysis and Solutions of Crack Problems. Vol.I. Mechanics of Fracture (G.C. Sih, Ed.). Noordhoff Int.Pub., 21–45.
Theocaris, P.S. and C.P. Spyropoulos (1983). Photoelastic determination of complex stress intensity factors for slant cracks under biaxial loading with higher-order term effects. Acta Mechanica, 48, 57–70.
Williams, J.G. and P.D. Ewing (1972). Fracture under complex stress – The angled crack problem. Int. J. Fract. Mechs., 8, 441–446.

A Comparison of Finite Element Techniques for Contact Stress Analysis

S. K. PASCOE, J. E. MOTTERSHEAD and T. K. HELLEN*

Department of Mechanical Engineering, University of Liverpool and
*Central Electricity Generating Board, Berkeley Nuclear Laboratories

ABSTRACT

This paper is intended to give background information about the major displacement finite element contact methods in current use and to expose their relative strengths and weaknesses. It is shown that the choice of algorithms should depend largely upon the various features of the particular contact problem. To obtain greatest advantage, the user should be aware of the different methods available, the mechanisms by which they identify contact and how these effects are incorporated in the system of finite element equations. On a computational level this is important, as the contact problem is generally a changing boundary condition problem and different methods require quite different CPU times.

KEYWORDS

Finite element; contact stress analysis; penalty method; Lagrange multiplier method; transformation matrix method.

INTRODUCTION

The finite element method has advanced over the past twenty years to the extent that it is now the major numerical tool in structural stress analysis. However contact type problems, where two or more objects come together resulting in a contact zone, have presented considerable difficulties in finite element modelling. The reason for this is that often no a priori information is known about the contact conditions. For example, in the case of an unloaded sphere resting on a flat surface, contact will initially be at a point. However loading of the sphere results in the region of contact increasing. What in this circumstance is the resulting contact area? What is the force distribution in the contact zone? What is the contact state (i.e. is sticking or sliding occurring at the interface)? These are all conditions that a useful finite element contact algorithm should be able to determine.

In this paper the authors wish to explain and compare some of the currently applied finite element methods for solving contact problems. To allow comparison, key phases in the solution process are identified and discussed. Essentially three phases are shown to exist, these are (a) load application, (b) identification of contact conditions, and (c) implementation of contact constraints in the finite element method.

By studying these phases in the solution process it is possible to assess the accuracy, reliability, computational efficiency and overall suitability of a contact algorithm.

PROBLEMS PECULIAR TO CONTACT ANALYSIS

When two or more bodies due to externally applied loading come into contact, a region of mutual contact will result. In some cases this region is well defined in that very little change in the shape of the interface zone occurs, for example contact between two rigid flat slabs. In other cases however, like the sphere on a flat surface, the contact region may increase or decrease with load. This latter case represents a changing boundary problem which may be solved by a non-linear contact analysis procedure, whereas the former may possibly be solved using a linear algorithm.

Upon identifying the size of the contact region what condition should then be applied? There are three different states which could possibly occur,

(i)   sticking contact,
(ii)   frictionless sliding contact,
(iii)   frictional sliding contact.

It is also possible that combinations of these states could be occurring in a single contact problem at different regions on the contact surface. Hence the identification of the size of the contact region, the decision on the appropriate contact state and how the various contact states are included within the framework of existing finite element packages are the prime reasons for the difficulty in modelling contact stress problems.

CONTACT MODELLING

In the previous section general difficulties associated with contact were identified. In implementing a contact algorithm using finite elements an ideal method should be capable of handling a variety of problems associated with contact modelling,

(a)   to be valid in both 2-D and 3-D analyses - however extra difficulties are encountered in 3-D problems as explained later in the paper,
(b)   the position where contact may occur not known in advance,
(c)   possibly relatively few nodes in the contact region - characteristic of non-conforming contact,
(d)   possibly many nodes contacting - characteristic of conforming contact,
(e)   sticking or sliding contact including frictional effects permitted,
(f)   non-node-on-node contact,
(g)   large amount of relative sliding,

(h) sliding over curved surfaces,
(i) reliable results.

To be completely general the following advanced features should be included,

(j) dynamic effects,
(k) material non-linearity including plasticity,
(l) geometric non-linearity.

However the above features are ideal and in practice few (if any) contact algorithms can claim to be fully general and to correctly include all the above requirements. In fact it may not be necessary to attempt to model every form of contact problem with one general method, but to use the method most suited to a particular contact application. As a consequence, methods have been developed in the past which are only applicable to a certain class of contact problem. It is important for the user that the limitations of a particular method are known so as to allow him to make the most appropriate selection.

It is an obvious requirement that the algorithm should be computationally efficient. This point is particularly important with contact problems due to their inherent non-linearity which must be treated using an incremental solution, an iterative solution or a combination of both. Hence any reduction in the number of increments, iterations or the number of finite element equations can yield considerable savings in CPU time.

FINITE ELEMENT CONTACT ALGORITHMS

In all finite element contact methods, the algorithm or solution process can be basically divided into three separate stages. The understanding of how these stages are tackled can be helpful to the user who has to identify accurate and efficient algorithms. These three stages are,

(a) load application,
(b) gap monitoring and decision on contact state,
(c) implementation of the contact constraints in the finite element method.

A more detailed evaluation of the stages follows.

Stage (a) Load Application

There are primarily three different methods of applying the external loads, these are,

(i)   fully loading,
(ii) incrementally loading in defined load steps,
(iii) loading in stages of one boundary condition change per load step.

If the fully loaded approach is used, then the loading history of the structure is obscured because iterations are performed until force equilibrium and displacement compatibility are satisfied simultaneously. This is effectively linearising a non-linear problem into one load stage and as such creates errors when modelling contact with closing gaps.

However modelling of non-varying contact areas using this approach is acceptable (Hitchings, 1989; Stadler and Weiss, 1979).

The other extreme in loading is to only allow one extra boundary condition to be included per load stage (Okamoto and Nakazawa, 1979; Torstenfelt, 1983; Pascoe and Mottershead, 1988, 1989). This can be achieved by applying the full external loads, and then scaling the result such that just one new extra boundary condition is included in the next solution. This process yields an accurate build-up and history of the contact zone. However the computational time may become excessive in problems where many contact nodes are present.

An alternative to the two approaches above, is to apply load increments as in standard non-linear analyses. Iteration is then undertaken to obtain convergence of nodal states, displacement and contact forces within each increment. This approach has been used by Bathé and Chaudhary (1985), Chaudhary and Bathé (1986), Mazurkiewicz and Ostachowicz (1983) and Hellen (1988). This is probably the most attractive approach to problems involving many nodes in the contact region, in that it allows reasonable tracking of the load history and is not restricted to one boundary condition change per load step.

Stage (b) Gap Monitoring and Decision on the Contact State

Irrespective of the method of load application it is necessary to know what the current contact conditions are, so that they can be implemented into the system of finite element equations. Within the finite element scheme several distinctly different methods of monitoring the contact conditions have been developed. The main ones are based on monitoring at the element level, nodal level or on Gauss point monitoring as suggested recently by Hitchings (1989).

The purpose of monitoring the interface is to decide the size of the current contact area, i.e. to which nodes or elements contact conditions apply and whether sticking or sliding constraints are to be imposed. The size of the contact area can be obtained by updating the geometry after each solution, the resulting deformation profiles indicate the region of contact with previously defined nodes or elements just touching and new nodes or elements overlapping. With contact methods where gap elements are used, the normal forces generated within the element can be used as a contact indicator such that negative forces indicate mesh overlap. However such a force is generated by the product of stiffness and displacement, and it is again the overlap of displacements which is indirectly the contact indicator.

In some early methods only purely sticking or purely sliding contact could be imposed (Francavilla and Zienkiewicz, 1975) which required the user to know in advance the type of contact. This is satisfactory for many types of problem and considerably reduces the computational effort. However, the user may often be unaware of the contact conditions in advance and in this case general purpose algorithms are required which can automatically apply the correct constraints.

In these general purpose methods, new contacting zones are initially defined as sticking with sticking constraints included in the system of equations. From the solution it must then be decided whether sticking was correctly defined or whether sliding should have been imposed and its frictional effects also included. This check on the state of contact is undertaken by the calculation of interface forces on elements, at nodes or at Gauss points. Which of these is used is an important characteristic of the contact algorithm.

Almost all algorithms are nodal based which results in nodal forces being calculated. Use of nodal forces is acceptable in defining the contact state when the force distribution for an element face, under uniform loading, yields compressive loads at each surface node. However in elements such as the 20 noded, 3-D brick element, uniform loading implies tensile forces at the corners when the shape functions are used to distribute the loading at the nodes. This would incorrectly indicate the release of a node. Hence in these problems, decisions on the contact state, based on either element or Gauss point information would be efficacious. In the element approach, the nodal forces previously calculated are converted into equivalent surface pressures in the normal and tangential directions. These pressures are then used to calculate total normal and tangential element forces. From these forces the element's state of sticking, sliding or release can be defined. The nodal state is then given by comparison of adjacent element states (Bathé and Chaudhary, 1985).

The Gauss point method of monitoring has been developed by Hitching (1989) and is particularly useful in identifying the contact state. This is because forces are sampled not at nodes or elements, but at the Gauss points. Since Gauss points are the positions at which the most accurate stresses are available then sampling forces at the Gauss points provides a direct indication of state. This method can be imposed by preventing Gauss point overlap rather the usual nodal or element overlap. This may result in slight mesh overlap occurring elsewhere although the important values of stresses and hence contact forces will be most accurate.

Stage (c) Implementation of the Contact Conditions

The method by which the contact conditions are included within the system of finite element equations has an important effect on the CPU time. There are predominantly three main methods of doing this, i.e.

(i)   Penalty methods
(ii) Lagrange multipliers
(iii) Transformation matrix method.

All of these methods involve different techniques for applying displacement constraints to the stiffness matrix. The penalty method is an approximate method and as such iteration for a particular contact stage may be necessary. The Lagrange multiplier and transformation matrix methods can apply the displacement constraint exactly in one stage.

An important feature of contact algorithms is their iterative nature, hence any reductions in the iteration scheme can considerably improve the solution time. For example in elastic contact analysis, the structures being analysed can have their internal degrees of freedom statically condensed out, leaving a superelement for which just the boundary nodes remain. This dramatically reduces the size of matrix to be solved. Further savings can be made if the same matrix is used for each iteration (usually possible when the contact zone is initially well defined) making re-inversion unnecessary with iteration affecting only the force vector.

A more detailed discussion and comparison of the three methods is provided in the following section.

IMPLEMENTATION OF CONSTRAINTS

Penalty Method

In this method displacement constraints defining the contact zone are imposed by the addition of extra stiffness terms into the stiffness matrix. Physically this can be interpreted as special gap elements being introduced between the contacting bodies, which upon identification of contact become very stiff, effectively joining the meshes together. Line elements predominate rather than area elements between the contacting surfaces in contemporary finite element codes. This is primarily because of their ease of implementation, with nodal pairs which are coupled together with a line element defined across a potential contact region.

The most common method of implementation involves defining a low stiffness for the gap element if contact is not occurring and definition of a high value of it is. This change in stiffness results in changes to the overall system of equations requiring re-inversion. The effects of sticking contact are included by the use of line elements of equal (high) stiffness in the normal and tangential directions. Frictionless sliding is modelled by removing the tangential line elements, allowing sliding to an equilibrium frictionless position to occur. The effects of friction are usually included by initially imposing frictionless sliding, then calculating the resulting product of the normal reaction force and a coefficient of friction (Coulomb's Law), which is then included in the right hand side force vector of the next solution. An alternative approach is to insert tangential line elements with a stiffness equal to the coefficient of friction times the normal stiffness in the initial formulation of the stiffness matrix (Hitchings, 1989 and Mazurkiewicz et al., 1983). However this method requires many iterations with new force terms and can be quite inefficient.

The main advantage of the penalty method is that it is relatively straightforward to implement into existing finite element schemes. This results in no change to the size or bandedness of the system of equations, which is of advantage with certain equation solvers.

A main disadvantage is that generally the gap stiffnesses must be pre-defined by the user. Too low a value results in a contact displacement constraint not being strictly applied, allowing overlapping of the meshes to occur which requires extra iterations to remove. Also problems can occur if the stiffness is too high which may result in a singular matrix. Furthermore the concept and use of gap elements between surfaces prevents large amounts of relative sliding from being accurately modelled.

Lagrange Multipliers

This is a method which allows displacement constraints to be applied exactly to a system of equations. The overall effect on the system is that extra variables are introduced which represent the normal and tangential forces to enforce these displacement constraints. These constraints are included as extra (usually symmetric) rows and columns in the stiffness matrix. To apply sticking conditions for a node, extra rows and columns are added representing the normal and tangential nodal displacement constraint. For frictionless sliding the tangential constraint is removed, allowing the node to slide to its equilibrium position. For frictional sliding two methods are available. In the first method frictionless sliding is initially imposed, the normal forces generated from this solution are then used to evaluate the missing frictional forces (usually by Coulomb's Law) which are included in the force vector of the next solution (Bathé and Chaudhary, 1985). The second method includes friction effects directly after the sticking solution, by including the dynamic coefficient of friction term within the force Lagrange terms of the stiffness (Pascoe and Mottershead, 1988). This method reduces the number of required solutions to obtain a converged frictional solution, although it results in an unsymmetric matrix.

A main advantage of the Lagrange multiplier method is that no special gap elements are necessary, hence node on node contact is not a requirement. This feature allows large amounts of relative sliding to be modelled (using incremental loading) and different types of element to be in contact. Also if allowing one boundary condition change per iteration, then the contact nodal forces are directly obtained as the Lagrange multipliers in the solution vector. Additionally no re-iteration for a particular contact condition is necessary as the displacement constraints are imposed exactly.

A disadvantage of the method is that the order of the system of equations increases and the banded nature of the equations may be lost. However, the maintenance of bandedness is not important using modern finite element equation solvers. A further disadvantage is that extra coding is necessary to monitor the 'housekeeping' of touching nodes, their relative positions and their normals and tangents. However the extra computational burden imposed by this housekeeping is insignificant in comparison with the necessary matrix inversions.

Transformation Matrix Method

In this method the displacement constraint to be imposed on a system is formed into a transformation matrix. By pre- and post-multiplying the stiffness matrix by the transformation matrix, the order of the stiffness matrix may be reduced with the displacement constraint imposed exactly. The effects of friction can be included as in other methods by applying initially a frictionless sliding constraint, and then including frictional forces in subsequent iterations (Torstenfelt, 1983).

Advantages with this method are that the order of the matrix to be solved may be reduced and that the displacement constraint is imposed exactly. A disadvantage is that a complete change in the reduced stiffness matrix is necessary, for each change of contact condition. Thus a full re-inversion is required. Additionally, unlike the Lagrange multiplier method, displacement constraints involving overlaps cannot be imposed exactly because the elements of the transformation matrix must all be coefficients of the nodal variables.

In Table 1 a review of some of the contact methods is shown along with the advantages and disadvantages of each technique.

When comparing the different methods using particular contact examples, the particular problem being solved has a major influence on which method is considered 'best'. This point is quite important as generally similar results can be obtained by the different methods, but the CPU times can vary considerably. For some types of problem certain methods are not recommended.

A range of different contact problems are described in Table 2 with their particular simplifying and difficult features pointed out. The suitability of the six methods shown in Table 1 is also discussed. In incremental solution procedures two important factors are the number of pre-defined load increments and the convergence criteria applied. The influence of these factors on the solution time and accuracy requires further detailed study. Additionally the effects of variation of penalty number needs investigation. Hence it becomes apparent that an exact comparison between the different methods is not possible because of different ways in which the above features are treated in various algorithms. Thus a general description of their suitability is provided in the Table.

CONCLUSIONS

This paper has shown that there are a variety of different methods by which contact conditions may be imposed in a finite element system of equations. If it is critical that the extra coding should be minimal in implementation then the gap element approach is usually best. However large relative deformations between the contacting bodies cannot be modelled with this method, nor can problems where the contact region is not well known in advance. To be able to include these desirable features either the transformation matrix or Lagrange multipliers need to be used. In problems where relatively few contact nodes are present then one change in boundary condition per load step is an acceptable approach. However problems involving many contact nodes are best suited to an incremental solution process with pre-defined load steps.

The application of nodal constraints is satisfactory in most cases, however decisions based on the nodal forces cannot be relied upon in certain 3-D elements. These elements require either element or Gauss point forces to be used to obtain a reliable estimate of the contact state. The Gauss point approach appears to be the best, since the stresses are monitored at the most favourable positions.

ACKNOWLEDGEMENT

S. K. Pascoe is supported by an SERC (CASE) award in collaboration with the Central Electricity Generating Board.

REFERENCES

Bathé, K-J. and A. Chaudhary (1985). A solution method for planar and axisymmetric contact problems. Int. J. Numer. Meth. Engng., 21, 65-88.

Chaudhary, A. and K-J. Bathé (1986). A solution method for static and dynamic analysis of three-dimensional contact problems with friction. Comput. Struct., 24, 855-873.

Francavilla, A. and O. C. Zienkiewicz (1975). A note on numerical computation of elastic contact problems. Int. J. Numer. Meth. Engng., 9, 913-924.

Hellen, T. K. (1988). A gap element facility in BERSAFE. CEGB Report TPRD/B/ 1043/R88.

Hitchings, D. (1988). Contact analysis using finite elements. Institute of Physics Conference - Contact Stress Analysis I, 7 December 1988, I.O.P. Publishing Ltd., in press.

Mazurkiewicz, M. and W. Ostachowicz (1983). Theory of finite element method for elastic contact problems of solid bodies. Comput. Struct., 17, 51-59.

Okamoto, N. and M. Nakazawa (1979). Finite element incremental contact analysis with various frictional conditions. Int. J. Numer. Meth. Engng., 14, 337-357.

Pascoe, S. K. and J. E. Mottershead (1988). Linear elastic contact problems using curved elements and including dynamic friction. Int. J. Numer. Meth. Engng., 26, 1631-1643.

Pascoe, S. K. and J. E. Mottershead (1989). Two new finite element contact algorithms. Comput. Struct., in press.

Stadter, S. J. and R. O. Weiss (1979). Analysis of contact through finite element gaps. Comput. Struct., 10, 867-873.

Torstenfelt, B. (1983). Contact problems with friction in general purpose finite element computer programs. Comput. Struct., 16, 487-493.

Table 1. A Summary of Different Contact Algorithms

Method	Description
A. Penalty method using line elements between contacting nodes, with incremental loading. (Mazurkiewicz et al., 1983; Hellen, 1988)	Line elements are inserted across the interface which upon identification of contact attain a high stiffness. Any overlap is eliminated by including a term equal to the product of stiffness and overlap in the force vector.
B. Penalty method using area elements between the contacting bodies. (Stadter et al., 1979)	Area elements are inserted across the interface. Upon identification of contact a high stiffness is defined. Overlaps are eliminated in the Stadler and Weiss method by redefinition of the stiffness value. Eventually contacting elements have a very high stiffness and non-contacting elements having a very low value. This method essentially linearises the contact problem into one load stage.
C. Penalty method using line elements between Gauss points of adjacent bodies, full loading. (Hitchings, 1988)	Line elements having normal and tangential stiffness are inserted across the interface. Full loads are applied with overlaps corrected by modification of the r.h.s. force vector until force equilibrium is obtained.
D. Lagrange multipliers to apply the constraints between touching bodies with one boundary condition change per load step. (Pascoe et al., 1988)	Lagrange multipliers used to apply the constraints. After each solution scaling is undertaken until the next boundary condition becomes operative. The remainder of load is applied with the new boundary condition included. Summation of each scaled solution is computed to obtain the overall solution.
E. Lagrange multipliers to apply linear constraints with incremental loading. (Bathé et al., 1985)	Lagrange multipliers are used to apply the constraints. After each solution the contact state is defined by converting the residual loads into equivalent element pressures. This may result in changes of state which results in a new force vector. Each stage represents a solution, summation of every stage giving the final result. Iteration is undertaken for each load increment until the residual terms are sufficient small.
F. Transformation matrix to apply constraints with one boundary condition change per step. (Torstenfelt, 1983)	The contact displacement constraint is imposed by forming a transformation matrix which pre and post multiplies the stiffness matrix. Scaling is undertaken to allow just one boundary condition change per load step.

Table 1 (Cont.) A Summary of Different Contact Algorithms

Advantages	Disadvantages
A. 1) Relatively simple to implement and monitor. 2) Size and bandedness of the matrix is maintained.	1) Restricted to node on node contact problems, hence the contact zone must be known in advance with limited amounts of sliding. 2) Possible poor convergence if too low a penalty number is selected and possible singular matrix or ill-conditioning if too high. 3) Cannot be used between 3-D, 20 node brick elements as incorrect separation is suggested by negative forces at the corner nodes.
B. 1) Relatively simple to implement.	1) Linearising into one load stage causes problems if large gaps have closed. 2) Restricted to small amounts of movement and known contact zones. 3) The whole element is in contact or released. This can cause convergence problems as nodal control is not permitted.
C. 1) No change in the contact conditions is allowed. Thus the stiffness matrix remains unchanged and re-solution is performed with new force vectors. 2) Monitoring and application of constraints is applied at the Gauss points. Higher order elements can be included in the contact model.	1) Linearising the problem into one load stage creates errors when modelling closing gaps - hence best suited for non-varying contact area problems. 2) Since the stiffness matrix is not reformed, change of state from sticking to sliding may cause difficulties. 3) Sliding with friction requires many iterations.
D. 1) Contact conditions can be defined from the Lagrange multipliers. 2) Accurate build-up of the load history. 3) Non node on node contact. 4) Sliding around curved surfaces can be allowed by definition of an average normal/tangential direction	1) Many solution stages are required if the number of contacting nodes is large. 2) Since nodal information is used to define the contact state 20-noded brick elements cannot be included.

Table 1 (Cont.) A Summary of Different Contact Algorithms

Advantages	Disadvantages
D. 5) The displacement constraints are applied exactly.	
E. 1) Reasonable build-up of the load history. 2) Reasonable number of stages to obtain the overall solution. 3) Non node on node contact. 4) The displacement constraints are applied exactly.	1) Linear constraints are used restricting this method to linear elements. 2) Friction forces are averaged over each element – this can cause convergence problems.
F. 1) Good build-up of the load history. 2) The displacement constraint is applied exactly. 3) The size of matrix to be solved is reduced. 4) No major changes in F.E. coding.	1) Many solution stages are required.

Table 2. Analysis of Different Contact Problems

F.E. Model	'Easy' Features	'Difficult' Features	Recommended Methods
	1. Single node contact (A on B). 2. Node on node contact. 3. Rigid base. 4. Frictionless contact.		A,D,E & F These methods are recommended because they apply nodal constraints. Methods D & F are the most efficient requiring just two re-inversions.
	1. Node on node contact. 2. Non varying contact area.	1. Deformable base. 2. Multi-node contact.	A,B,C,D,E & F All the methods work well as there is no change in the contact area, although incremental methods may require a large number of iterations.
	1. Sliding over a relatively flat surface. 2. Non varying contact area.	1. Non node on node contact. 2. Multi-node contact. 3. Large amounts of relative sliding.	D & E These methods are recommended as they can cope with non node on node contact and large amounts of relative sliding.

Table 2 (Cont.). Analysis of Different Contact Problems

F.E. Model	'Easy' Features	'Difficult' Features	Recommended Methods
	1. Non varying contact area.	1. Non node on node contact. 2. Sliding over a curved surface. 3. Multi-node contact. 4. Large amounts of relative sliding. 5. Frictional effects.	D This method is recommended because node on node contact is not a requirement and sliding over a curved boundary is allowed.
	1. Node on node contact to occur in the contact region.	1. Varying contact area. 2. Frictional effects.	A,B,C,D,E & F However B & C may not attain the same accuracy due to linearisation errors.
		1. Varying contact area. 2. Large relative sliding causing non-node on node contact to occur. 3. Frictional effects. 4. Multi-node contact.	D & E These methods allow non-node on node contact. If many nodes come into contact then the incremental approach (E) will be faster.

A Boundary Element Computer Program for Practical Contact Problems

A. A. BECKER

Department of Mechanical Engineering,
Imperial College of Science, Technology and Medicine,
London SW7 2BX, UK

ABSTRACT

A computer program based on the Boundary Element Method (BEM) formulation is applied to several two-dimensional and axisymmetric practical contact problems. A brief review of the BEM formulation is presented followed by a description of the automatic iterative process used by the program. The BEM solutions show excellent agreement with other analytical solutions.

KEYWORDS

Boundary Element Method, Boundary Integral Equation, contact, stress analysis

INTRODUCTION

The stress analysis of contact problems is of major concern in many applications such as ball bearings, gears, rollers, mechanical seals and pressure vessel attachments. The numerical modelling of practical contact problems requires special attention because the actual contact area between the contacting bodies is not known in advance and if friction is present, the behaviour may be dependent on load history. Thus, the correct solution of such problems must be determined by an iterative and/or incremental procedure.

The Boundary Element Method (**BEM**), also known as the Boundary Integral Equation (**BIE**), is now well established as an accurate computational stress analysis tool. The method is very suitable for contact problems for the following reasons.
1- Due to the fact that the BE approximations are confined to the surface (boundary), there is a much higher resolution of stresses than other "domain" techniques such as the Finite Element Method (**FEM**).
2- The surface tractions are calculated to the same degree of accuracy as the displacements. This enables accurate coupling of contact variables.
3- Re-meshing a specific region on the boundary is relatively an easy task. In most practical contact problems, re-meshing an initial mesh is usually required before accurate results can be determined.

313

Analytical solutions of contact problems were first established by Hertz (1896) for frictionless contact problems. Contact stress analysis theories and numerical algorithms were later developed by several researchers such as Mindlin (1949), Conry and Seireg (1971), and Johnson (1982) . The rapid development of the FEM in the early seventies inevitably lead to its application to contact problems. Wilson and Parson (1970) were one of the first to apply the FEM to two-dimensional frictionless contact problems using constant strain elements, while Francavilla and Zienkiewicz (1975) used parabolic elements. The range of applications were later extended to include friction and non-linear problems (see, for example, Okamoto and Nakazawa ,1979 , Torstenfelt ,1984 and Bathe and Chaudhary ,1985).

The Boundary Element Method (BEM) has been continually developed since it was first applied to elastostatic problems by Rizzo (1967). It was first applied to contact problems by Andersson et al (1980) who used constant elements in two-dimensional frictionless problems, and later extended the applications to include friction and parabolic elements (Andersson and Allan-Persson, 1983). Isoparametric quadratic elements were used by Karami (1983) , Abdul-Mihsein et al(1985) and Becker and Plant(1987) to model two-dimensional and axisymmetric contact problems.

The aim of this paper is to describe how a general purpose BEM computer code can be adapted to solve practical contact problems with minimum interaction from the user. A brief review of the BEM formulation is described where isoparametric quadratic elements are used to model the boundary curves. The contact variables are coupled directly to produce a set of simultaneous linear equations with a unique solution. Several types of contact interface conditions are covered in this paper including heat conduction, frictionless, Coulomb frictional contact, and interference problems. The contact computer program (**BEACON**) is applied to several two-dimensional and axisymmetric contact problems. These problems include thermal stresses in compound spheres, contact of cylindrical rollers, receding contact of closely conforming bodies, shrink-fit cylinders and contact in a pressure vessel assembly.

REVIEW OF THE BEM FORMULATION

The formulation and theoretical background of the BEM has been dealt with adequately in the literature (see, for example the textbooks by Banerjee and Butterfield, 1981 and Brebbia et al, 1984), and will not be repeated here. It suffices to mention that the partial differential equations of elasticity are transformed into integral equations applicable over the boundary relating the displacement at an interior (load) point p to the displacements and tractions at a boundary (field) point Q . In standard tensor notation this boundary integral equation (BIE) for 2D and 3D problems can be written as follows.

$$u_i(p) = - \int_S T_{ij}(p,Q) \; u_j(Q) \; dS(Q) \quad + \quad \int_S U_{ij}(p,Q) \; t_j(Q) \; dS(Q) \qquad (1)$$

where u and t are the displacement and traction vectors, S is the surface (boundary) of the solution domain, while U_{ij} and T_{ij} are the displacement and traction kernels. The axisymmetric version of equation (1) can be derived by integrating the kernels in a ring path around the axis of symmetry (Becker, 1986).

A set of linear algebraic equations is then constructed by taking each point on the boundary in turn as the field point p and performing the integrations (using Gaussian quadrature formulae) over each element on the boundary.

$$[A] \; [u] = [B] \; [t] \qquad\qquad (2)$$

where the matrices [A] and [B] contain the integrals of T_{ij} and U_{ij}, respectively. Up to this stage, the boundary conditions are not applied and the matrices [A] and [B] are simply functions of geometry and material properties only. The simplest boundary conditions take the form of either prescribed displacements or prescribed tractions . The program assumes displacements to be the unknown variables except at nodes where displacements are prescribed. For these nodes, the tractions become the unknown variables and are moved to the left hand side. Therefore, after the application of all the boundary conditions, the solution matrix becomes :

$$[C] \ [x] \ = \ [d] \tag{3}$$

where [C] is the solution matrix, and [d] contains all known quantities. The equations are then solved by Gaussian elimination since [C] is fully populated.

In contact problems, however, the boundary conditions are more complex since the elements on the contact interface do not have prescribed displacements or tractions. Instead, they must satisfy continuity of displacements and equilibrium conditions (equal and opposite tractions). Each domain is treated separately to form equations (2), and the resulting matrices [A] and [B] are coupled together according to the relevant contact conditions, with the number of unknowns remaining equal to the number of equations. Thus, the system of linear equations has a unique solution. Some of the contact interface conditions are listed below.

Heat Conduction

In perfect thermal conduction between two contacting solids, (a) and (b), continuity of temperatures and heat fluxes must be maintained at contacting nodes.

$$T^{(a)} = T^{(b)} + C$$
$$k^{(a)} \ dT/dn \ ^{(a)} = k^{(b)} \ dT/dn \ ^{(b)} \tag{4}$$

where T and dT/dn are temperature and normal temperature gradient, respectively, while k is the thermal conductivity. C is a prescribed step in temperature that can be caused by a heat source in the interface, e.g. a thin squeezed film of high temperature. Once all temperatures and fluxes are calculated for each node they are then fed into the elastic analysis as body forces (see Becker and Fenner, 1983).

No-Slip Contact

This occurs when the contact surfaces are glued together and they are not allowed to slip in any direction. Continuity of displacements and equilibrium conditions must be maintained resulting in the following four relationships.

$$u_x^{(a)} + X^{(a)} = u_x^{(b)} + X^{(b)}$$
$$u_y^{(a)} + Y^{(a)} = u_y^{(b)} + Y^{(b)}$$
$$t_x^{(a)} = - t_x^{(b)}$$
$$t_y^{(a)} = - t_y^{(b)} \tag{5}$$

where X and Y are the coordinates of the nodal points. Note that it is important to allow for a gap between the contacting nodes, i.e. a different starting position of each node (see the example of two cylindrical rollers).

Slip Contact

This occurs when the interface is either frictionless or the ratio of tangential to normal traction exceeds the value of the Coulomb coefficient of friction, μ. Referring to the local components of tractions and displacements, the following relationships satisfy the continuity of displacements in the normal direction as well as equilibrium conditions

$$t_t = \pm \mu \ t_n$$
$$u_n^{(a)} = u_n^{(b)}$$
$$u_t^{(a)} = u_t^{(b)} + \delta_t \tag{6}$$

where the subscripts n and t refer to normal and tangential directions, respectively, and δ_t is the amount of slip in the tangential direction. Since the final system of linear equations is expressed in terms of global coordinates, it is necessary to convert local components into global components using the contact angle θ, resulting in the following four relationships.

$$u_y{}^{(b)} = u_y{}^{(a)} + Y^{(a)} - Y^{(b)} + \tan\theta \,[u_x{}^{(b)} - u_x{}^{(a)} + x^{(b)} - x^{(a)}\,]$$

$$t_x{}^{(a)} = t_y{}^{(a)}\,[\pm\mu\cos\theta + \sin\theta\,]/[-\cos\theta \pm \mu\sin\theta\,]$$

$$t_x{}^{(b)} = -t_x{}^{(a)} \tag{7}$$

$$t_y{}^{(b)} = -t_y{}^{(a)}$$

Interference (shrink-fit)

This condition applies when a cylinder or shaft sleeve is heated and then fitted on another cylinder and allowed to cool creating favourable compressive hoop stresses. The coupling of the contact variables is the same as above except that there is and extra component , δ_n, to allow for the amount of interference or clearance. i.e.

$$u_n{}^{(a)} = u_n{}^{(b)} + \delta_n \tag{8}$$

THE COMPUTER PROGRAM "BEACON"

The main complication which arises in most practical contact problems is that the actual contact area is *a priori* unknown, and the problem is nonlinear because the contact area is not linearly dependent on the applied load. In the absence of friction, a reasonable initial contact area is assumed and the contact variables are examined to satisfy two conditions; no geometrical overlap just outside the contact area, and no tensile stresses anywhere in the contact area (Becker and Plant, 1987).

The above BEM formulation is implemented in the computer program **BEACON** (**B**oundary **E**lement **A**nalysis of **CON**tact) which is derived from the general purpose program **BEETS** (Becker,1988). The program uses isoparametric quadratic elements to discretise the boundary and performs automatic iterations until the correct contact area is arrived at. After each iteration, the program performs the following three checks.

1- Overlap Check : The displacements of the pairs of elements just outside the assumed contact area are checked to determine whether overlap has occured. If overlap is detected, the program automatically includes these pairs of elements in the contact area in the next iteration.

2- Tensile Stress Check : If tensile stress is detected in any element pair in the contact area, then the program releases these element pairs in the next iteration.

3- Friction Slip Check : If friction is present, the program checks that the ratio of the tangential to normal tractions does not exceed the value of μ. It it does, it allows the relevant element pairs to slip in the next iteration and imposes the Coulomb friction slip condition on the tangential traction.

The program automatically terminates the iterations when all three conditions are met. However, although it performs these iterations automatically, a certain amount of engineering judgment is necessary to minimise the number of iterations. This is particularly important in complex geometries, because the edge of the contact area is unlikely to terminate exactly at an element edge (resulting in overlap detection in one iteration and tensile stress detection in the next). Furthermore, if the initial guess of contact area is very different from the actual one, it may become necessary to re-mesh the contact area to obtain better accuracy .

It should be mentioned that the matrices [A] and [B] remain unchanged throughout the iteration process since the kernels T_{ij} and U_{ij} are functions of the nodal positions and material properties (not boundary conditions). This means that no further integrations are necessary after the first iteration, resulting in modest CPU time used in subsequent iterations.

CONTACT APPLICATIONS

The program BEACON is applied to several practical contact problems to establish its reliability and accuracy. The results are compared to analytical solutions where applicable.

Contact of Two Cylindrical Rollers

Consider two identical cylindrical rollers made of steel and compressed together by a radial force P under plane strain conditions. The contact interface is assumed frictionless. The BEM mesh of 62 elements used to model a symmetrical half of this problem is shown in Fig. 1 together with a plot of the contact pressure distribution which is clearly in good agreement with the well established Hertz analytical solution (Lipson and Juvinall, 1963).

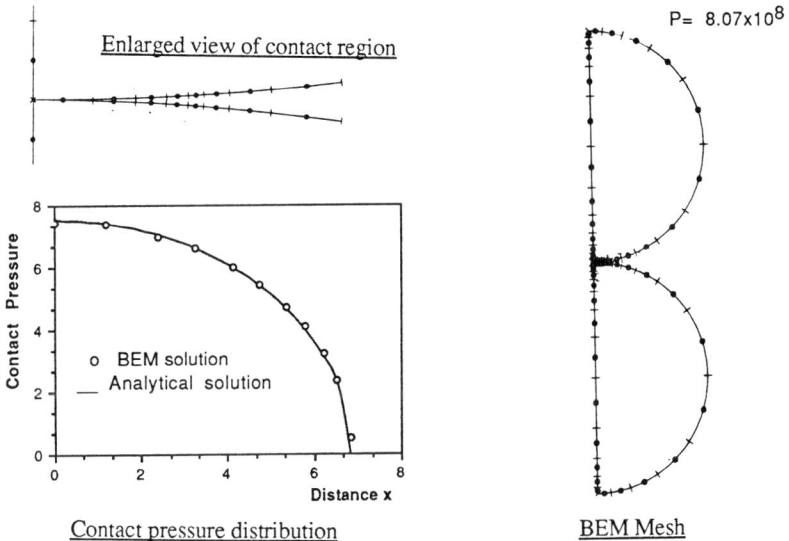

Fig. 1. Cylindrical rollers problem

Shrink-fit cylinders

This application involves the shrink-fit stresses resulting from fitting two dissimilar cylinders with a prescribed interference, δ, and subjected to internal pressure P_1 and external pressure P_2. The BEM mesh of 24 elements and a plot of the hoop stress distribution are shown in Fig. 2 where good agreement with the analytical solution (Timoshenko and Goodier, 1970) is obtained.

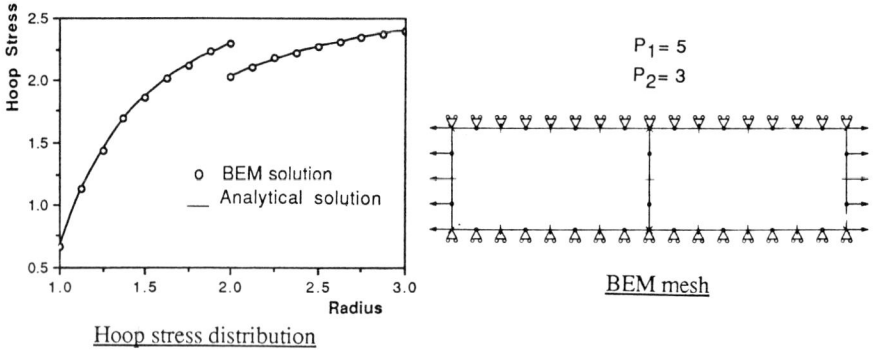

Hoop stress distribution

Fig.2. **Shrink-fit problem**

Separation of Inclusions (Receding Contact)

This application represents a receding contact of closely conforming bodies in which the original contact area decreases as the load is increased. An interesting feature of receding contact problems is that the extent of contact is independent of the magnitude of the applied load (Keer *et al*, 1972).

Consider a spherical inclusion embedded in a homogeneous infinite matrix of the same material. A remote tensile stress is applied to the matrix causing the inclusion to separate from it . The contact interface is assumed frictionless. The BEM of 88 elements is shown in Fig. 3 together with the contact pressure distribution where it shows good agreement with the corresponding analytical solution of Wilson and Goree (1967). By applying different values of the tensile load, it was easily verified that the contact area remained the same.

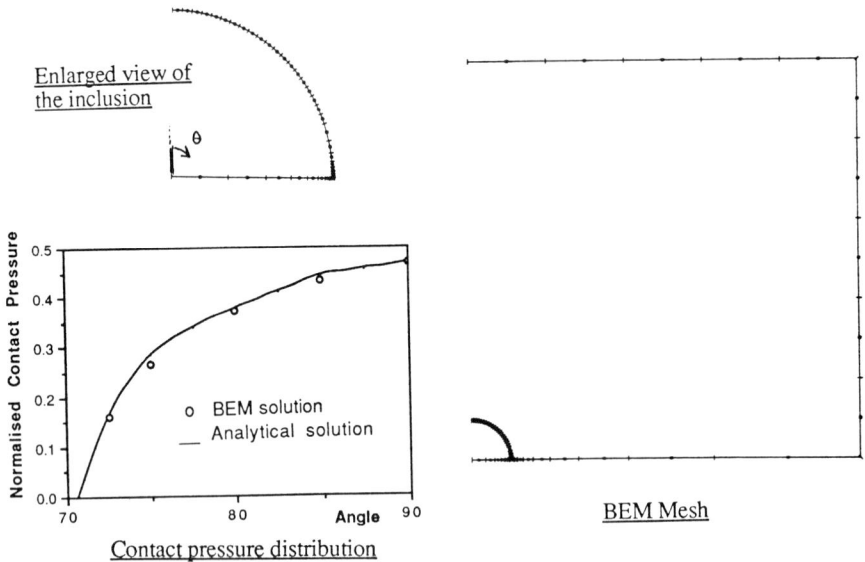

Contact pressure distribution

Fig. 3. **Separation of inclusions problem**

Thermal Stresses in Compound Spheres

Consider three dissimilar thick-walled hollow spheres in perfect contact with each other, with the inner surface subjected to temperature T_1 and pressure P_1 and the external surface subjected to temperature T_2 and pressure P_2. The BEM mesh of 51 elements is shown Fig. 4 as well as a plot of the calculated Von Mises stress which shows excellent agreement with the corresponding analytical solution (Timoshenko and Goodier, 1970).

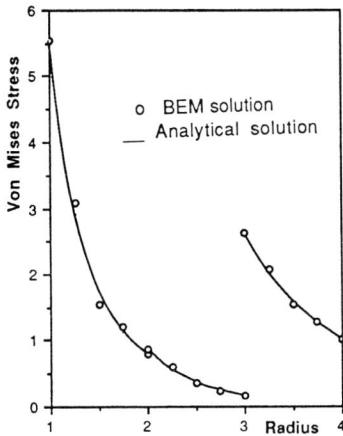

Von Mises stress distribution

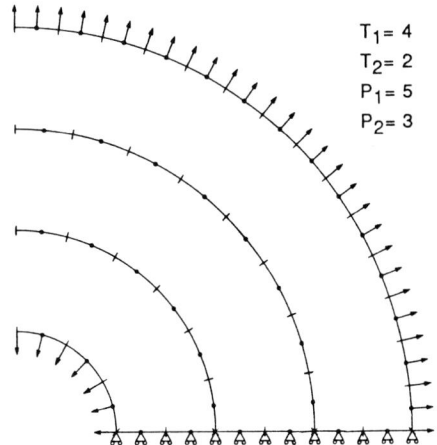

BEM Mesh

$T_1= 4$
$T_2= 2$
$P_1= 5$
$P_2= 3$

Fig. 4. Thermal stress problem

Pressure Vessel Assembly

This problem demonstrates the application of BEACON to an industrial problem of no known analytical solution. The application involves the contact that occurs in a pressure vessel assembly between the top cover of the vessel and the clamp that holds the top cover to the vessel body. The geometry is assumed axisymmetric, and the contact areas are assumed frictionless. The BEM mesh of 126 elements is carefully designed to allow accurate modelling of high stress gradient regions, as shown in Fig. 5. The results are presented in terms of Von Mises equivalent stresses plotted as vectors normal to the nodes they belong to. High stress concentrations occur in the groove area near the contact region as well as on the top hole of the top cover due to bending.

CONCLUSIONS

The BEM formulation is implemented in a computer program for practical contact problems. The automatic iterative process works equally well for both advancing and receding contact problems. The accuracy of the BEM results is impressive in view of the relatively small number of elements needed to obtain good agreement with other available solutions. The isoparametric quadratic elements demonstrate excellent modelling capabilities both in modelling complex geometries and , more significantly, rapidly varying stress gradients. This paper demonstrates that the BEM is very suitable for practical elastic contact problems.

320

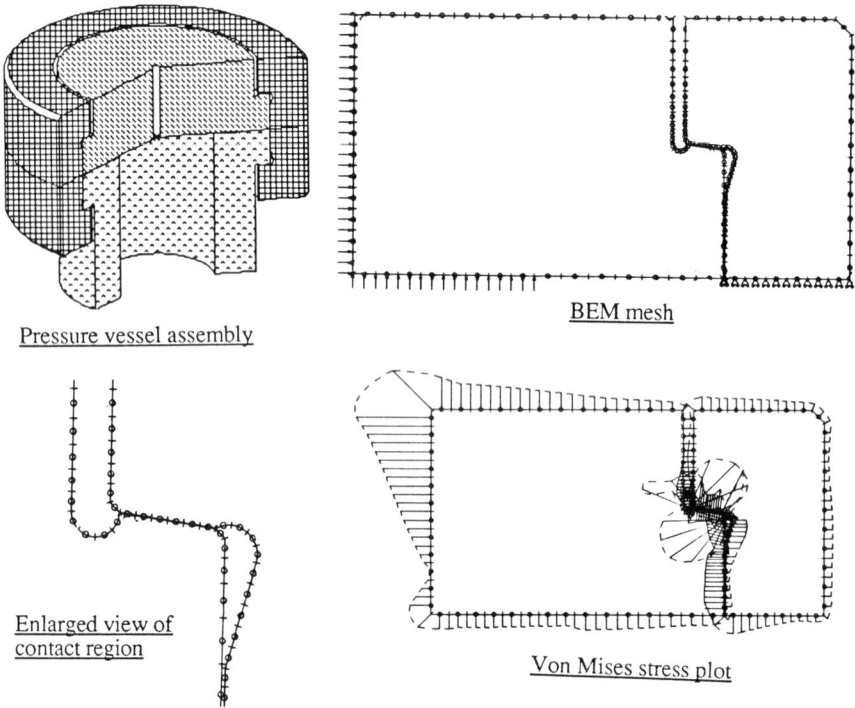

Pressure vessel assembly

BEM mesh

Enlarged view of
contact region

Von Mises stress plot

Fig. 5.   Pressure vessel problem

REFERENCES

Abdul-Mihsein, M.J., A.A. Becker and A.P. Parker (1986), A boundary integral equation method for axisymmetric elastic contact problems , *Computers and Structures*, 23, 787-793.

Andersson, T. , B. Fredriksson and B.G. Allan-Persson (1980), The boundary element method applied to two-dimensional contact problems, *New Developments in Boundary Element Methods* (Edited by C.A. Brebbia), CML Publications, Southampton.

Andersson, T. and B.G. Allan-Persson (1983), The boundary element method applied to two-dimensional contact problems , *Progress in Boundary Element Methods- Vol 2*, (edited by C.A. Brebbia), Pentech Press, London.

Banerjee, P.K. and R. Butterfield (1981), *Boundary Element Methods in Engineering Science*, McGraw-Hill , London.

Bathe, K.J. and A. Chaudhary (1985), A solution method for planar and axisymmetric contact problems, *Int. J. Numerical Methods Eng.* , 21, 65-88.

Becker (Bakr) A.A. (1986), *The Boundary Integral Equation Method in Axisymmetric Stress Analysis Problems*, Springer-Verlag, Berlin.

Becker, A.A. (1988), *BEETS Manual (version 9) - Boundary Elements in Engineering Thermal and Stress Problems*, Department of Mechanical Engineering, Imperial College, London.

Becker, A. A. and R.T. Fenner (1983), Boundary integral equation analysis of axisymmetric thermoelastic problems, *J. Strain Analysis*, 18, 239-251.

Becker, A.A. and R.C.A. Plant (1987), Contact mechanics using the boundary element method , *Proceedings of the International Conference on Tribology, Friction, Lubrication and Wear*, 975-980, The Institution of Mechanical Engineers, London.

Brebbia, C.A., J.C.F. Telles and L.C. Wrobel (1984), *Boundary Element Techniques - Theory and Applications in Engineering* , Springer-Verlag, Berlin.

Conry, T.F. and A. Seireg (1971), A mathematical programming method for the design of elastic bodies in contact, *Trans. ASME, J. Applied Mechanics*, 94, 387-392.

Francavilla, A. and O.C. Zienkiewicz (1975), A note on numerical computation of elastic contact problems, *Int. J. Numerical Methods Eng.*, 9, 913-924.

Green, D.J. (1983), Microcracking mechanisms in ceramics, *Fracture Mechanics of Ceramics - Volume 5* ,(edited by R.C. Bradt, A.G. Evans, D.P.H. Hasselman and F.F. Lange), Plenum Press, New York.

Hertz, H. (1896), *Miscellaneous Papers on the Contact of Elastic Solids*, (translated by D.E. Jones), Macmillan, London.

Johnson, K.L. (1982), One hundred years of Hertz contact, *Proc. I. Mech. E.*, 196, 363-378.

Karami, G. (1983), *A Boundary Integral Equation Method for Two-Dimensional Elastic Contact Problems* , PhD Thesis, Imperial College, University of London.

Keer, L.M. , J. Dunders and K.C. Tsai (1972), "Problems involving a receding contact between a layer and a half space", *Trans ASME , J. Applied Mechanics*, 39, 1115-1120.

Lipson, C. and R.C. Juvinall(1963), *Handbook of Stress and Strength*, Macmillan, New York.

Mindlin, R.D.(1949), Compliance of elastic bodies in contact, *Trans. ASME, J. Applied Mechanics*, 17, 259-268.

Okamoto, N. and M. Nakazawa (1979), Finite element incremental contact analysis with various frictional Conditions, *Int. J. Numerical Methods Eng.*, 14, 337-357.

Rizzo, F.J. (1967), An integral equation approach to boundary value problems of classical elastostatics, Q. Applied Mathematics, 25, 83-95.

Timoshenko, S.P. and J.N. Goodier (1970), *Theory of Elasticity - Third Edition*, McGraw-Hill, Tokyo.

Torstenfelt, B. (1984), An automatic incrementation technique for contact problems with friction, *Computers and Structures*, 19, 393-400.

Wilson, H.B. and J.G. Goree (1967), Axisymmetric contact stresses about a smooth elastic sphere in an infinite solid stressed uniformly at infinity, *Trans ASME, J. Applied Mechanics*, 34, 960-966.

Elastoplastic Stress Analysis by the Boundary Element Method

M. R. LACEY*, D. A. YORK* and J. A. BLAND**

*Department of Mechanical Engineering, Trent Polytechnic,
Nottingham NG1 4BU, UK
**Department of Mathematics, Statistics and Operational Research,
Trent Polytechnic, Nottingham NG1 4BU, UK

ABSTRACT

This paper describes the implementation of a 2-dimensional elastoplastic stress analysis program on an IBM compatible microcomputer. The program utilises the boundary element (BE) or boundary integral equation method of analysis; this provides, compared with existing domain methods, for a much reduced data input requirement and in many instances a much improved accuracy for stresses and deflections.

The program utilises constant, linear or isoparametric quadratic order elements for the boundary with isoparametric quadratic triangular or quadrilateral cells for those parts of the domain where plastic deformation occurs.
An initial strain approach has been taken to allow both perfectly plastic and strain hardening materials to be modelled. Within an incremental load stepping algorithm an iterative procedure is used at each step to evaluate the unknown plastic strain rates. The stress and strain rates at the boundary are readily derived from boundary traction and displacement rates, those in the interior require further computational effort and this is achieved by direct evaluation of the integral identities.

The paper describes some typical applications of boundary elements in elastoplasticity, outlining some advantages of its use. It is concluded that the BE method, although requiring elemental segmentation within the domain for plastic analysis, is computationally very accurate and efficient for elastoplastic stress analysis.

KEYWORDS

Boundary elements, Elastoplastic, initial strain, quadratic.

323

INTRODUCTION

All numerical solutions to the integral equations of elastoplasticity entail the discretisation of both boundary and domain. However, unlike other numerical techniques the Boundary Element Method (BEM) requires only the region where plasticity is likely to occur to be discretised.Therefore, although the reduction in dimensionality by one for elasticity is not retained the application of the method continues to be more attractive then existing techniques that require full domain discretisation.

The first BEM formulation for elastoplastic problems was by Swedlow and Cruse (1971), although the paper presented was only theoretical and had no internal solution for stress or strain. Ricardella (1973) expanded this work to the first 2D elastoplastic analysis, using piecewise constant interpolation of the plastic strain rates. Further implementations were presented (Mukherjee, 1977; Bui, 1978) improving the above techniques but all were limited by the use of constant elements making solution times prohibitive.

Telles and Brebbia (1979) presented a complete linear formulation for 2D and 3D bodies showing how different numerical procedures could be implemented with major emphasis on accuracy and efficiency. At about the same time Banerjee and Cathie(1980) developed a numerical formulation using constant elements for both the boundary and domain. A Finite difference approximation was employed for the stress-strain equations which is intrinsically less accurate than the above but reduces the computational time.

The above papers and many more of that era may be described as the first generation of BEM for elastoplasticity where the formulations, with the exception of a paper by Faria et al (1981), are limited to constant or linear elements, and no use of forward extrapolation or acceleration of iteration was used.

The second generation began with papers such as Cathie and Banerjee(1982) where a new Visco-plastic approach was described. This allowed for the inclusion of strain-softening, strain-hardening and perfect plasticity with quadratic boundary elements and a constant variation of problem variable over the internal cells.
More recently (Lee and Fenner, 1986; Banerjee and Raveendra, 1986; Li H-B and Han G-M, 1985) the search has been for efficient isoparametric models with accelerated convergence of iteration, the investigation of new methods for evaluating singular integrals and methods such as particular integrals (Henry and Banerjee, 1988)and ´Dual Reciprocity´,(Brebbia and Telles, 1988) to allow the domain integrals to be written at the boundary.
This paper presents the development of a Fortran Computer code BESS (*Boundary Element Solution Scheme*) to model 2D elastoplastic stress analysis using the most advanced techniques described above.

GOVERNING EQUATIONS

The case considered is a Two-dimensional homogeneous, isotropic body where the plastic deformation is considered time-independent.

The equilibrium conditions, ignoring body forces are

$$\dot{\sigma}_{ij,i} = 0 \tag{1}$$

The constitutive equations may be written as

$$\dot{\sigma}_{ij} = 2G \left(\dot{\epsilon}_{ij} - \dot{\epsilon}_{ij}^{P} \right) + \frac{2G\nu}{1-2\nu} \dot{\epsilon}_{kk} \delta_{ij} \tag{2}$$

If the material is assumed to obey the von-Mises yield criterion the incremental elastoplastic flow rules may be shown (Lee and Fenner, 1986), to give the plastic strain rates as

$$\dot{\epsilon}_{ij}^{P} = - \frac{\overline{S}_{kl} \dot{\sigma}_{kl} + \overline{S}_{33} \dot{\sigma}_{33}}{H' \overline{\sigma}_{eq}^{2}} \overline{S}_{ij} \tag{3}$$

where $\overline{S}_{ij}$ and $\overline{\sigma}_{eq}$ are the current deviatoric stress tensor and equivalent stress respectively. H' represents the plastic modulus (value from the stress plastic strain curve during uniaxial tension) with the mixed hardening model proposed by Axelsson and Samuelsson being employed. This model breaks the plastic strain rate down into its Isotropic and Kinematic parts as follows

$$\dot{\epsilon}_{ij}^{P(i)} = M \dot{\epsilon}_{ij}^{P} \tag{4}$$

$$\dot{\epsilon}_{ij}^{P(k)} = (1-M) \dot{\epsilon}_{ij}^{P}$$

where M is called the mixed hardening parameter for which

$$
\begin{matrix}
M = 1 & \text{Isotropic hardening} \\
M = 0 & \text{Kinematic hardening} \\
M = -1 & \text{Isotropic softening}
\end{matrix} \tag{5}
$$

An initial strain approach has been adopted whereby the plastic strain rates are considered as initial strain rates. Hence the starting equation for the plane strain problem is given by

$$c_{ij} \dot{U}_{j} = \int_{\Gamma} U_{ij}^{*} \dot{t}_{j} \, d\Gamma - \int_{\Gamma} T_{ij}^{*} \dot{u}_{j} \, d\Gamma + \int_{\Omega} \Sigma_{kij}^{*} \dot{\epsilon}_{jk}^{P} \, d\Omega \tag{6}$$

where U_{ij}^{*} and T_{ij}^{*} are the second order displacement and traction tensors and Σ_{kij}^{*} the third order tensor for plastic strain terms. Γ represents the boundary of the domain and Ω its interior.

FUNDAMENTAL SOLUTIONS

The fundamental singular solutions to Kelvins problem in two dimensions can be written as

$$U_{ij}^* = - \frac{1}{8\pi G(1-\nu)} \left\{ (3-4\nu) \ln(r) \delta_{ij} - r,_i r,_j \right\} \tag{7}$$

$$T_{ij}^* = - \frac{1}{4\pi r(1-\nu)} \left\{ [(1-2\nu)\delta_{ij} + 2 r,_i r,_j] \frac{\delta r}{\delta n} \right.$$
$$\left. - (1-2\nu) (r,_i n,_j - r,_j n,_i) \right\} \tag{8}$$

In the above expressions U_{ij}^* and T_{ij}^* represent the displacement and traction tensors in the j direction due to a unit force in the i direction. Also r is the distance between the point of application of the load (P) to the point under consideration and n_i is the component in the i direction of the outward normal.

The stresses at any point within the body are given by

$$\Sigma_{Kij}^* = - \frac{1}{4\pi r(1-\nu)} \left\{ [(1-2\nu)\delta_{ij} + 2 r,_i r,_j] \frac{\delta r}{\delta n} \right.$$
$$\left. - (1-2\nu) (r,_i n,_j - r,_j n,_i) \right\} \tag{9}$$

NUMERICAL IMPLEMENTATION

If the boundary is split into m isoparametric quadratic elements, i.e 2m nodes and the interior where plasticity is likely to occur into n internal cells, then equation (6) may be rewritten

$$c_{ij} \dot{u}_j + \sum_1^m \sum_1^3 \dot{u}_i \int_{S_b} T_{ki} M^\alpha J d\xi = \sum_1^m \sum_1^3 \dot{t}_i \int_{S_b} U_{ki}^\alpha M J d\xi$$
$$+ \sum_1^n \sum_1^\sigma \epsilon_{ij}^P \int_{S_I} \Sigma_{kij} N^b J d\xi \tag{10}$$

Where S_b denotes the bth boundary element and S_I the Ith internal cell.

Boundary

The geometry of a boundary element is defined in terms of quadratic shape functions of intrinsic coordinate ξ which can vary from -1 to +1 along the element.

The Cartesian coordinates of a point on an element may be defined,

$$x_i(\xi) = M^a(\xi)\, x_i^a \qquad a = 1 \text{ to } 3 \qquad (11)$$

M^a are the shape functions and x^a the coordinates of the ath node.

The Jacobian of transformation is given by

$$J(\xi) = \left(\frac{\delta x_i}{\delta \xi} \; \frac{\delta x_i}{\delta \xi} \right)^{1/2} \qquad (12)$$

Domain

The domain can be discretised into two types of internal cell, namely 6 noded triangles or 8 noded quadrilaterals.

Quadrilateral Cells. The geometry is defined in terms of two intrinsic coordinates ξ_1 and ξ_2 (see fig 1.) both of which vary from -1 to +1.

The Cartesian coordinates are defined as

$$x_i(\xi) = N^b(\xi)\, x^b \qquad b = 1 \text{ to } 8 \qquad (13)$$

where N^b are the shape functions.

The Jacobian of transformation is

$$J(\xi) = \left| \frac{\delta x_1}{\delta \xi_1} \frac{\delta x_2}{\delta \xi_2} - \frac{\delta x_1}{\delta \xi_2} \frac{\delta x_2}{\delta \xi_1} \right| \qquad (14)$$

Triangular Cells. The geometry is defined in terms of quadratic shape functions of three intrinsic coordinates ξ_1, ξ_2 and ξ_3 (see fig 1) all varying from 0 to +1 over each element. Only two are independent, the third comes from the relation

$$\xi_1 + \xi_2 + \xi_3 = 1 \qquad (15)$$

and the cartesian coordinates are defined

$$x_i(\xi) = N^c(\xi)\, x^c \qquad c = 1 \text{ to } 8 \qquad (16)$$

If ξ_1 and ξ_2 are taken as the independent coordinates the above Jacobian (14) is also used.

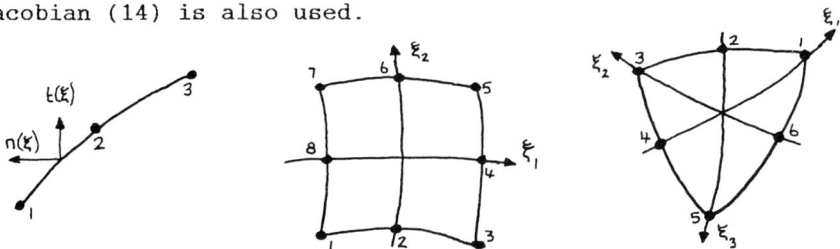

Fig.1 Diagram of Boundary Elements and internal cells

EVALUATION OF MATRIX COEFFICIENTS

Three different situations require consideration :-

(1) P does not belong to either S_b or S_I in which case case the integrals do not possess any singularities and may be evaluated using standard Gaussian quadrature.

(2) P is a node of S_b. The code allows one of two different methods to be applied.

    (a) Force the $\ln\left[\frac{1}{r(\xi)}\right]$ term to a logarithmic quadrature format.

    (b) Split the term into two parts (singular logarithmic and non-singular logarithmic) and solve with explicit definition of $r(\xi)$.

(3) P is a node of S_I. The internal cell is first subdivided depending on type of cell and whether P is a corner or mid-side node. (see fig.2).

Quadrilateral   Triangle

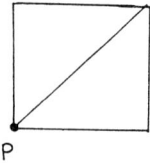

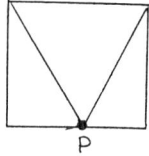

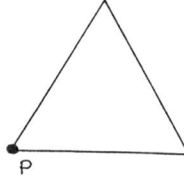

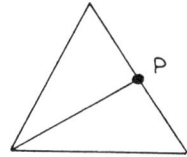

Fig.2 Sub-division of Internal cells for singular Integration.

The sub-elements are then treated as degenerate quadrilaterals and solved using Gaussian quadrature.

SOLUTION OF SYSTEM OF EQUATIONS

Equation (10) is non-linear because the interior values for the plastic strain rates depend on the rate of stress as well.

Therefore an iterative procedure is required with successively better approximations to the unknown plastic strain rates achieved.

The model is set up such that at each node either the displacement or traction rates are known and equation (10) reduces to a set of 4m equations for 4m unknowns and 3h unknown domain nodal values where h is the number of domain nodes.

Expressing this in matrix form

$$[A] \ \dot{z} = \dot{b} + [\Sigma] \ \dot{\epsilon}^{P}$$ (17)

where [A]   Coefficient matrix of unknowns   (4m x 4m)
$\dot{z}$ Vector consisting of the unknown rates (4m x 1)
$\dot{b}$ Second member containing products of known nodal
      values with matrix coefficients (4m x 1)
[Σ]   Matrix plastic strain rate coefficients (4m x 3h)
$\dot{\epsilon}^{P}$ Vector of plastic strain rates (3h x 1)

Procedure

(1) Evaluate all coefficients in equation (10) and store as
    they will not change for the rest of the analysis.
(2) Apply boundary conditions to get second member $\dot{b}$.
(3) Solve system (17) to achieve initial elastic solution.
(4) Scale highest stressed node to yield and re-compute.
(5) Obtain set of approximations for $\dot{\epsilon}^{P}$ by applying (3)
(6) Update Right Hand side.. Find new approx. for $\dot{z}$
(7) Check for convergence. If not go to 5.

Repeat within loading increment routine until failure, end of
load cycle or unloading.

        INTERNAL SOLUTION

The displacement rates may be found from an integral identity
written for the interior and then numerically differentiated
for the stress and strain rates. However this often proves
computationally inefficient and so other techniques have been
developed whereby the integral identities for the strain rates
are utilised to obtain the strain at the interior nodes. The
stress rates then follow from the stress-strain relations.

Special care has to be taken when dealing with the case of P
belonging to an internal cell as the domain integrals exist
only the sense of Cauchy principal   values. To overcome
this problem the two methods most widely employed are either
(a)semi-analytical   methods   often   requiring   total
discretisation of the domain or (b) techniques involving the
derivation of additional integrals for this special case.

The method utilised in BESS follows the derivation of a
technique by Lee and Fenner (1986) which suffers from none of
the above drawbacks. The derivation is lengthy and therefore
not detailed here, but for completeness it involves the
creation of a secondary boundary over the discretised part of
the domain at which the integral identities are written. The
displacements and tractions at this secondary boundary are
then required for evaluation of the stress rates at the
interior.

THE BESS COMPUTER CODE

The FORTRAN code developed (BESS) includes all the above and the following procedures to improve accuracy and computational efficiency wherever possible.

To accurately model corners where ambiguities in the definition of the outward normals exist, the concept of double noding is employed.(see fig.3)

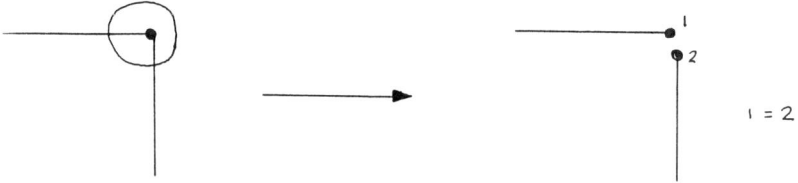

Fig.3 Double Nodes at a corner

The above allows the following combinations of boundary condition to be prescribed.

  (1) Traction at 1 and 2.
  (2) Traction at 1, displacement at 2.
  (3) Displacement at 1, Traction at 2.

The use of convergence accelerators has been investigated to minimise the computational time for iteration.Also Four different types of convergence criteria are employed with the user having the ability to utilise 1,2 or all.

TEST PROBLEMS

To examine the accuracy of this BEM formulation results for some simple examples are given.

In fig.4 the displacements on the outer surface of the thick cylinder show excellent agreement with the analytical solution. In this example 96 internal cells were used. The plastic front is at a position r=1.6a.

For the perforated strip problem fig.5 the computed stresses at the root show good agreement with exerimental data and compare favourably with other numerical solutions, Telles and Brebbia (1981).

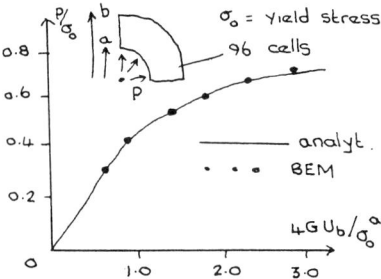

Fig.4

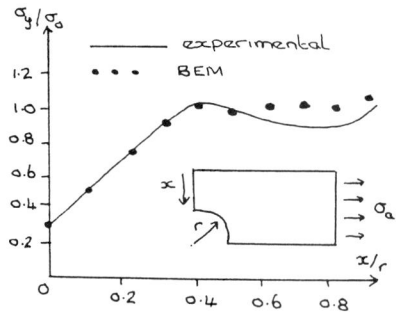

Fig.5

CONCLUSIONS AND FURTHER WORK

A background to the numerical formulation of the boundary element method in elastoplasticity is described together with the implementation of BESS, a two dimensional elastoplastic stress analysis code. It has been demonstrated that the initial results are both accurate and computationally efficient. The future effort will now be to fully test the mixed strain hardening capabilities and to extend the model to enable contact problems to be solved.

REFERENCES

Banerjee,P.K. and Cathie D.N.(1980). A direct formulation and numerical implementation of the BEM for 2D problems of elasto-plasticity.*Int.J. Mech Sciences.22.* 233-245.

Banerjee,P.K. and Raveendra,S.T.(1986).Advanced BE analysis of 2D and 3D problems of Elastoplasticity.*Int.J. for NM in Eng.23.* 985-1002.

Brebbia,C.A. and Telles,J.C.F.(1988). A generalised approach to transfer the domain integrals onto boundary ones for potential problems in BEM. *Boundary Elements IX, Proc.9th Int. Seminar on recent adv. in BEM.* 99-116.

Bui,H.D.(1978). Some remarks about the formulation of 3D thermo elastoplastic problems by integral equations. *Int. J. of solids and structures.14.* 935-939.

Cathie,D.N. and Banerjee,P.K.(1986). Boundary element methods for plasticity and creep including a visco-plastic approach. *Research mechanica.4.* 3-22.

Faria,L.M.,Soares,C.,Pereira,M. and Brebbia C.A.(1981). BE in 2D plasticity using quadratic shape functions. *Applied math. modelling.5.* 371-375.

Henry,D.P. and Banerjee,P.K.(1988). A new BEM formulation for 2D and 3D elastoplasticity using particular integrals.*Int.J. for NM in engineering.26.* 2079-2096.

Lee,K.H. and Fenner,R.T.(1986). A quadratic formulation for 2D elastoplastic analysis using BIEM. *J. of Strain analysis,*Vol 21. no 3, 159-175.

Li,H-B. and Han,G-M.(1985). A new method for evaluating singular integrals in stress analysis of solids by the direct BEM. *Int.J. for NM in Eng.21.* 2071-2098.

Mukherjee,S.(1977).Corrected BIE in planar thermoplasticity. *Int.J. Solids and structures.13.* 331-335.

Ricardella,P.(1973). An implementation of the boundary integral technique for plane problems in elasticity and elasto-plasticity.*PhD thesis, Carnegie-Mellon university.*

Swedlow,J.L. and Cruse,T.A.(1971). Formulation of boundary integral equations for three dimensional elastoplastic flow. *Int.J. solids and structures.7.* 1673-1683.

Telles,J.C.F. and Brebbia,C.A.(1980). The boundary element method in plasticity. *Proc. second int.seminar on recent advances in BEM.(Pentech press,Plymouth).* 295-317.

Consistency Analysis and Optimization Approach for Multivariable Finite Elements

CHANG-CHUN WU

*Institut für Mechanik (Bauwesen), Universität Stuttgart,
Pfaffenwaldring 7, 7000 Stuttgart 80, FRG*

ABSTRACT

A general analysis of the consistency of multivariable finite elements based on incompatible trial functions is presented. The convergence and optimization conditions for the discrete solutions of a nonlinear system are set up. The relevant optimization approach for the hybrid element is suggested and applied to the three dimensional problem in elasticity.

KEYWORDS

Multivariable; Incompatible; Consistency condition; Element optimization condition; Optimization stress pattern.

INTRODUCTION

In recent years one has found that the numerical performance of the finite element with multivariables could be effectively improved by introducing incompatible displacements. Some relevant researches have been presented by T. H. H. Pian et al since 1982. This kind of elements bring us some new problems to be considered. Besides the convergency and stability of discrete solutions, the further improvement of element behaviour is also worth considering. A new concept of the optimization design of hybrid modes has been presented (Wu et al, 1987), and a more ideal plane stress hybrid element has been obtained in terms of an optimization condition. But how can we get the condition? why it may be used to optimize the element numerical performance? and what about the nonlinear forms of the convergence and optimization conditions? All of the questions should be discussed systematically. Another task of the paper is to extend the optimization design into 3-dimensional problem. An optimization stress pattern of the hybrid solid element will be constructed explicitly.

ENERGY CONSISTENCY CONDITION

We start from a discrete version of the Hellinger-Reissner principle in nonlinear elasticity (Washizu, 1975; Bufler, 1979). If the displacement boundary condition is satisfied and the prescribed external boundary force $\bar{T}_i = 0$, the functional of a discrete system takes the form

$$\Pi_R = \sum_e \int_{v^e} \left[\frac{1}{2} (u_{i,j} + u_{j,i} + u_{m,i} u_{m,j}) \sigma_{ij} - B(\sigma_{ij}) \right] dv \tag{1}$$

where the complementary function is

$$B(\sigma_{ij}) = \frac{1}{2} S_{ijkl} \sigma_{ij} \sigma_{kl}$$

$\sigma_{ij} = \sigma_{ji}$ means the 2nd Piola-Kirchoff stress tensor and the symbol v^e indicates the volume of an individual element.

Usually the displacement u_i in (1) is required to be compatible over the system region $v = Uv^e$, namely $u_i \in C^0(v)$. On the other hand the stress σ_{ij} is allowed to be piecewise continuous. The question is now that what about the case of $u_i \notin C^0(v)$? Here an incompatible displacement is introduced into the functional Π_R defined by (1). In this case the strain

$$\varepsilon_{ij} = \frac{1}{2} (u_{i,j} + u_{j,i} + u_{m,i} u_{m,j})$$

will exhibit a singular distribution on the interfaces of elements due to the discontinuity of u_i. Similar to the analysis of incompatible finite elements (Strang and Fix, 1973; Ciarlet, 1978), ignoring the discontinuity between elements, the energy functional of an incompatible discrete system is now be defined as a sum of an individual one, Π^e, directly:

$$\Pi = \sum_e \Pi^o = \sum_e \int_{v^e} \left[\frac{1}{2} (u_{i,j} + u_{j,i} + u_{m,i} u_{m,j}) \sigma_{ij} - B(\sigma_{ij}) \right] dv \tag{2}$$

Our goal is to find a condition under which the finite element trial function set (σ_{ij}, u_i) converges to the true solution under a variational meaning. Obviously, the desired condition must conform to the stationary condition of the functional Π, Such that virtual work equation $\delta\Pi = 0$ holds. An incompatible displacement u_i is here expressed as a sum of the compatible part u_i^o and the discontinuous one u_i^Δ:

$$u_i = u_i^o + u_i^\Delta \tag{3}$$

where u_i^o and u_i^Δ are linearly independent of each other. Substituting (3) into (2) and taking a functional variation with σ_{ij}, u_i^o and u_i^Δ, we have

$$\delta\Pi(\sigma_{ij}, u_i^o, u_i^\Delta) = \sum_e \int_{v^e} \left\{ \left[\frac{1}{2} (u_{i,j} + u_{j,i} + u_{m,i} u_{m,j}) - \frac{\partial}{\partial \sigma_{ij}} B(\sigma_{ij}) \right] \delta\sigma_{ij} \right.$$

$$-(\sigma_{mi} + \sigma_{ij} u_{m,j})_{,i} (\delta u_m^o + \delta u_m^\Delta) \Big\} dv$$

$$+ \sum_e (\oint_{\partial v^e} T_i \delta u_i^o ds + \oint_{\partial v^e} T_i \delta u_i^\Delta ds) \tag{4}$$

Here the surface traction is

$$T_i = n_j \sigma_{mj} (\delta_{im} + u_{i,m}) \qquad \text{on } \partial v^e. \tag{5}$$

Denoting the element boundaries with prescribed external force and displacement as s_σ^e and s_u^e respectively, thus $\partial v^e = s_\sigma^e \cup s_u^e \cup s_{ab}$, where $s_{ab} = v_{(a)}^e \cap v_{(b)}^e$ is the interface of the neighbouring element (a) and (b). Observing the fact that u_i^Δ is usually non-conforming with prescribed displacement $\bar{u}_i$ on s_u^e, so the displacement boundary condition is always

implemented by the compatible one, i.e. $u_i^o = \dot{u}_i$ on s^e.
Thus we have $\delta u^o|_{su}^e = 0$. By means of the above relations and the continuity of u_i^o on s_{ab}, Eq.(4) can be rewritten as

$$\delta\Pi = \sum_e \int_{v^e} \left\{ \left[\frac{1}{2}(u_{i,j} + u_{j,i} + u_{m,i}u_{m,j}) - S_{ijkl}\sigma_{kl} \right]\delta\sigma_{ij} \right.$$
$$- \left[\sigma_{ij}(\delta_{mj} + u_{m,j}) \right]_{,i}(\delta u_m^o + \delta u_m^\Delta) \bigg\} \, dv$$
$$+ \sum_{ab} \int_{s_{ab}} (T_i^{(a)} + T_i^{(b)})\delta u_i^o \, ds$$
$$+ \sum_e \int_{s_\sigma^e} T_i \delta u_i^o \, ds$$
$$+ \sum_e \oint_{\partial v^e} T_i \delta u_i^\Delta \, ds \qquad (6)$$

where $T_i^{(a)}$ and $T_i^{(b)}$ are a pair of tractions on the interface s_{ab}. Considering the functional stationary condition $\delta\Pi = 0$, we obtain the equilibrium equations in v^e and on s_{ab}, the u_i-σ_{ij} relationship and the boundary condition $T_i = 0$ on s_σ^e, But also another kind of condition

$$\sum_e \oint_{\partial v^e} T_i \delta u_i^\Delta ds \equiv \sum_e \oint_{\partial v^e} n_j \sigma_{mj}(\delta_{im} + u_{i,m})\delta u_i^\Delta ds = 0 \qquad (7)$$

This means that the sum of virtual work done by the element surface traction T_i to the discontinuous virtual displacement δu_i^Δ along ∂v^e must vanish. For an incompatible discrete system, (7) is a necessary condition for $\delta\Pi = 0$, so it is also a necessary condition to obtain a rational approximate solution. (7) shown us an energy consistency requirement on the multivariable finite elements with incompatible displacements would be called the energy consistency condition.

For the multivariable elements based on the Hu-Washizu principle, it can be shown that the condition (7) is also necessary for us to obtain a correct solution when some incompatible displacement trials are adopted.

CONVERGENCE AND OPTIMIZATION CONDITIONS

If we consider only the convergence of discrete solutions, the above energy consistency condition will be reduced to another version. In the limit situation of refining element meshes, i.e. the mesh measure parameter $h \to 0$, the stress σ_{ij} will tend to a constant state, denoted by σ_{ij}^c, within the individual element. Thus the condition (7) takes the form

$$\sum_e \oint_{\partial v^e} n_j \sigma_{mj}^c(\delta_{im} + u_{i,m})\delta u_i^\Delta \, ds = 0 \qquad (8)$$

In order to guarantee the convergence of incompatible finite elements, the convergence criteria (8) must be satisfied. In the case of linear elasticity (8) is of the fome

$$\sum_e \oint_{v^e} n_j \sigma_{ij}^c \delta u_i^\Delta \, ds = 0 \qquad (9)$$

This is just an expression of Irons' patch test (Strang et al, 1973). Now let us consider a more practical situation where the discrete system is composed of the element meshes with finite size, i.e. $h \nrightarrow 0$. Obviously, the consistency condition (7) cannot be completely realized by only the convergence criteria (8), and there exsists another condition that should be satisfied simultaneously. If we define an element higher order stress as

$$\sigma^h_{ij} = \sigma_{ij} - \sigma^c_{ij} \tag{10}$$

the desired supplementary condition should be

$$\sum_e \oint_{\partial v_e} n_j \sigma^h_{mj} (\delta_{im} + u_{i,m}) \delta u^\Delta_i \, ds = 0 \tag{11}$$

The condition (11) is depedent on the mesh division and the combination style of elements, so it is not convenient to be used. Instead of (11) we take its strong form for an individual element as follows,

$$\oint_{\partial v_e} n_j \sigma^h_{mj} (\delta_{im} + u_{i,m}) \delta u^\Delta_i \, ds = 0 \tag{12}$$

Under (8), the Eq.(12) makes energy consistency condition (7) be satisfied exactly even the element meshes are finite and rough so that the element behaviour could be improved or optimized. We term (12) the optimization condition of multivariable finite element with incompatible displacements.

OPTIMIZATION APPROACH FOR HYBRID ELEMENTS

The above convergence and optimization conditions take a simplified manner for an linear elastic element:

$$\oint_{\partial v_e} n_j \sigma^c_{ij} \delta u^\Delta_i \, ds = 0 \quad \text{(or equivalently} \quad \oint_{\partial v_e} n_i \delta u^\Delta_i \, ds = 0) \tag{13}$$

and

$$\oint_{\partial v_e} n_j \sigma^h_{ij} \delta u^\Delta_i \, ds = 0 \tag{14}$$

respectively. In the following (13) and (14) will be applied to the model design of hybrid elements, and the element optimization approach will be presented. Taking a matrix expression, the element displacement is now defined as

$$u = u_q + u_\lambda = N_q q + N_\lambda \lambda \tag{15}$$

Here q is the node displacement parameter of the compatible displacement u_q, and λ is the element inner parameter of the incompatible displacement u_λ. Usually the trial $u_\lambda = N_\lambda \lambda$ does not satisfy the condition (13), but it may be modified as another form, which satisfied (13), by means of the virtual parameter method (Wu et al, 1987):

$$u^*_\lambda = N^*_\lambda \lambda \tag{16}$$

On the other hand the element optimization condition (14) can be absorbed by an optimization stress pattern, σ^*, as you see in the next. An initially assumed stress trial is defined as

$$\sigma = \phi \, \beta, \quad (\beta = \text{element stress parameter}) \tag{17}$$

We rewrite it as a sum of constant stress σ_c and higher order one σ_h:

$$\sigma = \sigma_c + \sigma_h = \phi_c \beta_c + \phi_h \beta_h = \beta_c + \begin{bmatrix} \phi_I & \phi_{II} \end{bmatrix} \begin{bmatrix} \beta_I \\ \beta_{II} \end{bmatrix} \tag{18}$$

where it is required that $\dim(\beta_{II}) = \dim(\lambda)$. Corresponding to the trial (16) and (18), the element optimization condition (14) takes a discrete version:

$$\oint_{\partial v^e} \delta u_i^{\Delta} n_j \sigma_{ij}^h \; ds \equiv \oint_{\partial v^e} \delta u_\lambda^{*^T} n \sigma_h \; ds = \delta \lambda^T M \, \beta_h = 0 \tag{19a}$$

$$M = \oint_{\partial v^e} N_\lambda^{*^T} n \left[\phi_I \; \phi_{II} \right] ds = \left[M_I \; M_{II} \right] \tag{19b}$$

Then a restriction equation on the higher order stress parameter β_h is obtained by (19), i.e.

$$M \, \beta_h = \left[M_I \; M_{II} \right] \begin{bmatrix} \beta_I \\ \beta_{II} \end{bmatrix} = 0 \tag{20}$$

If the determinant $\left| M_{II} \right| \neq 0$, β_{II} can be expressed by β_I and eliminated from (18). Finally, the desired optimization stress pattern may be formulated as follows,

$$\sigma^* = \sigma_c + \sigma_h^* = \left[I \; \phi_h^* \right] \begin{bmatrix} \beta_c \\ \beta_I \end{bmatrix} = \phi^* \beta^* \,, \tag{21}$$

$$\phi_h^* = \phi_I - \phi_{II} M_{II}^{-1} M_I$$

In accordance with the improved element trial u_λ^* and σ^*, in which the condition (13) and (14) has been absorbed respectively, the individual element functional should be of the form

$$\Pi^e(\sigma^*, \, u_q, \, u_\lambda^*) = \int_{v^e} \left[\sigma^{*^T}(Du_q + Du_\lambda^*) - \frac{1}{2} \sigma^{*^T} S \, \sigma^* \right] dv$$

$$= \int_{v^e} \left[\sigma^{*^T}(Du_q) + \sigma_h^{*^T}(Du_\lambda^*) - \frac{1}{2} \sigma^{*^T} S \, \sigma^* \right] dv \tag{22}$$

where D = linear strain operator and S = elastic compliance matrix.
There exists another possibility in the element optimization design, where, instead of u_λ^* which passed the patch test, the incompatible displacement u_λ can be introduced into element formulations directly. But, in the present case, the element boundery integral

$$\oint_{\partial v^e} (n\sigma_c)^T u_\lambda \; ds \neq 0 \tag{23}$$

and the resulted hybrid element does not pass the patch test. In order to guarantee the convergence, the non-zero integral in (23) must be cancelled from the energy functional, such that we have a modified one:

$$\Pi^e(\sigma^*, \, u_q, \, u_\lambda) = \int_{v^e} \left[\sigma^{*^T}(Du_q + Du_\lambda) - \frac{1}{2} \sigma^{*^T} S \, \sigma^* \right] ds - \oint_{\partial v^e} (n\sigma_c)^T u_\lambda \; ds$$

$$= \int_{v^e} \left[\sigma^{*^T}(Du_q) + \sigma_h^{*^T}(Du_\lambda) - \frac{1}{2} \sigma^{*^T} S \, \sigma^* \right] dv \tag{24}$$

In which the optimization stress pattern σ^* is still defined by (21) provided that N_λ takes the place of N_λ^* in (19b). We see that for the second optimization approach, indeed, a constant stress multiplier trick is used to relax the energy restriction (13). It is notable that the inner displacement u_λ in (24) and u_λ^* in (22) should be of higher order in comparision to u_q, and σ^* is the higher order part in σ^*. Therefore the energy integral terms with incompatible displacements are the higher order small quantities in comparision to the first term in (24)/(22), and which may be neglected from (24) and (22) respectively. Thus we obtain a simplified functional formulation without inner displacement u_λ:

$$\Pi_s^e(\sigma^*, \, u_q) = \int_{v^e} \left[-\frac{1}{2} \sigma^{*^T} S \, \sigma^* + \sigma_h^{*^T}(Du_q) \right] dv \tag{25}$$

So far the optimization design of hybrid elements is concluded as only one problem, i.e. how to establish σ^*.

APPLICATION TO 3-D SOLID ELEMENT

Now we apply the element optimization approach to the 3-dimensional elastic analysis. and the σ^* patern of a hybrid solid element will be established. Considering a 8-node hexahedron element shown in Fig.1, as a normal isoparametric element, the compatible displacement u_q is defined as a set of trilinear interpolations in the isoparametric coordinate ξ, η and ζ.

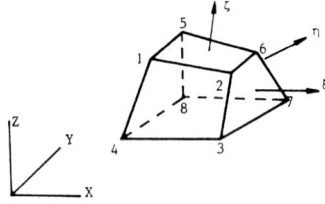

Fig.1   8-node hexahedral element

The Jacobi matrix for the isoparametric transformation is then

$$
J = \begin{bmatrix} \dfrac{\partial x}{\partial \xi} & \dfrac{\partial y}{\partial \xi} & \dfrac{\partial z}{\partial \xi} \\[2mm] \dfrac{\partial x}{\partial \eta} & \dfrac{\partial y}{\partial \eta} & \dfrac{\partial z}{\partial \eta} \\[2mm] \dfrac{\partial x}{\partial \zeta} & \dfrac{\partial y}{\partial \zeta} & \dfrac{\partial z}{\partial \zeta} \end{bmatrix} = \begin{bmatrix} a_1 + a_4\eta + a_5\zeta + a_7\eta\zeta & : & : \\ a_2 + a_4\xi + a_6\zeta + a_7\xi\zeta & :(a_i \rightarrow b_i) & :(b_i \rightarrow c_i) \\ a_3 + a_5\xi + a_6\eta + a_7\xi\eta & : & : \end{bmatrix} \qquad (26)
$$

where the coefficients are given by

$$
\begin{bmatrix} a_1 & b_1 & c_1 \\ \vdots & \vdots & \vdots \\ a_7 & b_7 & c_7 \end{bmatrix} = \frac{1}{8} \begin{bmatrix} -1 & 1 & 1 & -1 & -1 & 1 & 1 & -1 \\ -1 & -1 & -1 & -1 & 1 & 1 & 1 & 1 \\ 1 & 1 & -1 & -1 & 1 & 1 & -1 & -1 \\ 1 & -1 & 1 & -1 & 1 & -1 & 1 & -1 \\ -1 & 1 & -1 & 1 & -1 & 1 & -1 & 1 \\ -1 & -1 & 1 & 1 & 1 & 1 & -1 & -1 \\ 1 & -1 & 1 & -1 & -1 & 1 & -1 & 1 \end{bmatrix} \begin{bmatrix} x_1 & y_1 & z_1 \\ \vdots & \vdots & \vdots \\ x_8 & y_8 & z_8 \end{bmatrix} \qquad (27)
$$

The element inner displacement is defined as

$$
u_\lambda = \begin{bmatrix} u_\lambda \\ v_\lambda \\ w_\lambda \end{bmatrix} = \begin{bmatrix} \xi^2 & \eta^2 & \zeta^2 & & & & & & \\ & & & \xi^2 & \eta^2 & \zeta^2 & & & \\ & & & & & & \xi^2 & \eta^2 & \zeta^2 \end{bmatrix} \begin{bmatrix} \lambda_1 \\ \vdots \\ \lambda_9 \end{bmatrix} = N_\lambda \lambda \qquad (28)
$$

such that $u_q + u_\lambda$ consist of a complete quadratic element function.
On the other hand, based on a set of complete linear functions, the initial stress trial may be defined as

$$
\sigma \equiv \begin{bmatrix} \sigma_x & \sigma_y & \sigma_z & \tau_{yz} & \tau_{zx} & \tau_{xy} \end{bmatrix}^T = \sigma_c + \sigma_h = \beta_c + \begin{bmatrix} \phi_I & \phi_{II} \end{bmatrix} \begin{bmatrix} \beta_I \\ \beta_{II} \end{bmatrix} \qquad (29)
$$

$$
\beta_c = \begin{bmatrix} \beta_1 \\ \vdots \\ \beta_6 \end{bmatrix}, \qquad \beta_I = \begin{bmatrix} \beta_7 \\ \vdots \\ \beta_{18} \end{bmatrix} \quad \text{and} \quad \beta_{II} = \begin{bmatrix} \beta_{19} \\ \vdots \\ \beta_{27} \end{bmatrix};
$$

$$
[\phi_I | \phi_{II}] = \begin{bmatrix}
\eta & \zeta & \eta\zeta & o & o & o & o & o & o & o & o & o & \bigg| & \xi & o & o & o & o & o & o & o & o \\
o & o & o & \zeta & \xi & \zeta\xi & o & o & o & o & o & o & \bigg| & o & o & o & o & \eta & o & o & o & o \\
o & o & o & o & o & o & \xi & \eta & \xi\eta & o & o & o & \bigg| & o & o & o & o & o & o & o & o & \zeta \\
o & o & o & o & o & o & o & o & o & \xi & o & \bigg| & o & o & o & o & o & \zeta & o & \eta & o \\
o & o & o & o & o & o & o & o & o & o & \eta & \bigg| & o & o & \zeta & o & o & o & \xi & o & o \\
o & o & o & o & o & o & o & o & o & \zeta & o & o & \bigg| & o & \eta & o & \xi & o & o & o & o & o
\end{bmatrix}
$$

Here three bilinear term $\eta\zeta$, $\zeta\xi$ and $\xi\eta$ are added into σ in order to suppress the element zero energy deformations (Pian and Chen, 1983).

For the 3-D solid element it is convenient to calculate the matrix M in (19) in terms of solid integrals, and it can be rewritten as

$$
M = \int_{v_e} \left((DN_\lambda)^T [\phi_I | \phi_{II}] + N_\lambda^T [D^T\phi_I | D^T\phi_{II}] \right) \, dv = [M_I | M_{II}] \tag{30}
$$

It has been shown that $|M_{II}| \neq 0$ provided that u_λ and σ_h are defined by (28) and (29) respectively. Thus the optimization stress pattern can be formulated by means of (21). Unfortunately, it is difficult to obtain an explicit scheme of σ^* since the integrand in (30) is very complex. But if the value of Jaconi matrix (26) is taken at only the element center point, i.e.

$$
J(\xi, \eta, \zeta) = J_c(o, o, o) = \begin{bmatrix}
a_1 & b_1 & c_1 \\
a_2 & b_2 & c_2 \\
a_3 & b_3 & c_3
\end{bmatrix} \tag{31}
$$

This is a constant matrix, then the calculation of M is easily finished. Under the assumption (31), it can be shown that the incompatible displacement defined by (28) has passed the patch test condition (13). From (31),

$$
J_c^{-1} = \frac{1}{|J_c|} \begin{bmatrix}
j_{11} & j_{12} & j_{13} \\
j_{21} & j_{22} & j_{23} \\
j_{31} & j_{32} & j_{33}
\end{bmatrix} = \frac{1}{|J_c|} \begin{bmatrix}
b_2c_3 - b_3c_2 & b_3c_1 - b_1c_3 & b_1c_2 - b_2c_1 \\
c_2a_3 - c_3a_2 & c_3a_1 - c_1a_3 & c_1a_2 - c_2a_1 \\
a_2b_3 - a_3b_2 & a_3b_1 - a_1b_3 & a_1b_2 - a_2b_1
\end{bmatrix} \tag{32}
$$

Now M_I, M_{II} and M_{II}^{-1} are easily calculated, and the desired optimization stress pattern in accordance with (21) is formulated as follows,

$$
\sigma^* = \phi^* \begin{bmatrix} \beta_1 \\ \cdot \\ \cdot \\ \cdot \\ \cdot \\ \cdot \\ \beta_{18} \end{bmatrix} = \begin{bmatrix} I \\ (6\times6) \end{bmatrix}
\begin{bmatrix}
j_{22}^2\eta & j_{33}^2\zeta & \eta\zeta & o & j_{21}^2\xi & o \\
j_{12}^2\eta & o & o & j_{33}^2\zeta & j_{11}^2\xi & \zeta\xi \\
o & j_{13}^2\zeta & o & j_{23}^2\zeta & o & o \\
o & o & o & -j_{23}j_{33}\zeta & o & o \\
o & -j_{13}j_{33}\zeta & o & o & o & o \\
-j_{12}j_{22}\eta & o & o & o & -j_{21}j_{11}\xi & o
\end{bmatrix}
$$

$$
\begin{bmatrix}
j_{31}^2\xi & o & o & o & 2j_{21}j_{31}\xi & o \\
o & j_{32}^2\eta & o & o & o & 2j_{12}j_{23}\eta \\
j_{11}^2\xi & j_{22}^2\eta & \xi\eta & 2j_{13}j_{23}\zeta & o & o \\
o & -j_{32}j_{22}\eta & o & -j_{13}j_{33}\zeta & j_{11}^2\xi & -j_{12}j_{22}\eta \\
j_{31}j_{11}\xi & o & o & -j_{23}j_{33}\zeta & -j_{21}j_{11}\xi & j_{22}^2\eta \\
o & o & o & j_{33}^2\zeta & -j_{31}j_{11}\xi & -j_{32}j_{22}\eta
\end{bmatrix}
\begin{bmatrix} \beta_1 \\ \cdot \\ \cdot \\ \cdot \\ \cdot \\ \cdot \\ \beta_{18} \end{bmatrix} \tag{33}
$$

When the element is a brick hexahedron, and its local coordinate system $\xi-\eta-\zeta$ is parallel to the structural coordinate system X-Y-Z, the coefficient $j_{kl} = o$ $(k \neq l)$ such that the stress pattern (33) will be reduced to the result of Loikkanen and Irons (1984).

For the resulted hybrid element based on the trial u_q and σ^* in (33) and the functional (25) some numerical examples are presented here. The example 1 is a cantilever beam of constant section acted on by two sets of loads. where, as shown in Fig.2, five irregular solid elements are used. another example, shown in Fig.3, is a thick sphere under the internal pressure q. The present solutions of the optimization element are compared with that of the normal 8-node isoparametric element DM8.

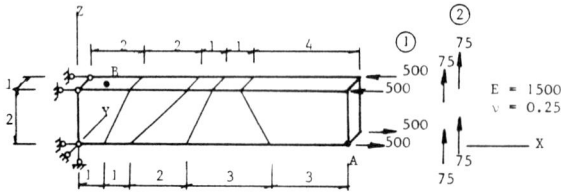

Element	Case 1		Case 2	
	w_A	σ_{xB}	w_A	σ_{xB}
DM8	43.24	-1732	48.43	-2412
Present	96.05	-3016	98.01	-4076
Exact	100	-3000	102.6	-4050

Fig.2 Cantilever beam under end loads

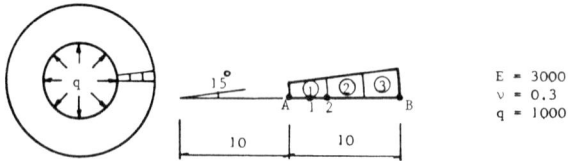

Element	u_{rA}	u_{rB}	σ_{r1}	$\sigma_{\phi1}=\sigma_{\theta1}$	σ_{r2}	$\sigma_{\phi2}=\sigma_{\theta2}$
DM8	25.80	9.88	-512.6	587.6	-425.4	334.9
Present	26.27	9.99	-545.4	528.0	-365.0	329.3
Exact	26.67	10	-576.8	502.7	-339.3	383.9

Fig.3 Thick sphere with internal pressure

In the following we consider the variation of solutions when an element distortion takes place. In the numerical test shown as Fig.4(a,b,c) and Fig.5(a,b,c), a cantilever with two silod elements is considered. The degree of distortion is marked by the dimension "a", and LO8:7-APC and LO8:7-APR are two hybrid elements established by means of the symmetry group theory (Punch and Atluri,1984). We see that the present optimization hybrid element is less sensitive to the geometric distortion and provides the most excellent results.

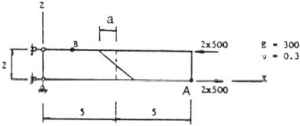

Fig.4a Element distortion test I

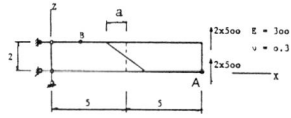

Fig.5a Element distortion test II

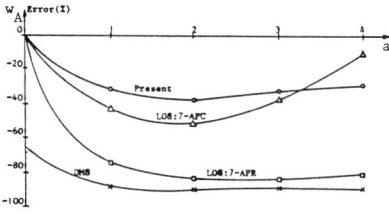

Fig.4b

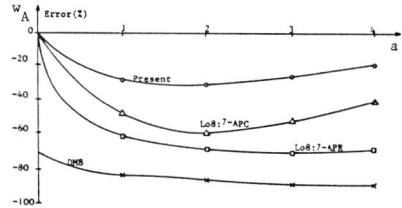

Fig.5b

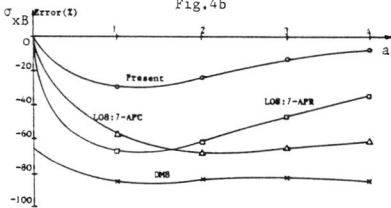

Fig.4c

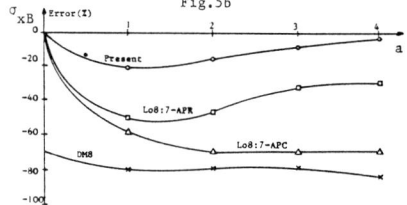

Fig.5c

ACKNOWLEDGEMENT The work was undertaken with the support of Alexander von Humboldt Foundation. The kind invitation and many helps of Prof. Dr.-Ing. H. Bufler of Universität Stuttgart are gratefully acknowledged.

REFERENCES

Bufler, H.(1979). Generalized variational principles with relaxed continuity requirements for certain nonlinear problems with an application to nonlinear elasticity. Comp. Meth. Appl. Mech. Eng. 19, 235-255.

Loikkanen, M.J.and B.M. Irons (1984). A 8-node brick finite element. Int. J. Num. Meth. Eng. 20, 523-528.

Pian, T.H.H. and D.P. Chen (1983). On the suppression of zero energy deformation modes. Int. J. Num. Meth. Eng. 19, 1741-1752.

Punch, E.F. and S.N. Atluri (1984), Applications of isoparametric three dimensional hybrid-stress finite elements with least order stress fields. Computers & Structures 19, 409-430.

Strang, G. and G.J.Fix (1973). A Analysis of the Finite Element Method. Prentice-Hall, Englewood Cliffs, NJ.

Washizu, K. (1975). Variational Methods in Elasticity and Plasticity, 2nd ed. Pergamon Press.

Wu, C.C., S.L. Di and M.G. Huang (1987).Optimizatuion design of hybrid elements. Kexue Tongbao 32, 1236-1239.

Wu, C.C., M.G. Huang and T.H.H. Pian (1987). Consistency condition and convergence criteria of incompatible elements: General formulation of incompatible functions and its application, Computers & Structures. 27, 639-644.